Endocytosis

Frontiers in Molecular Biology

SERIES EDITORS

B. D. Hames

Department of Biochemistry and Molecular Biology University of Leeds, Leeds LS2 9JT, UK

D. M. Glover

Department of Genetics, University of Cambridge, UK

TITLES IN THE SERIES

1. **Human Retroviruses** — *Bryan R. Cullen*
2. **Steroid Hormone Action** — *Malcolm G. Parker*
3. **Mechanisms of Protein Folding** — *Roger H. Pain*
4. **Molecular Glycobiology** — *Minoru Fukuda and Ole Hindsgaul*
5. **Protein Kinases** — *Jim Woodgett*
6. **RNA–Protein Interactions** — *Kyoshi Nagai and Iain W. Mattaj*
7. **DNA–Protein: Structural Interactions** — *David M. J. Lilley*
8. **Mobile Genetic Elements** — *David J. Sherratt*
9. **Chromatin Structure and Gene Expression** — *Sarah C. R. Elgin*
10. **Cell Cycle Control** — *Chris Hutchinson and D. M. Glover*
11. **Molecular Immunology (Second Edition)** — *B. David Hames and David M. Glover*
12. **Eukaryotic Gene Transcription** — *Stephen Goodbourn*
13. **Molecular Biology of Parasitic Protozoa** — *Deborah F. Smith and Marilyn Parsons*
14. **Molecular Genetics of Photosynthesis** — *Bertil Andersson, A. Hugh Salter, and James Barber*
15. **Eukaryotic DNA Replication** — *J. Julian Blow*
16. **Protein Targeting** — *Stella M. Hurtley*
17. **Eukaryotic mRNA Processing** — *Adrian Krainer*
18. **Genomic Imprinting** — *Wolf Reik and Azim Surani*
19. **Oncogenes and Tumour Suppressors** — *Gordon Peters and Karen Vousden*
20. **Dynamics of Cell Division** — *Sharyn A. Endow and David M. Glover*
21. **Prokaryotic Gene Expression** — *Simon Baumberg*
22. **DNA Recombination and Repair** — *Paul J. Smith and Christopher J. Jones*
23. **Neurotransmitter Release** — *Hugo J. Bellen*
24. **GTPases** — *Alan Hall*
25. **The Cytokine Network** — *Fran Balkwill*
26. **DNA Virus Replication** — *Alan J. Cann*
27. **Biology of Phosphoinositides** — *Shamshad Cockcroft*
28. **Cell Polarity** — *David G. Drubin*
29. **Protein Kinase Functions** — *Jim Woodgett*
30. **Molecular and Cellular Glycobiology** — *Minoru Fukuda and Ole Hindsgaul*
31. **Protein–Protein Recognition** — *Colin Kleanthous*
32. **Mechanisms of Protein Folding (Second Edn)** — *Roger Pain*
33. **The Yeast Nucleus** — *Peter Fantes and Jean Beggs*
34. **RNA Editing** — *Brenda L. Bass*
35. **Chromatin Structure and Gene Expression (Second Edn)** — *Roger Pain*
36. **Endocytosis** — *Mark Marsh*

Endocytosis

EDITED BY

Mark Marsh

MRC Laboratory for Molecular Cell Biology, UCL, Gower Street, London WC1E 6BT

OXFORD
UNIVERSITY PRESS

Great Clarendon Street, Oxford OX2 6DP

Oxford University Press is a department of the University of Oxford
It furthers the University's objective of excellence in research, scholarship,
and education by publishing worldwide in

Oxford New York
Athens Auckland Bangkok Bogotá Buenos Aires Calcutta
Cape Town Chennai Dar es Salaam Delhi Florence Hong Kong Istanbul
Karachi Kuala Lumpur Madrid Melbourne Mexico City Mumbai
Nairobi Paris São Paulo Singapore Taipei Tokyo Toronto Warsaw

with associated companies in
Berlin Ibadan

Oxford is a registered trade mark of Oxford University Press
in the UK and in certain other countries

Published in the United States
by Oxford University Press Inc., New York

First published 2001

A catalogue record for this title is available from the British Library

Library of Congress Cataloging in Publication Data

Endocytosis / edited by Mark Marsh.
p. cm. (Frontiers in molecular biology)
Includes bibliographical references and index.
1. Endocytosis I. Marsh, Mark. II. Series.

QH634 .E532 2001 571.6′55–dc21 00-053076

ISBN 0 19 963852 7 (Hbk.) ISBN 0 19 963851 9 (Pbk.)

1 3 5 7 9 10 8 6 4 2

Typeset by Footnote Graphics, Warminster, Wilts

Printed in Great Britain on acid free paper by
Bookcraft (Bath) Ltd, Midsomer Norton, Avon

Preface

Endocytosis has been at the forefront of cell biology for much of the last century. Since Metchnikoff observed phagocytosis in amoeboid cells in the mid-1800s, investigators have sought to understand the functions of endocytosis and latterly the molecular mechanisms that underlie these processes. Now, at the beginning of the third millennium, we are in the position of having multiple endocytic pathways clearly established, molecular details understood for most of these routes, a plethora of functions ascribed to each, and still a multitude of questions to ask and details to establish.

For me two things have made studies of endocytosis particularly appealing. First, endocytosis is required for a vast number of functions that are essential for the well being of the cell and, in the case of metazoans, the organism. In protozoa endocytosis is vital for providing nutrients. Bacterial prey is internalized by phagocytosis and the fluid environment by macropinocytosis. In multicellular organisms these mechanisms have been adapted and developed for the benefit of the whole organism, to enhance communication between cells, to maintain the composition of the extracellular environment, to provide protective functions such as immunity, to remove dead and dying cells, and to modulate the composition of the plasma membrane. In turn, pathogens have learned to exploit the opportunities these pathways offer as a means to gain entry to the cell and have themselves become exquisite tools for studies of endocytosis.

The second reason for the appeal of endocytosis is that these pathways offer special opportunities to study the fundamental cellular processes involved in membrane traffic and membrane protein and lipid sorting. The endocytic pathway gave the first insights into the molecular mechanisms that underlie membrane protein sorting in cells and still today endocytic clathrin-coated vesicles remain the best characterized vesicular transport intermediates. Molecular biology is uncovering a vast array of proteins and lipids that function together in the formation and regulation of endocytic vesicles. Structural studies are providing views of the machinery at atomic resolution, and high-resolution morphological techniques are providing detailed insights to the structure of assembled endocytic machines. Endocytosis has proven to be particularly amenable to genetic analysis. Initial studies in yeast have identified a vast array of proteins and interactions that regulate traffic through the endocytic pathway, while studies in *Dictyostelium*, *Drosophila*, *Caenorhabditis elegans*, and the mouse are beginning to probe the functions of endocytic pathways and specific endocytic proteins in whole organisms.

Currently, we are able to describe endocytosis in reasonably precise morphological, biochemical, and kinetic terms. Before the advent of electron microscopy and molecular biology, endocytic phenomena were largely defined by the nature of the

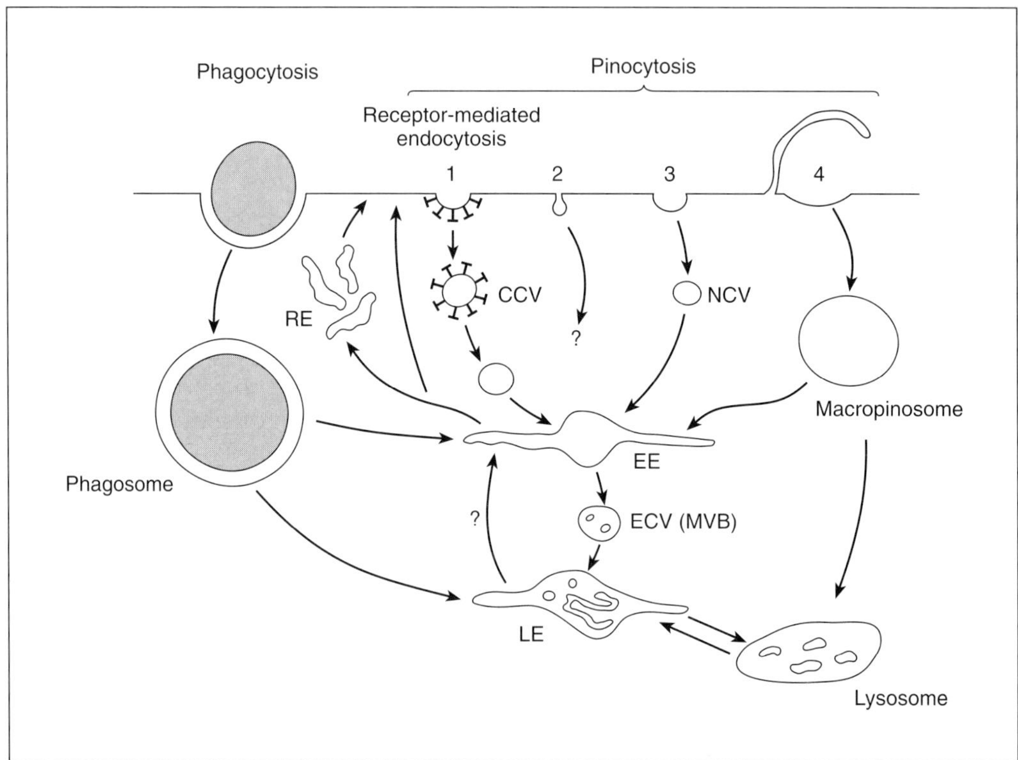

Fig. 1 Cartoon of the principal endocytic mechanisms. Four different mechanisms for pinocytosis are illustrated: (1) via clathrin-coated pits and vesicles (CCV), (2) via caveolae, (3) via non-clathrin coated vesicles (NCV), (4) via macropinosomes. The CCV pathway is also frequently associated with receptor-mediated endocytosis. In addition, the phagocytic pathway responsible for internalization of large particles (> 200 nm in diameter) but not fluid phase is illustrated. The organelles of the endocytic pathway are also illustrated, together with the principal pathways linking these organelles. These include early endosomes (EE), recycling endosomes (RE), late endosomes (LE), endosome carrier vesicles (ECV) or multivesicular bodies (MVB), and lysosomes.

material taken up by cells. The terms 'phagocytosis' and 'pinocytosis' were adopted to describe the uptake of large particles (cell eating) and extracellular medium (cell drinking), respectively. More recently studies on the uptake of low density lipoproteins (LDL) pointed to the process of receptor-mediated endocytosis (or adsorptive endocytosis) in which cells efficiently recruit extracellular ligands to cell surface receptors that then undergo endocytosis. The observation that LDL and many other ligands are internalized through clathrin-coated pits has led to receptor-mediated endocytosis and clathrin-mediated endocytosis being considered as one and the same. However, this is not always the case. Examples of receptor-mediated internalization that is not clathrin-dependent are increasing (Chapter 2); indeed phagocytosis is receptor-mediated but clathrin-independent. Conversely, clathrin-coated vesicles (CCVs) internalize fluid phase (a vesicle cannot form without content) and can therefore effect both pinocytosis (fluid phase endocytosis) and receptor-mediated endocytosis. The relative contribution of different endocytic mechanisms to total cellular endocytic activity is still, in many cases, unclear. In many tissue culture cell lines CCVs are responsible for the bulk of pinocytosis, but in immature dendritic cells the bulk of fluid uptake can occur through macropinocytosis. In HeLa cells at least an alternative pathway can be activated when CCV formation is blocked (Chapter 2). Thus endocytic activity is plastic: different mechanisms operate to a greater or lesser extent in different cell types and the balance of these mechanisms may change as cells differentiate, undergo activation, or adopt specific functional roles. Currently, the mechanisms through which different endocytic processes are monitored and controlled is not understood, though information is beginning to emerge for at least some systems.

The pathways discussed in this book are defined as follows (see Fig. 1):

(a) **Clathrin-dependent endocytosis**. Endocytosis mediated by small (approx. 100 nm diameter) vesicles that have a morphologically characteristic bristle coat made up of a complex of proteins that includes clathrin. Clathrin-coated vesicles (CCVs) are found in virtually all cells and form from domains of the plasma membrane termed clathrin-coated pits (Chapter 1). Coated pits can concentrate a large number of different receptors responsible for the receptor-mediated endocytosis of ligands, e.g. LDL, transferrin, growth factors, antibodies, and many others.

(b) **Caveolae**. Caveolae are small (approx. 50 nm diameter) flask-shaped invaginations of the plasma membrane seen in many, but not all cell types (Chapter 2). Caveolae require the cholesterol-binding protein caveolin (Vip21) for formation and are associated with plasma membrane glycolipid-rich microdomains. Caveolae have been implicated in the endocytosis of the SV40 virus (Chapters 2 and 11) and in transcytosis, but the kinetics of their uptake and contribution to total cellular endocytic activity is not well established.

(c) **Macropinocytosis**—the formation of vesicles (0.5–5 μm in diameter) that internalize large volumes of fluid (equivalent to 10^3 to 10^6 CCVs; Chapter 4). Prominent in *Dictyostelium discoideum* and in some antigen presenting cells such as dendritic cells. Involves formation of membrane ruffles and lamellaepodia, is regulated by Rho family small GTPases, and is inhibited by agents that depolymerize actin.

(d) **Phagocytosis** (Chapter 3). Well established process involved in the uptake of large particular ligands including bacteria and apoptotic cells. Requires ligand binding and receptor signalling, as with macropinocytosis involves redistribution of actin under the control of Rho family GTPases.

(e) **Dynamin- and clathrin-independent endocytosis** (Chapter 2). There has been increasing evidence for a pinocytic activity that is independent of clathrin. This uptake could occur through caveolae or macropinocytosis. Recent studies suggest that endocytosis through CCVs, caveolae, and macropinosomes is dependent on dynamin, but pinocytosis can continue in cells expressing dominant negative dynamin indicating a dynamin- and clathrin-independent mechanism. It is unclear what contribution this pathway normally makes to cellular endocytosis or whether it is a single mechanism.

These distinct vesiculation mechanisms are responsible for the uptake of extracellular fluid, macromolecular and cellular material from the cell surface. Once formed, the vesicles for the most part undergo fusion with organelles that collectively make up the endocytic pathway. Within the endocytic organelles the internalized membrane and content are processed. Some components, such as particular types of receptors and lipids, are sorted for recycling to the cell surface; others are selected for delivery to late endocytic compartments where they may be degraded. Alternatively, components may be sorted for delivery to other cellular destinations such as the Golgi apparatus, TGN, or a transcytotic route.

The principle components of the endocytic pathway are (Fig. 1):

(a) **Early endosomes**—the first station on the endocytic pathway. Early endosomes are often located in the periphery of the cell and receive most types of vesicles coming in from the cell surface. They have a characteristic tubulo-vesicular morphology (vesicles up to 1 μm in diameter with connected tubules of approx. 50 nm diameter) and a mildly acid pH. They are principally sorting organelles that select components for recycling to the cell surface (via tubules) or for transport to late compartments (via the vesicular component; note that this vesicular component can form multivesicular bodies (MVBs) or endosomal carrier vesicles (ECVs) that are involved in transport to late endosomes). Early endosomes are now thought to comprise at least two subcompartments—sorting endosomes and recycling endosomes. Sorting endosomes receive and redistribute material coming from the cell surface. Recycling endosomes harbour recycling receptors such as the transferrin receptor and are comprised of clusters of small (50 nm diameter) vesicles and tubules. Recent evidence has pointed to the markers EEA1 and Rab5 being located on sorting endosomes, while Rab11 is preferentially associated with recycling endosomes (Chapter 5). However, the precise relationship between these organelles is still to be established.

(b) **Late endosomes**—receive internalized material en route to lysosomes, usually from early endosomes or the TGN (Chapter 6). Late endosomes often contain many membrane vesicles or membrane lamellae (Chapter 8) and proteins characteristic of lysosomes, including lysosomal membrane glycoproteins and acid hydrolases. They are acidic (approx. pH 5.5), carry Rab7, and are part of the trafficking itinerary of mannose-6-phosphate receptors. Late endosomes are thought to mediate a final set of sorting events prior to delivery of material to lysosomes. Sorting in late endosomes involves both the invagination of membrane vesicles into the lumen of organelles, or the formation of transport vesicles that return MPR and other proteins to the Golgi. The internal vesicles and membranes seen in late endosomes are enriched in the lipid lysobisphosphatidic acid and in proteins such as CD63.

(c) **Lysosomes** (see Chapter 6). Lysosomes are the last compartment of the endocytic pathway. They are acidic (approx. pH 4.8) and by EM usually appear as large vacuoles (1–2 μm in diameter) containing electron dense material. They have a high content of lysosomal membrane proteins and active lysosomal hydrolases, but no mannose-6-phosphate receptor. They are generally regarded as the principle hydrolytic compartment of the cell, though recent work suggests that their functions are more heterogeneous. For example, lysosomes, or subpopulations of lysosomes, make up the secretory organelles of haematopoietic cells including the granules of cytotoxic T cells, and the pigment containing granules of melanocytes.

In this volume the contributors, all experts in their respective fields, have focused on recent studies that have begun to dissect the molecular mechanisms that underlie endocytosis. The first four chapters discuss the various modes of endocytic vesicle uptake from the cell surface. The following chapters discuss the properties of the organelles that make up the endocytic pathway, the nature of signals that direct proteins through these different compartments, and the way in which yeast genetics has been used to dissect aspects of endocytosis. The final three chapters discuss specific roles of endocytosis, its function in antigen presentation, its role in synaptic vesicle biogenesis and cycling, and its exploitation in pathogen entry. In selecting the contributions for this volume I have tried to present a broad view of endocytosis. Nevertheless, the coverage is not complete. Aspects of the dynamics and plasticity of endocytic traffic and endocytic organelles have not been covered, nor is the emerging information on lipid trafficking and interactions with the cytoskeleton discussed. These will be topics for future volumes. In the mean time, we have tried to point to the appropriate literature.

Endocytosis has been a prominent area for research, its unquestionable importance in cell function will ensure this interest continues. The current integration of genetics, structural biology, morphology, molecular biology, and biochemistry is providing remarkable new insights into the molecular mechanisms of endocytosis. Progress will continue at a rapid pace and the coming years will be exciting times for the field.

London M. M.
August, 2000

Contents

Contributors

FRANCES M. BRODSKY
The G W Hooper Foundation, Department of Microbiology and Immunology, School of Medicine and Department of Biopharmaceutical Sciences, School of Pharmacy, University of California, San Francisco, CA 94143–0552, USA.

EMMANUELLE CARON
MRC Laboratory for Molecular Cell Biology, University College London, Gower Street, London WC1E 6BT, UK.

ALICE DAUTRY-VERSAT
Unité de Biologies des Interactions Cellulaires, URA CNRS 1960, Institut Pasteur, 25 Rue du Dr Roux, 75724 Paris Cedex 15, France.

SYLVIE FRIANT-MICHEL
Biozentrum of the University of Basel, Klingenbergstr. 70, CH-4056 Basel, Switzerland.

ALAN HALL
MRC Laboratory for Molecular Cell Biology, University College London, Gower Street, London WC1E 6BT, UK.

STEFAN HÖNING
Georg-August University, Heinrich-Düker-Weg 12, Göttingen D-37073, Germany.

MONIQUE J. KLEIJMEER
Department of Cell Biology, Institute of Biomembranes, UMC Utrecht, AZU GO2.525 Heidelberglaan 100, 3584 CX Utrecht, The Netherlands.

SHU-HUI LIU
Connectics Corporation, 3400 West Bayshore Road, Palo Alto, CA 94303, USA.

RUBEN LOMBARDI
Biozentrum of the University of Basel, Klingenbergstr. 70, CH-4056 Basel, Switzerland.

J. PAUL LUZIO
Wellcome Trust for Molecular Mechanisms in Disease, University of Cambridge, Wellcome Trust / MRC Building, Hills Road, Cambridge CB2 2XY, UK.

WILLIAM G. MALLET
Genesoft, Inc., Two Corporate Drive, Suite 100, South San Francisco, CA 94080, USA.

MARKUS MANIAK
Abt. Zellbiologie, Universität GhK, Heinrich-Plett-Str. 40, D-34132 Kassel, Germany.

MARK MARSH
MRC-Laboratory for Molecular Cell Biology, University College London, Gower Street, London WC1E 6BT, UK.

HARVEY T. MCMAHON
Medical Research Council, Hills Road, Cambridge CB2 2QH, UK.

GRAÇA RAPASO
Compartimentation et Dynamique Cellulaire, Institut Curie, CNRS UMR 144, Paris, France.

HOWARD RIEZMAN
Biozentrum of the University of Basel, Klingenbergstr. 70, CH-4056 Basel, Switzerland.

DAVID G. RUSSELL
Department of Microbiology and Immunology, College of Veterinary Medicine, Cornell University, Ithaca, NY 14853, USA.

MATTHEW N. J. SEAMAN
Wellcome Trust for Molecular Mechanisms in Disease, University of Cambridge, Wellcome Trust/MRC Building, Hills Road, Cambridge CB2 2XY, UK.

HARALD STENMARK
Department of Biochemistry, Norwegian Radium Hospital, Montebello, N-0310 Oslo, Norway.

PATRICK WIGGE
Medical Research Council, Hills Road, Cambridge CB2 2QH, UK.

MARINO ZERIAL
European Molecular Biology Laboratory, Meyerhofstrasse 1, D-69012 Heidelberg, Germany.

Abbreviations

# indicates that the attached acronym is followed by a number indicating a specific protein or family member.

ADP	adenine diphosphate
AEP	asparagine-specific cysteine endopeptidase
AP	adaptor protein complexes
APC	antigen presenting cell
ATP	adenine triphosphate
BCR	B cell receptor
Cat	cathepsin
CCV	clathrin-coated vesicle
CD#	cluster of differentiation antigen number
CD-MPR	cation-dependent mannose-6-phosphate receptor
CHO	Chinese hamster ovary cells
CIIV	(MHC) class II-containing vesicle
CI-MPR	cation-independent mannose-6-phosphate receptor
CLIP	class II-associated invariant chain peptide
COP	coat protein complex
CPS	carboxypeptidase S
CPY	carboxypeptidase Y
CR#	complement receptor number
CURL	compartment of uncoupling of receptor and ligand
DAG	diacylglycerol
DAMP	3-(2,4-dinitroanilino)-3′-amino-*N*-methyldipropylamine
DC	dendritic cell
DIG	detergent insoluble glycolipid-rich domain (also termed DIMs or DRMs)
DIM	detergent insoluble membrane
DRM	detergent-resistant membrane
ECV	endosome carrier vesicle
EE	early endosome
EEA1	early endosomal antigen 1
EGF	epidermal growth factor
EH domain	Eps15-homology domain
EM	electron microscopy
End	endocytosis mutant (yeast)
Eps	EGF receptor phosphorylated substrate protein
ER	endoplasmic reticulum

EV	endocytic vesicle
FcR	receptor for immunoglobulin Fc domains
FDC	follicular dendritic cell
FFE	free flow electrophoresis
GAP	GTPase-activating protein
GDI	GDP dissociation inhibitor
GDP	guanine diphosphate
GEF	GDP/GTP exchange factor
GFP	green fluorescent protein
GILT	gamma-interferon-inducible lysosomal thiol reductase
GPCR	heterotrimeric GTP-binding protein-coupled receptor
GPI	glycophosphatidyl inositol
GTP	guanine triphosphate
GTPase	guanine triphosphatase
HEL	hen egg lysozyme
HLA	human leukocyte antigen
HRP	horse radish peroxidase
HRV	human rhinovirus
HSC	heat shock protein
IEM	immuno-electron microscopy
Ii	invariant chain
IL-2	interleukin 2
ITAM	immune receptor tyrosine-base activation motif
ITIM	immune receptor tyrosine-base inhibitory motif
LAMP#	lysosome-associated membrane protein
LAP	lysosomal acid phosphatase
LDL	low density lipoprotein
LDLR	low density lipoprotein receptor
LE	late endosome
LGP	lysosomal membrane glycoprotein
MHC I	major histocompatibility complex antigen class I
MHC II	major histocompatibility complex antigen class II
MIIC	MHC class II-containing compartment
MPR	mannose-6-phosphate receptor
MR	mannose receptor
MVB	multivesicular body
NSF	*N*-ethyl maleimide-sensitive fusion protein
PH domain	pleckstrin homology domain
PI	phosphoinositides
PI3K	phosphatidylinositol 3-kinase
PI3-kinase	phosphatidylinositol 3-kinase
PP2A	protein phosphatase 2A
PrA	proteinase A
PtdIns	phosphatidylinositol

PtdIns(3)P	phosphatidylinositol-3-phosphate
PtdIns(4,5)P2	phosphatidylinositol-4,5-bisphosphate
PtdIns(3,4,5)P3	phosphatidylinositol-3,4,5-trisphosphate
RE	recycling endosome
R-ME	receptor-mediated endocytosis
SFV	Semliki Forest virus
SH domain	Src homology domain
SNAP	soluble NSF attachment protein
SNARE	SNAP receptor
TAP	transporter associated with antigen processing
TCR	T cell receptor
TfR	transferrin receptor
TGN	*trans*-Golgi network
t-SNARE	target membrane-associated SNARE
Vam	vacuolar acidification mutant (yeast)
V-H^+ATPase	vacuolar proton ATPase
Vps	vacuolar protein sorting mutant (yeast)
v-SNARE	vesicle-associated SNARE
VSV	vesicular stomatitis virus

1 | Clathrin-mediated endocytosis

SHU-HUI LIU, WILLIAM G. MALLET, and FRANCES M. BRODSKY

1. Introduction

Clathrin-mediated endocytosis from the plasma membrane allows cells to internalize proteins and other biomolecules from their environment via specific receptors. By this mechanism, cells take up nutrients, modulate the expression of cell surface molecules, and control their responses to external stimuli. For example, endocytosis allows the clearance of lipoproteins from the blood stream upon binding to cognate cell surface receptors. Also, endocytosis of foreign antigens and subsequent processing and presentation is essential for generation of the cellular immune response. Receptors are endocytosed by their capture in clathrin-coated vesicles (CCVs) budding from the plasma membrane. The interactions of clathrin, receptors, and associated structural and regulatory molecules during this process is the focus of this chapter. The functional role of polymerization of clathrin and associated adaptor proteins into a cytoplasmic vesicle coat to facilitate receptor sorting and internalization is one of the most widely accepted and best-characterized concepts in membrane transport. Over the last twenty years, steady progress has yielded an increasingly sophisticated description of clathrin-mediated endocytosis at the molecular level. From early work demonstrating the existence of endocytic vesicles surrounded by a protein coat to current efforts aimed at understanding the regulation and specificity of endocytosis, a battery of approaches has been brought to bear on this important problem. The result is a mechanism requiring a large array of protein and membrane interactions, exquisitely co-ordinated in space and in time. In this chapter, we will describe how our model of the endocytic mechanism has been successively refined and matured, and we will highlight areas where recent work has been particularly illuminating. We will develop this model over the following sequence of topics:

(a) Section 2: History and biochemistry of the clathrin-coated vesicle.
(b) Section 3: Mechanism of CCV formation during receptor-mediated endocytosis.
(c) Section 4: Regulation of clathrin-mediated endocytosis.
(d) Section 5: Recent advances in the resolution of coat protein structures.
(e) Section 6: Methods for studying clathrin function *in vivo*.

2. History and biochemistry

Internalization involves the formation of specialized protein 'coats' at the cytoplasmic face of the plasma membrane (Fig. 1). The coats comprise polymers of the cytoplasmic protein clathrin and several associated proteins, as will be discussed below. The clathrin coats assemble as the membrane is inwardly deformed ('invaginated'). Progressive invagination eventually promotes the budding of a coated vesicle within the cytoplasm, enclosing receptors, proteins, and other substrates from the extracellular milieu. Whether membrane-bound or soluble, these substrates must be delivered to the appropriate intracellular destinations for degradation, recycling, or other fates. The delivery requires the endocytosed vesicle to fuse with other membrane-bound organelles, a process that can proceed only upon the removal ('uncoating') of the clathrin lattice from the membrane. After uncoating, the clathrin subunits are then available to mediate another round of endocytosis from the plasma membrane.

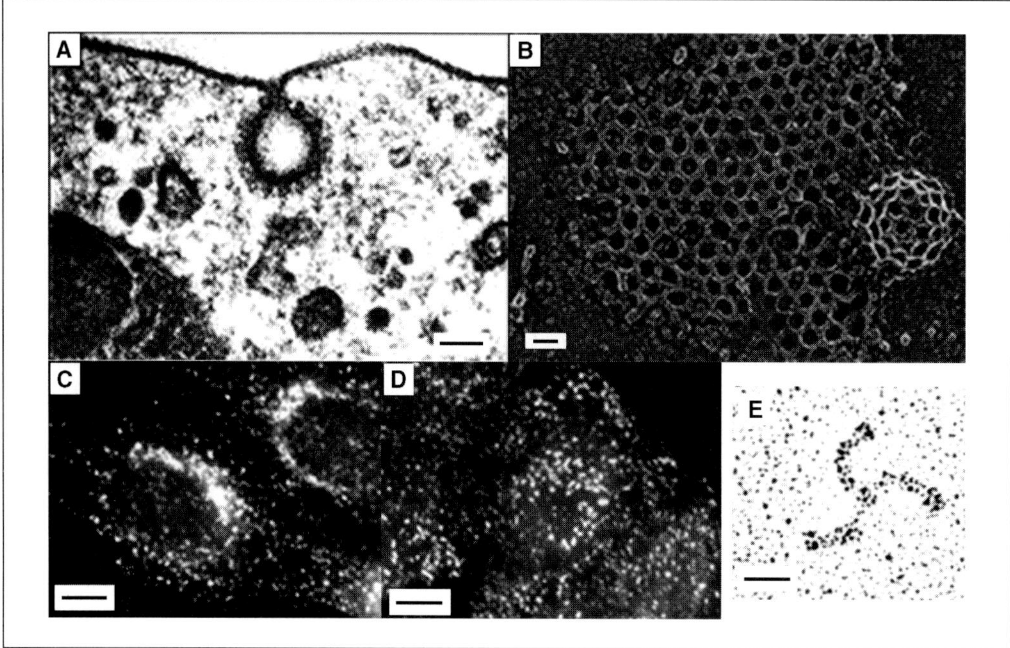

Fig. 1 Morphology of clathrin and clathrin-coated vesicles. (A) A clathrin-coated pit internalizing from the plasma membrane, visualized by transmission electron microscopy (92). (B) A clathrin lattice with a budding clathrin-coated pit on the cytoplasmic face of the plasma membrane, visualized by freeze-etch electron microscopy (124). (C) The distribution of clathrin-coated pits and vesicles at the plasma membrane and in the *trans*-Golgi network region in a cell stained with the anti-clathrin heavy chain monoclonal antibody X22 (125), visualized by immunofluorescent labelling. (D) The distribution of endocytic clathrin-coated pits and vesicles containing the AP-2 complex in a cell stained with the AP.6 monoclonal antibody (126), visualized by immunofluorescent labelling. (E) A clathrin triskelion, visualized by platinum shadowing. The bars indicate the following dimensions: (A) 50 nm, (B) 33 nm, (C) 5 μm, (D) 5 μm, (E) 20 nm. The images in A and B are reproduced with permission from reference (92) copyright 1999 American Association for the Advancement of Science and from reference (8) by copyright permission of the Rockefeller University Press, respectively. Panels C, D and E were generated in our laboratory.

Understanding of the function of clathrin emerged from a confluence of morphological, biochemical, and cell biological studies over more than two decades. In 1964, Roth and Porter described the uptake of yolk proteins into mosquito oocytes via 80 nm bristle-coated membrane vesicles (1). They observed the bristle coats by electron microscopy at both the oocyte membrane and on internal vesicular structures near the membrane. An example of a coated vesicle budding from the plasma membrane is shown in Fig. 1A. Further into the cytoplasm, they detected vesicles of similar size but lacking the bristle coat, suggesting that the association of the coat with the membrane may be transient. The coated vesicle intermediate has proved to be a general mechanism for uptake in a broad range of species and cell types, and coated vesicles were first purified in 1969 from pig brain (2). Their purification allowed these vesicles to be characterized in greater morphological detail, with the bristle coat revealed as a lattice of hexagons and pentagons surrounding the vesicle membrane (Fig. 1B). Pearse subsequently initiated the purification and characterization of the protein components of the coat with the landmark 1975 description of clathrin heavy chain component of coated vesicles (3), followed by the discovery of the clathrin light chain subunits in the early 1980s and the characterization of clathrin in solution (4, 5).

Clathrin purified from coated vesicles exists as a trimer of 192 kDa polypeptides ('heavy chains'), each associated with a 22–25 kDa 'light chain'. The heavy chain trimer appears as a three-legged pinwheel or triskelion, with legs composed of each heavy chain extending some 450 Å from the trimer hub (6) (Fig. 1E). The 'hub' of the triskelion extends from the centre, along the proximal segment of the heavy chain to a kink in the leg, followed by the distal segment and the globular terminal domain (Fig. 2A). Trimerization of the clathrin heavy chain is mediated by a short segment near the C-terminus, with the balance of the heavy chain distributed in thirds between the proximal and distal segments and the terminal domain (7, 8). In mammalian and avian clathrin there are two types of light chains, LCa and LCb, encoded by separate genes. The light chains are divided into a linear array of sequence motifs, and the central portions interact with the proximal leg segment of the clathrin heavy chain (reviewed in ref. 9). Also, neurons express splice variants of the light chain genes that have inserts of 12–18 amino acids with hydrophobic properties (10). LCa and LCb have characteristic levels of expression in different tissues (11). They compete with each other for heavy chain binding, creating four kinds of clathrin triskelia. Yeast and *Drosophila* express only one clathrin light chain, so their clathrin molecules are uniform (12, 13). The light chains have a regulatory function in clathrin assembly (14). The differential functions of LCa and LCb and their neuronal splicing variants have yet to be defined.

In the clathrin triskelion, the proximal and distal segments are available for binding to heavy chains from other triskelia with a defined geometry, and in this way clathrin triskelia can polymerize to form the regular polygonal lattice that is characteristic of coated vesicles (Plate 1A) (15). Keen and co-workers demonstrated that purified clathrin alone indeed could form polymeric lattices and basket-like spheres that in some ways resembled the vesicle coats (16). Although these baskets form only under non-physiological conditions of pH and salt and their dimensions differ from

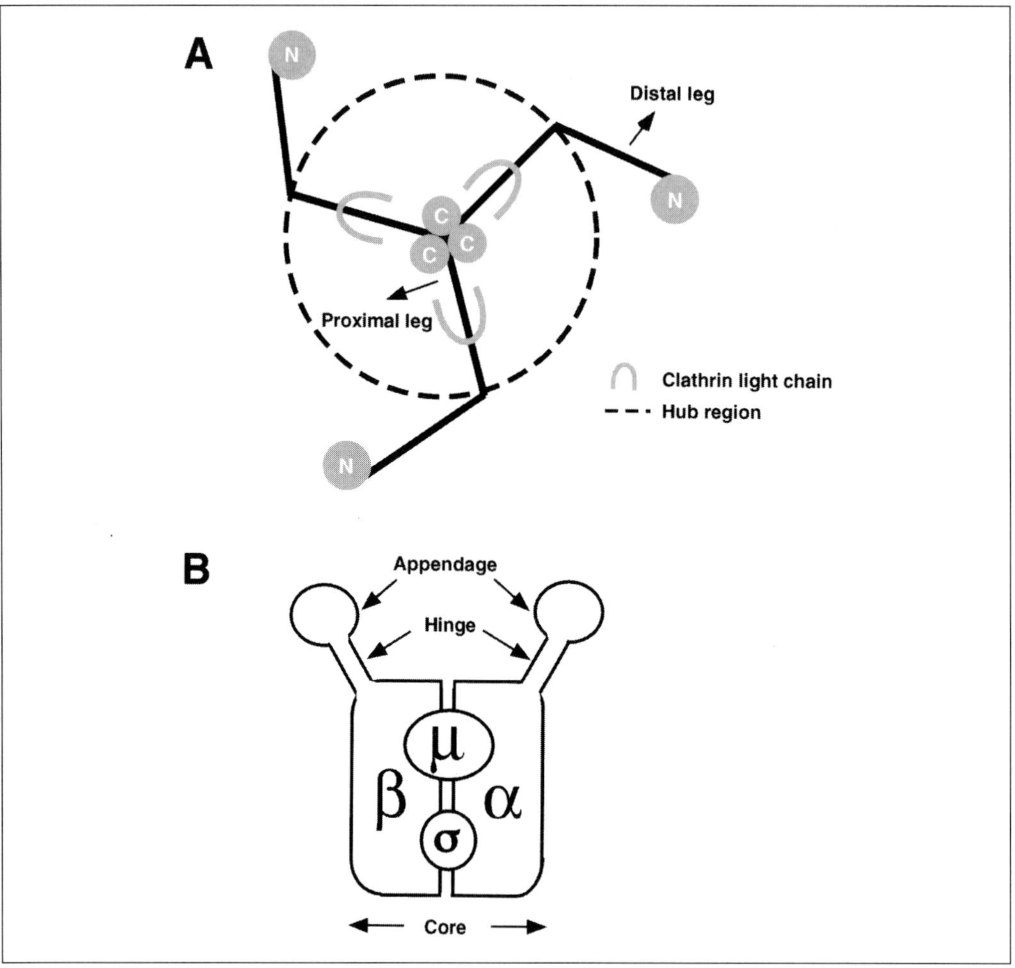

Fig. 2 Diagrams of the clathrin triskelion and the AP-2 complex. (A) The structural regions of the clathrin heavy chains and light chains and their distribution within the triskelion are indicated. Three clathrin heavy chains trimerize to form the basic triskelion structure. The globular terminal domain (N) comprises clathrin heavy chain residues 1–494, the distal leg comprises clathrin heavy chain residues 495–1073, and the hub is formed by clathrin heavy chain residues 1074–1521. The hub binds the light chain subunits and contains a region mediating trimerization near the C-terminus, residues 1522–1615 in the bovine clathrin heavy chain. The length of the C-terminal extension beyond the trimerization domain varies among the nine species for which the clathrin heavy chain sequence is known. Clathrin light chain variability is described in the text. The exact disposition of the light chains along the proximal leg is still to be determined, though there is biochemical evidence that they include a bend in the heavy chain binding region. The details of clathrin biochemistry have been reviewed (127). The crystallographic structures of some of these regions are shown in Plate 2. (B) The AP-2 complex is composed of four protein subunits of ~ 100 kDa (α subunit), ~ 100 kDa (β2 subunit), 50 kDa (μ2 subunit), and 17 kDa (σ2 subunit) and mediates the formation of endocytic clathrin-coated vesicles. The cartoon is based on the morphology observed in freeze-etch electron micrographs (128) and biochemical studies of subunit interactions (129). The appendage (ear) regions are the C-termini of the α and β subunits and the core region comprises their N-termini along with the other subunits. The hinge region of the β subunit interacts with clathrin, the μ subunit interacts with tyrosine-containing motifs in the cytoplasmic domains of receptors, and the core (head, trunk, or brick) region is responsible for intracellular localization of the AP complexes. The appendage (ear) region of the α subunit interacts with proteins involved in regulation of CCV formation. The crystallographic structures of portions of some of these subunits are shown in Plate 2.

those of purified coated vesicles, this finding led to the hypothesis that progressive clathrin polymerization at the plasma membrane could drive the invagination and budding of the vesicle through mechanical force. In addition, these initial studies demonstrated that clathrin assembly is regulated by additional proteins under physiological conditions in the cell. We now know that clathrin heavy chain trimers spontaneously assemble at physiological pH through formation of salt bridges. The light chain subunits interfere with this salt bridge formation through a conserved, negatively charged triplet of amino acids (17). This negative regulation ensures that physiological clathrin assembly depends on other cellular proteins, as described below.

At the time studies on the biochemistry of clathrin were initiated, Brown and Goldstein were establishing the mechanism of low density lipoprotein (LDL) uptake into cells, an effort that was recognized by the Nobel committee in 1986 (18). Over the course of these studies, the investigators showed that LDL is taken into cells by binding to receptors at the plasma membrane. The receptor–ligand complexes are highly clustered into coated membrane specializations and subsequently are transported into the cell in CCV. These findings firmly established CCV as a major route of internalization from the plasma membrane, and subsequent work has demonstrated the role of these structures in the uptake of a wide range of cell surface proteins, membrane constituents, and soluble factors. In some cases, the internalization proceeds without any apparent stimulus ('constitutive endocytosis'), whereas endocytosis of proteins such as growth factor receptors is modulated by ligand binding. The early studies pointed to the need for clathrin coats to select and accumulate the appropriate cargo within a coated vesicle, at the same time not enriching for or even excluding other plasma membrane components. The need for clathrin coats to provide selectivity and regulation in response to signals, in addition to their mechanical properties, further indicated that the coats are more complex than their morphology might suggest.

This hypothesis has been borne out by the discovery of a growing roster of coat-associated proteins since the late 1970s. The earliest additions to the coat protein family were made through the analysis of proteins isolated from purified coated vesicles in much the same way that clathrin itself was originally described. The presence of major coat components migrating at 100 kDa was revealed by chromatography and SDS–PAGE (4, 19). These polypeptides turned out to be the large subunits of another protein complex, termed the 'adaptor' or 'assembly' protein (AP). At least four distinct AP complexes have been described, designated AP-1, AP-2, AP-3, and AP-4, each with two large subunits of 100 kDa or more, one medium subunit of approximately 50 kDa, and a small 20 kDa subunit. Among the members of the family, one of the large subunits is distinct in sequence, whereas the other subunits are relatively highly conserved. By convention, the conserved subunits are named β (100 kDa), μ (50 kDa), and σ (20 kDa), and the variant subunit is given a unique Greek letter (γ for AP-1, α for AP-2, δ for AP-3, and ε for AP-4). Hence, the subunits of AP-2 are α, β2, μ2, and σ2. These subunits are arranged as a 'brick with ears': the 'brick' core is formed by the amino termini of the large subunits in complex with the μ and σ subunits; the carboxyl termini of the large chains form the globular appendages, 'ears'. The brick and ears are connected via flexible 'hinge' domains of the large subunits (Fig. 2B).

Although the organization of these APs is consistent across the family, they localize to different intracellular sites. The first two AP complexes identified, AP-1 and AP-2, are stoichiometric components of clathrin coats. AP-1 is associated with coats at the *trans*-Golgi network (TGN; the perinuclear clathrin population in Fig. 1C), while AP-2 is detected at the plasma membrane (Fig. 1D). Both AP-1 and AP-2 have the ability to trigger clathrin polymerization under physiological conditions. The extent to which AP-3 and AP-4 participate in CCV formation or whether they are components of novel types of coated vesicles is not yet clearly established. AP-4 lacks the clathrin-binding sequence identified for other AP complexes (20, 21), and there are variable reports of AP-3 binding to clathrin (22, 23). This chapter will focus on AP-2, the adaptor involved in clathrin-mediated endocytosis. Keen and others showed that purified AP-2 and clathrin associate *in vitro*, forming polymeric basket structures at physiological conditions with dimensions and morphology similar to those of CCV (19). Immuno-electron microscopy subsequently revealed the presence of adaptors in the interior of the vesicle coat, between the vesicle membrane and the surrounding clathrin lattice (24) (Plate 1B). Therefore, AP complexes are positioned to mediate the interaction of clathrin with the membrane and also to select the species for internalization, in addition to their biochemical function of stimulating coat assembly. Structural analyses indicate that the terminal domain of the clathrin heavy chain points into the interior of the coat, where it binds to the β2 hinge region of AP-2 (24–26). The AP-2 α subunit in turn is positioned to interact with the lipid membrane (27).

The enrichment of transmembrane proteins into CCV requires specific amino acid sequences in their cytoplasmic domains, discussed in the next section. The earliest illustration of an endocytosis signal was the description of the JD mutation of the LDL receptor, in which a critical tyrosine residue is replaced by a cysteine (28). Unlike the wild-type receptor, the JD mutant receptor is not clustered into plasma membrane coats and cannot mediate rapid clearance of LDL from the blood, resulting in a form of familial hypercholesterolemia. Similar tyrosine-containing internalization motifs have been identified in many transmembrane proteins, and other sequences are also believed to promote rapid internalization (see Chapter 7). A simple hypothesis is that these motifs are recognized by clathrin coat proteins, perhaps the AP complexes, so that the receptors will be concentrated into the growing clathrin coat lattice for internalization. During the 1980s, several attempts were made to demonstrate the interaction of AP-2 with transmembrane receptors, with a handful of suggestive reports. However, it was not until the applications of the yeast two-hybrid system and surface plasmon resonance in the mid-1990s that convincing *in vitro* evidence was produced showing the μ subunits of adaptors interacting with a class of internalization motifs in a specific manner (29). This work has been extended to identify several amino acid sequences that can serve as internalization signals, interacting with multiple sites on AP-2. While the binding of AP-2 to transmembrane proteins is accepted as a predominant mechanism to recruit cargo into coated vesicles, other coat components and possibly clathrin itself may fulfil this role in certain cases.

Additional cargo-recruiting proteins and an array of regulatory molecules involved in CCV formation, vesiculation, and uncoating have been identified over the past decade through the increasing use of recombinant techniques, the expansion of genetic databases, and the explosion of research on the formation and exocytosis of synaptic vesicles. The roles of these proteins are described in detail in the next section. Specifically, they include the arrestin protein family, which is involved in the recruitment of certain heterotrimeric G protein-coupled receptors (GPCR) into coated membranes (30) and the proteins dynamin and amphiphysin, which are involved in membrane invagination and/or vesicle budding in co-ordination with the assembly and growth of the clathrin lattice (31). Critical roles in clathrin-mediated endocytosis have also been demonstrated for intersectin (32), synaptojanin (33), Eps15 (34), and CALM/AP180 (formerly called AP-3) (35), which have all been implicated in clathrin coat

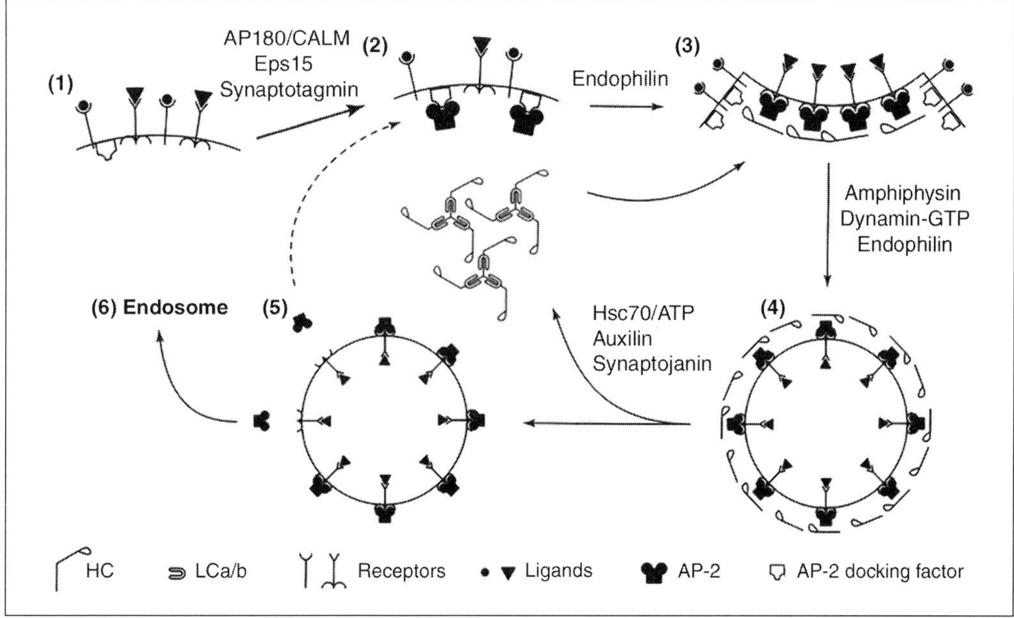

Fig. 3 Diagram of clathrin-mediated endocytosis. The sequence of events involved in clathrin-coated pit and vesicle formation is shown in numbered steps. The structural molecules involved are indicated in cartoon form, and the regulatory molecules are written in at the step where they function. The details of these processes and relevant references are discussed in the text and outlined, briefly, below. The diagram is modified from ref. 130, with permission copyright 1997 from Elsevier Science. (1) Receptors in the plasma membrane, with or without an AP-2 recognition motif. Uptake is generally ligand-independent for nutrient receptors and ligand-dependent for growth factor and G protein-coupled receptors. (2) AP-2 complexes interact with an unknown docking factor on the plasma membrane with high affinity. (3) AP-2 complexes interact with motifs in the cytoplasmic domains of receptors with low affinity and nucleate clathrin assembly. The AP-2 complexes in their membrane-bound form are dephosphorylated in their hinge region, so that they can bind clathrin. Clathrin assembly organizes the AP-associated receptors and sorts them from non-interactive receptors on the plasma membrane. Here the docking factor is shown as being excluded, but this is hypothetical. However, it must somehow remain primarily localized to the plasma membrane. (4) The fully polymerized clathrin-coated vesicle is excised from the membrane in a process requiring dynamin and associated regulatory molecules. (5) Clathrin and AP complexes are progressively uncoated from the internalized CCV, and the vesicle contents are fused into an early endosome. The AP complexes in the cytoplasm are phosphorylated at the clathrin-binding site, preventing non-productive clathrin assembly in the cytoplasm. The exact step at which AP-2 complex phosphorylation and dephosphorylation takes place is not yet established.

nucleation and recruitment of coat components. Recently, Schmidt *et al.* showed that the lysophosphatidic acid acyl transferase, endophilin I, is also required for vesicle formation, possibly as a consequence of alteration of the lipid bilayer geometry (36). After vesicle budding, the heat shock protein cognate hsc70 is believed to promote clathrin uncoating, in concert with the DnaJ homologue auxilin (37, 38). The uncoating factor hsc70 may also be required for coat assembly, serving the dual role of removing clathrin from endocytic vesicles and priming clathrin for another round of binding to the plasma membrane (39).

With so many new actors introduced during recent years, the mechanistic picture of endocytosis has become increasingly complex (Fig. 3). Endocytosis is not readily manipulated *in vivo*, so it is not straightforward to evaluate the hypotheses that have emerged from *in vitro* studies. Nevertheless, a reasonably detailed and plausible molecular model of internalization can be developed from the existing literature. We will now describe in greater depth the steps through which a CCV is generated, putting particular emphasis on the molecular mechanism of internalization and its regulation.

3. Mechanism of CCV formation during receptor-mediated endocytosis

Formation of endocytic CCV is a multi-step process orchestrated by a cohort of cellular proteins in a highly regulated manner. A great deal of knowledge regarding endocytosis comes from studying synaptic vesicle recycling, a process employed by the presynaptic neurons to ensure fast and efficient retrieval of membrane components for the biogenesis of new synaptic vesicles. Many neuron-specific proteins involved in this process have ubiquitous isoforms that may carry out similar functions in non-neuronal cells. In this section we will discuss the sequential events of CCV formation and incorporate recent findings that add on to the complexity of this process. The biogenesis of CCV during endocytosis can be divided into the following steps:

(a) Recruitment of coat components to the membrane, mediated through the interaction of AP-2 and accessory proteins with both lipids and membrane receptors.
(b) Assembly of clathrin coats to form deeply invaginated pits.
(c) Fission of vesicles by dynamin and other effector proteins.

Once released, the coated vesicles then go through a rapid uncoating process to disassemble the coat components. Each of the above steps requires an array of structural proteins and regulatory components that will be described in further detail below.

Multiple protein–lipid and protein–protein interactions required to recruit coat components to the membrane are being unveiled. In recent years, the importance of phosphoinositides in membrane trafficking has attracted much attention (40, 41). These phospholipids are suggested to play a critical role as recruitment signals for coat components and/or to modulate the interaction of coat proteins with receptor complexes. The major CCV component AP-2 has a high-affinity binding site for phosphoinositides in the N-terminal region of its α subunit crucial for its membrane targeting (42, 43). An attractive model is that modification of the plasma membrane to produce such phospholipids may create 'hot spots' for coat assembly. The existence of such hot spots is supported by recent studies from the Keen lab that illustrate the formation of clathrin-labelled structures repeatedly at the same regions of the plasma membrane (44). Contemporaneous work by De Camilli, Robinson, and others has suggested that the protein phosphatase synaptojanin and phospholipase D are important regulators of membrane composition during endocytosis (33, 45). Besides modification of membrane lipids, alterations in local cholesterol content may be another means by which the cell regulates coat assembly. In particular, plasma membranes depleted of cholesterol no longer support rapid uptake of transmembrane receptors, and the clathrin coats that form under these conditions appear 'frozen' in place (46, 47).

The specificity of cargo selection by CCV is attributed to the endocytic signals present on the cytoplasmic domains of the receptors (see Chapter 7). Several classes of these sorting signals have been extensively studied, including the NPXY, tyrosine, and di-leucine motifs (reviewed in refs 48, 49). Through yeast two-hybrid screening (29) and other biochemical approaches (50, 51), it is now widely accepted that the YXXØ is recognized by the μ subunit of adaptors. The di-leucine motif (in some cases one of the leucines is substituted by either methionine or isoleucine), was first identified in the cytoplasmic tail of CD3γ subunit of the T cell receptor (52) and was later found in many trafficking proteins such as the mannose-6-phosphate receptor, Limp-II, GLUT4, CD4, and HIV Nef. Binding of different adaptor complexes to the LL motif is more diverse and may not be generalized to a single subunit on all adaptors (53).

Interactions of AP-2 with receptors are relatively low-affinity. There is considerable evidence that targeting of AP-2 to specific sites on the plasma membrane requires another membrane component, a so-called 'docking protein' (27, 54, 55). Synaptotagmin, a neuron-specific transmembrane protein that binds to AP-2 through its C2B domain (56), was shown to serve this purpose during synaptic vesicle recycling. Peptides containing the tyrosine-based endocytic motif stimulate binding of AP-2 to synaptotagmin and enhance AP-2 recruitment to the plasma membrane (57). Because synaptotagmin isoforms are widely expressed in all tissues, it may be generally required for the recruitment of AP-2 to the plasma membrane.

Adaptor complexes are involved in multivalent protein–protein and lipid–protein interactions to initiate cargo recruitment and coated pit formation. There are other adaptor-like molecules that might also function to link cell surface receptors to CCVs. These proteins usually interact with clathrin or AP-2, and very frequently both. Some of them are receptor-specific, such as non-visual arrestins (β-arrestin and arrestin-3). Arrestins link β2-adrenergic receptors and some other GPCR to CCVs and are important for the endocyosis of these receptors (30, 58). They are hypothesized to work as molecular switches to co-ordinate receptor activation and internalization through phosphorylation/dephosphorylation events (59). AP180, a protein enriched in neuronal tissue, is capable of binding to clathrin and promotes cage assembly *in vitro* (35, 60). Mutation of the *Drosophila* AP180 homologue impairs synaptic vesicle endocytosis and alters the normal localization of clathrin in nerve terminals (61). The precise function of AP180 *in vivo* is still not clear, but it has been suggested that AP180 regulates the size of synaptic vesicles and also co-operates with AP-2 to achieve maximal efficiency to promote coat assembly (62). CALM (clathrin assembly lymphoid myeloid leukemia), a non-neuronal homologue of AP180 has been characterized (63). CALM interacts primarily with clathrin heavy chain, and overexpression of either full-length CALM or its clathrin-binding domain significantly inhibits receptor-mediated endocytosis (64). These results indicate that the AP180 family proteins might have a general function in clathrin-mediated endocytosis in all tissues. Synaptic vesicle biogenesis is a highly specialized process, and it remains to be determined whether CALM exerts the same effect on endocytosis in non-neuronal cells. Another AP-2 associated CCV component that has attracted much attention is Eps15. Eps15 was originally identified as a major cytosolic substrate for ligand-activated EGFR tyrosine kinase (65). Eps15 is organized in three distinct structural domains: an N-terminal region with the EH (Eps15 homology) domains (66), a central coiled-coil domain involved in Eps15 oligomerization (67), and a C-terminal AP-2 binding domain with multiple DPF repeats (68). Perturbation of Eps15 function in mammalian cells inhibits receptor-mediated endocytosis of both EGF and transferrin (34). The EH domain of Eps15 interacts with another protein called epsin (69). Although both Eps15 and epsin are crucial for endocytosis, the exact mechanism for their function remains to be determined. Together with epsin, Eps15 can either play a role in the dynamic rearrangement of coat assembly during bud invagination and fission (70) or assist AP-2 recruitment to specific sites on the plasma membrane (71). Because of its close association with pathways linked to cell proliferation, Eps15 could also regulate coated vesicle formation in response to various signalling events. This notion is supported by the potential of Eps15 to transform NIH-3T3 cells (65) and the fact that association of epsin and Eps15 with AP-2 is regulated by mitotic signal-induced phosphorylation events (72).

The ability of clathrin to assemble into closed basket-like structures is believed to be the driving force behind membrane deformation leading to vesicle budding. Under physiological conditions, the assembly reaction is triggered through the nucleation effect of adaptors. In the cell, the assembly reaction is restricted to the membrane by phosphorylation of the β subunit of AP complexes (73). When AP complexes are free in the cytoplasm, their β subunits are phosphorylated on serines in the clathrin-binding region, preventing their interaction with clathrin. When AP complexes are associated with the plasma membrane, the β subunits are dephosphorylated, by the action of a yet unidentified phosphatase, allowing AP complexes to stimulate assembly of the clathrin lattice. The clathrin lattice then undergoes an unknown molecular rearrangement to acquire the curvature that is essential for enclosing the vesicles. The protein dynamin subsequently mediates the release of invaginated coated pits from the membrane. Dynamin is a 100 kDa GTPase with at least three different isoforms and more than 25 splicing variants in vertebrates (74). The *Drosophila* dynamin mutant *Shibire* and cells expressing GTPase-defective dynamin isoforms exhibit endocytosis defects that correlate with a block in vesicle scission (75). Through its interaction with amphiphysin (which binds both clathrin and AP-2), dynamin is recruited to coated pits. *In vitro*, amphiphysin can transform spherical liposomes into narrow tubules, and co-assembly of amphiphysin and dynamin into ring-like structures promotes the liposome-fragmenting activity of dynamin (31). Dynamin has long been considered as a force generator for vesicle fission, and purified recombinant dynamin binds to acidic lipid vesicles to form helical tubes and constrict them upon the addition of GTP (76). Recently, however, the mechanism by which dynamin carries out the fission event *in vivo* has been re-evaluated. Instead of being the 'pinchase' to execute scission, dynamin may act as a switch to regulate downstream effectors that mediate the scission step (77). One possible candidate for this effector system is endophilin I, a presynaptic protein that binds to the proline-rich region of dynamin through its SH3 domain (78). Endophilin 1 exhibits lysophosphatidic acid acyl transferase activity and can change the membrane curvature by converting an inverted-cone-shaped lipid into a cone-shaped lipid in the cytoplasmic leaflet of the bilayer. Through this action, it is then hypothesized that endophilin 1 works with dynamin to mediate vesicle invagination and fission from the plasma membrane (36). Future experiments are needed to reconcile the above results and to clarify the molecular mechanism of dynamin-mediated fission.

Once they are released into the cytosol, the vesicles are quickly uncoated in an ATP-dependent manner by Hsc70 (37, 79, 80) and its associated protein auxilin (81). This step allows the recycling of coat components for multiple rounds of endocytosis, and enables nascent vesicles to fuse with endosomal compartments. Hsc70, a chaperone protein of the hsp70 family, uses ATP hydrolysis to remove the clathrin coat from CCVs. Besides its uncoating ATPase function, Hsc70 was recently suggested to have another role in priming free clathrin to interact with vesicle membranes (39). Auxilin contains a DnaJ domain and is required specifically to release clathrin from brain vesicles (81). Another protein that is suggested to participate in uncoating is synaptojanin. It is a poly-phosphoinositide phosphatase involved in regulating the interactions between coat proteins and the lipid bilayer and/or a pool of actin implicated in endocytosis. Synaptojanin knock-out mice have an increased number of CCVs in their nerve terminals and show symptoms of enhanced synaptic depression. This indicates that synaptojanin may function physiologically in the uncoating process (33).

4. Regulation of clathrin-mediated endocytosis

A complex process such as endocytosis will be regulated at many levels and many stages, and no single master regulator is anticipated. In this section, we will focus on the events and conditions that determine when and where coated vesicle formation will occur. Excepting mitosis, endocytosis seems to proceed efficiently irrespective of the environment and stimuli that the cell is experiencing. Nevertheless, there are signals that promote or modulate endocytic activity, particularly at the level of clathrin coat formation and cargo recruitment.

As mentioned previously, assembly of a clathrin coat with AP-2 occurs only at the plasma membrane. In fact, there appear to be specific loci of coat assembly, membrane patches that support coat formation repeatedly over multiple cycles (44). The exact composition and mechanisms behind these 'hot spots' is unknown. It is proposed that coat assembly is triggered by the binding of AP-2 to docking factors clustered in plasma membrane subdomains, although these clusters have not been demonstrated. The AP2 bound to these sites needs to be dephosphorylated to be able to trigger clathrin assembly, but whether dephosphorylation is a requirement for recruitment or occurs upon recruitment is not yet established (73). The binding of AP-2 to the docking factor would be an early event initiating coat assembly, after which the natural tendency of coat proteins to oligomerize would allow the growth of the coat protein lattice. By this mechanism, only a few copies of the docking factor would need to be active to promote the assembly of a complete coat.

Much recent work has focused on the relationship of coat assembly to the cytoskeleton and to the lipid bilayer. Phosphoinositides are important regulators of the actin cytoskeleton (82), and synaptojanin may also participate in controlling actin dynamics at the synapse. The effects of actin dynamics on endocytosis have been investigated most directly in yeast, and caution must be exercised before extending these observations to mammalian cells (83). In yeast, endocytosis is critically dependent on actin filaments, and many yeast gene products have dual functions for endocytosis and actin organization (84, 85). Cytoskeletal structures may help to organize the endocytic machinery at specific sites or 'hot spots' on the plasma membrane, to provide the force to drive membrane fission, or to promote vesicle movement into the cytoplasm. The interplay between endocytosis and the organization of the actin cytoskeleton and the microtubule network is only beginning to be explored. However, it seems very clear now that the dynamics of these networks must be regulated to allow vesicle formation and subsequent trafficking. Two possibilities are yet to be distinguished: the actin cytoskeleton may need to be rearranged to make room for clathrin coat formation; or cortical actin may provide the organization of the plasma membrane such that sites of coat formation are properly localized and maintained. Microtubules do not appear to have a role in these early stages of endocytosis, rather acting downstream during endosomal sorting (86).

We have pointed out previously the importance of the lipid bilayer in endocytosis; this is nowhere so clear as at the plasma membrane. If plasma membranes are depleted of cholesterol, assembled clathrin coats do not lead to productive vesiculation; instead, the coats remain apparently immobile over many seconds, as if they have been blocked in an intermediate stage of internalization (46, 47). Cholesterol is known to have profound effects on membrane fluidity, and the overall cholesterol content and distribution within a given cell type are tightly regulated. It is reasonable to suggest that membranes depleted of cholesterol are insufficiently fluid (or rigid) and cannot adopt the bilayer conformation necessary to proceed with endocytosis. It is well established *in vitro* that clathrin coat proteins and other endocytic factors have an affinity for acidic membrane phospholipids. However, it has not been directly demonstrated that this class of lipids is enriched at the sites of coat assembly or whether lipid composition at coat assembly sites is regulated.

The cargo molecules themselves may regulate coat formation in some circumstances. A long-standing question in endocytosis research is whether transmembrane proteins promote coat assembly, or if instead these proteins are recruited and immobilized within preformed coats (87). Many receptors, such as the transferrin receptor, are internalized efficiently irrespective of ligand binding, typically having a half-time at the plasma membrane of a few minutes. These receptors and similar transmembrane proteins are unlikely to exert any direct control over endocytosis. This would not contradict the high degree of enrichment of the transferrin receptor in clathrin coats. Given that many transmembrane proteins are dimeric at the cell surface, and given the extensive network of protein interactions that comprises a clathrin coat, even a low-affinity binding of a cargo protein with AP-2 should mediate the concentration of the cargo in the coat and its depletion from the rest of the membrane. For receptors that are internalized efficiently only in response to ligand binding, a more active role in coat assembly is proposed. There are two classes of receptors in this category, the receptor tyrosine kinases and the GPCR. The epidermal growth factor (EGF) receptor has served as a classic model system for internalization of receptor tyrosine kinases (88). This receptor carries at least two consensus internalization sequences that are cryptic in the unoccupied receptor. Upon growth factor binding, a phosphorylation cascade results in the exposure of the internalization sequences, Eps15 and AP-2 binding, and receptor internalization. In addition, EGF stimulates an increase in the proportion of clathrin bound to the membrane versus the cytosolic pool (89). Moreover, clathrin is a substrate for phosphorylation by Src kinase that is activated by EGF binding to its receptor. The site at which clathrin heavy chain is phosphorylated is in the light chain-binding region, so phosphorylation may influence clathrin assembly and/or recruitment. Inhibition of clathrin phosphorylation upon EGF stimulation blocks the recruitment of clathrin to the plasma membrane and changes the kinetics of EGF uptake (89). These data have led to the proposal that ligand binding and signal transduction allow the EGF receptor to initiate coat assembly and promote its own internalization. Some GPCR initiate their internalization upon ligand binding through non-visual arrestins, which rely on ligand-dependent phosphorylation of the receptor (59). As discussed above, these arrestins can potentially replace or enhance the clathrin-binding function of the AP-2 complex. It is not unreasonable that a growth factor or signalling receptor would have a mechanism to stimulate rapid uptake and subsequent down-regulation of its signalling activity. In contrast, nutrient receptors such as the transferrin receptor need not be controlled so tightly under normal circumstances, so their internalization can be dependent on the availability of 'pre-assembled' clathrin coats at the plasma membrane.

In vitro studies raise the possibility that other parts of the internalization machinery are regulated by phosphorylation. The protein interactions of dynamin, synaptojanin, and amphiphysin are all inhibited by their phosphorylation, indicating that these factors and the vesicle budding step may be a primary target of regulatory mechanisms (90). Quite possibly, the influence of phosphorylation on coat protein dynamics will be complex, with phosphorylation or dephosphorylation at distinct sites leading to varied or even opposing consequences.

A new player in the regulation of endocytosis is the 20 kDa GTPase, Rab5. Rab5 has been extensively studied in the context of endosomal trafficking, and it is well accepted that Rab5 is critical for the fusion of early endosomes with internalized vesicles and with one another. However, a recent report has demonstrated that Rab5 is also involved in clathrin coat assembly (91). In complex with its guanine nucleotide dissociation inhibitor (GDI), Rab5 appears to be required for membrane invagination. The target of Rab5-GDI at the clathrin-coated plasma membrane is unknown, and it remains to be shown that Rab5-GDI acts directly at this step in internalization.

5. Advances in the resolution of coat protein structures

Recently, an exciting development in the study of clathrin-mediated endocytosis has originated from the progress made in resolving the molecular structures of proteins involved in this process (reviewed in refs 92, 93) (Plate 2). CCV formation is a complex event that depends on and is regulated by a number of cytosolic factors that are recruited through various protein–protein and protein–lipid interactions. Specific structural determinants in each protein contribute to these intermolecular recognitions. The crystal structure of clathrin, the basic structural element of coated vesicles, is being revealed through a 'divide and conquer' approach (94). The first 330 amino acids of the N-terminal globular domain of clathrin heavy chain, which interacts with adaptor β subunits and a number of other adaptor-related proteins such as arrestins, forms a seven-bladed β-propeller (95) (Plate 2). The sequence that is responsible for arrestin binding, previously mapped by deletion analysis (96), lies on the margins of the grooves between blades 1 and 2. The clathrin-binding motif on AP-2 resembles that of arrestins, but further structural analysis on the AP-2 β subunit is required to provide a more definite picture of the interaction between clathrin and AP-2. A second breakthrough for resolving the clathrin structure comes from the crystallographic analysis of the proximal fragment of clathrin heavy chain (97) (Plate 2). This is the region involved in clathrin assembly and light chain binding (7, 8). The structure is an elongated right-handed superhelix made of short α helices linked by loops. Alignment analyses reveal a 145-residue domain, termed the CHCR (clathrin heavy chain repeat) motif, that is repeated seven times along the filamentous leg and appears in other proteins involved in vacuolar sorting in yeast. The structure indicates that the physical association of two proximal regions during clathrin self-assembly likely requires helix-face/helix-face contacts, facilitated by the complementary packing of helix ridges and grooves. While the balance of the structure of the clathrin molecule is under analysis, it is now feasible to begin to fit the known structures into the 21 Å resolution map created from image reconstruction of electron micrographs of assembled clathrin (Plate 1A) (15). This approach has been demonstrated by localization of the terminal domain structure within the lattice (98) and is being pursued for the filamentous legs to establish molecular interactions during clathrin assembly.

Recruitment of transmembrane proteins into clathrin coats is mediated by the interaction between AP-2 and the sorting motifs within the transmembrane protein cytoplasmic domains. For the YXXØ motif, the interaction is mediated by the μ2 chain of AP-2. Co-crystallization of the signal-binding portion of μ2 and several YXXØ containing peptides has provided a structural insight into this binding specificity (99)

(Plate 2). The μ2 fragment is an elongated banana-shaped β sheet structure. The YXXØ sequence binds to the surface of the two parallel β strands of μ2 such that the side chains of the Y and Ø residues are oriented into two hydrophobic pockets in the surface of μ2. The hydroxyl group of tyrosine contributes to multiple interactions with μ2 residues forming the hydrophobic pocket, explaining why substitution of the tyrosine with phenylalanine disrupts the endocytic signal (100). In addition to μ2, the crystal structure of another AP-2 domain, the appendage of the α subunit (aa 701–938), has been solved (101, 102) (Plate 2). Several proteins involved in endocytosis interact with this domain, including dynamin (103), amphiphysin (104), Eps15 (105), epsin (69), AP180, and auxilin (101). The structure contains an N-terminal nine-stranded β sandwich subdomain, which functions as a scaffold and spacer that supports a C-terminal five-stranded β sheet platform to which various proteins bind. The binding site apparently recognizes DPF/W sequences present in its binding partners. The number of DPF/W motifs present in each ligand is correlated with its affinity for the appendage domain. This is consistent with a model in which the α-appendage is a central co-ordinator of many of the protein–protein interactions necessary for the biogenesis of CCV during endocytosis. A single site for binding multiple ligands would allow for temporal and spatial regulation of coated vesicle assembly machinery.

In addition to the structural motifs of the CCV scaffold proteins, a variety of structural modules are used by proteins that bind to clathrin or adaptors to perform their functions in regulating coat assembly. The molecular structures of some of these domains involved in endocytosis have been characterized and are described in the following.

(a) **The PH (pleckstrin homology) domain**. This is a domain commonly used by cellular proteins to associate with phospholipids. The PH domain of dynamin is responsible for its membrane association (106) and is vital for dynamin function. The atomic structure of the lipid-binding PH domain of dynamin consists of seven β strands forming two roughly orthogonal antiparallel β sheets, and it is closed at one corner by a C-terminal α helix (107, 108). Co-crystallization of inositol triphosphate with a similar PH domain from phospholipase C further indicates that inositide-1,4,5-triphosphate interacts with the β1/β2 and β3/β4 loops of the PH domain (109).

(b) **The SH3 (src homology 3) domain**. SH3 domains are small modules found in a diverse array of proteins. The presence of an SH3 domain confers upon its resident protein the ability to interact with specific proline-rich domains (PRD) of other proteins. Several SH3 domain-containing proteins have been implicated in distinct steps during CCV formation, including amphiphysin, endophilin, intersectin, and syndapin (32). Amphiphysin binds to AP-2, dynamin, and clathrin simultaneously through different domains, and it has been implicated in recruiting dynamin and regulating its oligomerization (31). The SH3 domain of an amphiphysin isoform, amphiphysin II, contains an n-SRC loop that is at least partially responsible for the prevention of dynamin assembly (110).

(c) **The EH (Eps15 homology) domain**. This is another structure motif that is multiply linked to protein–protein interactions in CCV formation. The EH domain of Eps15 is composed of three 100 amino acid repeats located at the N-terminus. An Eps15 mutant lacking the EH domain no longer associates with the plasma membrane, and it impairs coated pit formation in transfected cells (71). NMR studies indicate that the EH domain consists of a pair of EF hand motifs, the second of which binds tightly to calcium (111). The EH domain recognizes the NPF sequence, which also exists in several proteins implicated in endocytosis, including epsin (69), CALM, and non-neuronal isoforms of synaptojanin (112). The NPF peptide is bound in a hydrophobic pocket between two α helices, and binding is mediated by a crucial aromatic interaction. Binding between the EH domain and the NPF peptide is of low affinity (Kd $\approx$ 560 $\pm$ 40 μM), but this interaction can be strengthened by multiple motifs found on both EH domain-containing proteins and their NPF-containing target proteins. It is also worth noting that it is common for one protein to have different motifs, thus increasing the complexity of the interaction network between these CCV-associated proteins. These multivalent associations could serve the purpose to fine-tune the endocytosis process to achieve greater efficiency and co-ordination.

6. Methods for studying clathrin-mediated endocytosis

The study of CCV was originated with morphological observations some thirty years ago, and followed by vast amounts of biochemical characterization of the coat components. Many aspects regarding the mechanism for CCV formation have been revealed; however, some key issues remain unanswered. Two common approaches have been developed to dissect the process of coated-vesicle formation *in vitro*:

(a) The use of isolated membrane fractions to study coat protein binding and recruitment (113).

(b) The use of semi-intact or perforated cells to reconstitute vesicle formation and budding (reviewed in ref. 114).

The principle of the first method is to reconstruct the coat assembly process by incubating purified membranes that have been stripped of endogenous coated pits with cytoplasm extracted from cultured cells. Through a series of experiments, this assay has established that both AP-2 and clathrin assemble from the cytosol onto saturable, high-affinity and protease-sensitive binding sites on the plasma membrane. Binding of coat proteins to the membrane can occur at 4 °C, but invagination is temperature-dependent. In turn, budding requires ATP and certain cytosolic factors. The second approach was developed to measure vesicle formation directly. Plasma membranes of the cells are ruptured by mechanical shear, and the sequestration of receptor-bound ligands into either constricted coated pits or fully-sealed coated vesicles is determined by the inaccessibility of the ligand to exogenously added probes (reviewed in ref. 114). This assay has facilitated dissection of early-stages of

CCV formation into biochemically distinct events. Another approach to reconstitute the coat assembly process *in vitro* is to use protein-free liposomes. In this scheme, liposomes made from defined lipid compositions are used as templates and purified coat components are then added to initiate vesiculation (76, 115). Since this system employs highly purified components, it can potentially be used to dissect both the lipid preference for coat assembly and the minimal requirement of protein components for each step. For example, purified recombinant dynamin can bind to liposomes made from anionic phospholipids and form helical tubular structures. Addition of GTP causes a rapid alteration in structure, resulting in constriction and fragmentation of the dynamin tubes. This result was used to support a previous notion that dynamin is a GTP-dependent force generator for vesicle formation. However, recent data using various dynamin mutants have challenged this theory. Instead of being the 'pinchase' to sever the vesicles, it is suggested that dynamin acts as a regulator for such enzymatic activity (77). The apparent disparities between these results highlight the importance to confirm the validity of *in vitro* data with *in vivo* experiments.

Researchers have employed the following approaches to study CCV formation *in vivo* over the years:

- genetic manipulation
- application of dominant negative mutants
- using fluorescently tagged molecules to trace assembly dynamics

Most clathrin-related coat proteins mentioned in this chapter have homologues in *Saccharomyces cerevisiae*, and the availability of various yeast deletion or temperature-sensitive mutants greatly contributes to the research of endocytosis (84, 85). There are also mutants related to CCV function developed in *Dictyostelium discoideum* (116), *Caenorhabditis elegans* (117), and *Drosophila melanogaster* (118). In mammals, however, such manipulation is usually restricted to proteins that are not involved in ubiquitous trafficking pathways since absence of such proteins is often lethal. To dissect the functions of these proteins in mammalian cells, another strategy is to introduce dominant negative mutants. One of the proteins subjected to this mutagenesis approach is dynamin. The dynamin K44A mutant, which is defective in its GTPase activity, is widely used to block vesicle budding *in vivo* (119). Because dynamin is also suggested to participate in similar fission events of internalization through caveolae (120, 121), additional mutants for clathrin-specific endocytosis are needed. To this end, transfection of the hub domain of clathrin into target cells produces a dominant negative clathrin phenotype. The hub domain is the centre portion of the clathrin triskelion, and it contains light chain binding and assembly domains (8). Overexpressed hub proteins sequester the regulatory light chain subunits from endogenous clathrin heavy chains and result in non-productive assembly of heavy chains on the membrane (122). Mutants with similar principles were later developed in μ2 (123), Eps15 (71), and epsin (72). The dynamics of coated pit formation has been a challenging topic in studying endocytosis. Recently, a technique combining green fluorescent protein (GFP)-tagged clathrin light chain and live cell microscopy was developed to observe clathrin coat assembly in real time (44). This approach has revealed an important relationship between the structural organization of clathrin-coated pits and the membrane skeleton, and should be useful in deciphering the temporal and spatial control of clathrin-mediated endocytosis in the future.

7. Conclusion

Over the last several decades, the study of clathrin-mediated endocytosis has evolved considerably. As we have described, the early focus of CCV research was the demonstration that these vesicles are intermediates in the internalization of biomolecules. Morphological and biochemical methods brought a simple and compelling model for the mechanism of coated vesicle formation: coat adaptors recruit receptors into coated areas of the plasma membrane, and the surrounding clathrin lattice forces the deformation of the membrane and eventually vesicle budding. The efficient self-association of the clathrin triskelion and the ability of AP-2 to bind to clathrin and to membrane components are consistent with the parameters of this model.

Initially, this picture seemed to satisfy the requirements for endocytosis. However, as the role of the coated vesicle in endocytosis became known in greater detail, the roster of cellular processes and proteins that depend on CCV steadily grew. Clathrin is now appreciated to be involved in the internalization of diverse biomolecules, in some cases through tightly regulated mechanisms. The need for CCV formation to be flexible and selectively controlled indicated that the early model would prove an incomplete one. Indeed, attempts to recapitulate the processes of coat assembly and vesiculation *in vitro* with purified components were unsuccessful. Clearly, these negative observations pointed to the importance of novel factors, as we have described.

Whereas clathrin and AP-2 are stoichiometric coat components that are readily purified and studied, the identification of new players has required new techniques in conjunction with protein biochemistry and cell biology. Neurobiology and signal transduction research contributed greatly to the characterization of such molecules as dynamin, Eps15, and endophilin. We now recognize the importance of many more factors in coated vesicle formation. However, these new players need to be placed correctly in the overall endocytic process. Current endocytosis research relies increasingly on describing the structures of these molecules to the atomic level. Also, recombinant techniques and genetic data bases have been applied to design mutants that disrupt endocytosis in specific ways, allowing researchers to place the mutated proteins and their interacting partners more accurately along the pathway.

The current model of how endocytosis is achieved still maintains the central roles of clathrin and AP-2, as proposed a generation ago. The additional factors allow CCV to accommodate a broad array of cargo molecules and to mediate endocytosis in response to specific stimuli. Echoing the importance of clathrin-mediated endocytosis in many areas of cell biology, coated vesicle research now depends heavily on the methods and concepts of all aspects of modern biology. Through these efforts, we now appreciate the complexity of the endocytic machinery, and we can anticipate the day when CCV formation will be thoroughly understood in molecular terms.

References

1. Roth, T. F. and Porter, K. R. (1964) Yolk protein uptake in the oocyte of the mosquito *Aedes aegypti. L. J. Cell Biol.*, **20**, 313.
2. Kaneseki, T. and Kadota, K. (1969) The 'vesicle in a basket'. A morphological study of the coated vesicle isolated from the nerve endings of the guinea pig brain, with special reference to the mechanism of membrane movements. *J. Cell Biol.*, **42**, 202.
3. Pearse, B. M. F. (1975) Coated vesicles from pig brain: purification and biochemical characterization. *J. Mol. Biol.*, **97**, 93.
4. Ungewickell, E. and Branton, D. (1981) Assembly units of clathrin coats. *Nature*, **289**, 420.
5. Kirchhausen, T. and Harrison, S. C. (1981) Protein organization in clathrin trimers. *Cell*, **23**, 755.
6. Heuser, J. E. and Kirchhausen, T. (1985) Deep-etch view of clathrin assembly. *J. Ultrastruct. Res.*, **92**, 1.
7. Näthke, I. S., Heuser, J., Lupas, A., Stock, J., Turck, C. W., and Brodsky, F. M. (1992) Folding and trimerization of clathrin subunits at the triskelion hub. *Cell*, **68**, 899.
8. Liu, S.-H., Wong, M. L., Craik, C. S., and Brodsky, F. M. (1995) Regulation of clathrin assembly and trimerization defined using recombinant triskelion hubs. *Cell*, **83**, 257.
9. Brodsky, F. M., Hill, B. L., Acton, S. L., Näthke, I., Wong, D. H., Ponnambalam, S., *et al.* (1991) Clathrin light chains: Arrays of protein motifs that regulate coated vesicle dynamics. *Trends Biol. Sci.*, **16**, 208.
10. Wong, D. H., Ignatius, M. J., Parosky, G., Parham, P., Trojanowski, J. Q., and Brodsky., F. M. (1990) Neuron-specific expression of high molecular weight clathrin light chain. *J. Neurosci.*, **10**, 3025.
11. Acton, S. and Brodsky, F. M. (1990) Predominance of clathrin light chain LCb correlates with the presence of a regulated secretory pathway. *J. Cell Biol.*, **111**, 1419.
12. Silveira, L. A., Wong, D. H., Masiarz, F. R., and Schekman, R. (1990) Yeast clathrin has a distinctive light chain that is important for cell growth. *J. Cell Biol.*, **111**, 1437.
13. Warner, T. S., Sinclair, D. A., Fitzpatrick, K. A., Singh, M., Devlin, R. H., and Honda, B. M. (1998) The light gene of *Drosophila melanogaster* encodes a homologue of VPS41, a yeast gene involved in cellular-protein trafficking. *Genome*, **41**, 236.
14. Ungewickell, E. and Ungewickell, H. (1991) Bovine brain clathrin light chains impede heavy chain assembly *in vitro*. *J. Biol. Chem.*, **266**, 12710.
15. Smith, C. J., Grigorieff, N., and Pearse, B. M. F. (1998) Clathrin coats at 21 Å resolution: A cellular assembly designed to recycle multiple membrane receptors. *EMBO J.*, **17**, 4943.
16. Keen, J. H., Willingham, M. C., and Pastan, I. H. (1979) Clathrin-coated vesicles: Isolation, dissociation, and factor-dependent reassociation of clathrin baskets. *Cell*, **16**, 303.
17. Ybe, J. A., Greene, B., Liu, S.-H., Pley, U., Parham, P., and Brodsky, F. M. (1998) Clathrin self-assembly is regulated by three light chain residues controlling the formation of critical salt bridges. *EMBO J.*, **17**, 1297.
18. Anderson, R. G. W., Brown, M. S., and Goldstein, J. L. (1977) Role of the coated endocytic vesicle in the uptake of receptor-bound low density lipoprotein in human fibroblasts. *Cell*, **10**, 351.
19. Zaremba, S. and Keen, J. H. (1983) Assembly polypeptides from coated vesicles mediate reassembly of unique clathrin coats. *J. Cell Biol.*, **97**, 1339.
20. Dell'Angelica, E. C., Mullins, C., and Bonifacino, J. S. (1999) AP-4, a novel protein complex related to clathrin adaptors. *J. Biol. Chem.*, **274**, 7278.

21. Hirst, J., Bright, N. A., Rous, B., and Robinson, M. S. (1999) Characterization of a fourth adaptor-related protein complex. *Mol. Biol. Cell*, **10**, 2787.
22. Dell'Angelica, E. C., Klumperman, J., Stoorvogel, W., and Bonifacino, J. S. (1998) Association of the AP-3 complex with clathrin. *Science*, **280**, 431.
23. Simpson, F., Peden, A. A., Christopoulou, L., and Robinson, M. S. (1997) Characterization of the adaptor-related protein complex, AP3. *J. Cell Biol.*, **137**, 835.
24. Vigers, G. P., Crowther, R. A., and Pearse, B. M. F. (1986) Location of the 100 kd-50 kd accessory proteins in clathrin coats. *EMBO J.*, **5**, 2079.
25. Shih, W., Gallusser, A., and Kirchhausen, T. (1995) A clathrin-binding site in the hinge of the β2 chain of mammalian AP-2 complexes. *J. Biol. Chem.*, **270**, 31083.
26. Greene, B., Liu, S.-H., Wilde, A., and Brodsky, F. M. (2000) Complete reconstitution of clathrin basket formation with recombinant protein fragments: Adaptor control of clathrin self-assembly. *Traffic*, **1**, 69.
27. Chang, M. P., Mallet, W. G., Mostov, K. E., and Brodsky, F. M. (1993) Adaptor self-aggregation, adaptor-receptor recognition and binding of α-adaptin subunits to the plasma membrane contribute to recruitment of adaptor (AP2) components of clathrin-coated pits. *EMBO J.*, **12**, 2169.
28. Anderson, R. G. W., Brown, M. S., and Goldstein, J. L. (1977) A mutation that impairs the ability of lipoprotein receptors to localize in coated pits on the cell surface of human fibroblasts. *Nature*, **270**, 695.
29. Ohno, H., Stewart, J., Fournier, M.-C., Bosshart, H., Rhee, I., Miyatake, S., *et al.* (1995) Interaction of tyrosine-based sorting signals with clathrin-associated proteins. *Science*, **269**, 1872.
30. Goodman, O. B., Jr., Krupnick, J. G., Santini, F., Gurevich, V. V., Penn, R. B., Gagnon, A. W., *et al.* (1996) β-arrestin acts as a clathrin adaptor in endocytosis of the β_2-adrenergic receptor. *Nature*, **383**, 447.
31. Takei, K., Slepnev, V. I., Haucke, V., and De Camilli, P. (1999) Functional partnership between amphiphysin and dynamin in clathrin-mediated endocytosis. *Nat. Cell Biol.*, **1**, 33.
32. Simpson, F., Hussain, N. K., Qualmann, B., Kelly, R. B., Kay, B. K., McPherson, P. S., *et al.* (1999) SH3-domain-containing proteins function at distinct steps in clathrin-coated vesicle formation. *Nat. Cell Biol.*, **1**, 119.
33. Cremona, O., Di Paolo, G., Wenk, M. R., Lüthi, A., Kim, W. T., Takei, K., *et al.* (1999) Essential role of phosphoinositide metabolism in synaptic vesicle recycling. *Cell*, **99**, 179.
34. Benmerah, A., Lamaze, C., Bègue, B., Schmid, S. L., Dautry-Varsat, A., and Cerf-Bensussan, N. (1998) AP-2/Eps15 interaction is required for receptor-mediated endocytosis. *J. Cell Biol.*, **140**, 1055.
35. McMahon, H. T. (1999) Endocytosis: an assembly protein for clathrin cages. *Curr. Biol.*, **9**, R332.
36. Schmidt, A., Wolde, M., Thiele, C., Fest, W., Kratzin, H., Podtelejnikov, A. V., *et al.* (1999) Endophilin I mediates synaptic vesicle formation by transfer of arachidonate to lysophosphatidic acid. *Nat. Cell Biol.*, **401**, 133.
37. Rothman, J. E. and Schmid, S. L. (1986) Enzymatic recycling of clathrin from coated vesicles. *Cell*, **49**, 5.
38. Ungewickell, E., Ungewickell, H., and Holstein, S. E. (1997) Functional interaction of the auxilin J domain with the nucleotide- and substrate-binding modules of Hsc70. *J. Biol. Chem.*, **272**, 19594.
39. Jiang, R., Gao, B., Prasad, K., Greene, L. E., and Eisenberg, E. (2000) Hsc70 chaperones clathrin and primes it to interact with vesicle membranes. *J. Biol. Chem.*, **275**, 8439.

40. De Camilli, P., Emr, S. D., McPherson, P. S., and Novick, P. (1996) Phosphoinositides as regulators in membrane traffic. *Science*, **271**, 1533.
41. Martin, T. F. J. (1997) Phosphoinositides as spatial regulators of membrane traffic. *Curr. Opin. Neurobiol.*, **7**, 331.
42. Gaidarov, I., Chen, Q., Falck, J. R., Reddy, K. K., and Keen, J. H. (1996) A functional phosphatidylinositol 3,4,5-trisphosphate/phosphoinositide binding domain in the clathrin adaptor AP-2 alpha subunit. Implications for the endocytic pathway. *J. Biol. Chem.*, **271**, 20922.
43. Gaidarov, I. and Keen, J. H. (1999) Phosphoinositide-AP-2 interactions required for targeting to plasma membrane clathrin-coated pits. *J. Cell Biol.*, **146**, 755.
44. Gaidarov, I., Santini, F., Warren, R. A., and Keen, J. H. (1999) Spatial control of coated-pit dynamics in living cells. *Nat. Cell Biol.*, **1**, 1.
45. West, M. A., Bright, N. A., and Robinson, M. S. (1997) The role of ADP-ribosylation factor and phospholipase D in adaptor recruitment. *J. Cell Biol.*, **138**, 1239.
46. Subtil, A., Daidarov, I., Kobylarz, K., Lampson, M. A., Keen, J. H., and McGraw, T. E. (1999) Acute cholesterol depletion inhibits clathrin-coated pit budding. *Proc. Natl. Acad. Sci. USA*, **96**, 6775.
47. Rodal, S. K., Skretting, G., Garred, O., Vilhardt, F., van Deurs, B., and Sandvig, K. (1999) Extraction of cholesterol with methyl-beta-cyclodextrin perturbs formation of clathrin-coated endocytic vesicles. *Mol. Biol. Cell*, **10**, 961.
48. Kirchhausen, T., Bonifacino, J. S., and Riezman, H. (1997) Linking cargo to vesicle formation: Receptor tail interactions with coat proteins. *Curr. Opin. Cell Biol.*, **9**, 488.
49. Marks, M. S., Ohno, H., Kirchhausen, T., and Bonifacino, J. S. (1997) Protein sorting by tyrosine based signals: adapting to the the Ys and wherefores. *Trends Cell Biol.*, **7**, 124.
50. Boll, W., Ohno, H., Songyang, Z., Rapoport, I., Cantley, L. C., Bonifacino, J. S., *et al.* (1996) Sequence requirements for the recognition of tyrosine-based endocytic signals by clathrin AP-2 complexes. *EMBO J.*, **15**, 5789.
51. Heilker, R., Manning-Krieg, U., Zuber, J.-F., and Spiess, M. (1996) *In vitro* binding of clathrin adaptors to sorting signals correlates with endocytosis and basolateral sorting. *EMBO J.*, **15**, 2893.
52. Letourneur, F. and Klausner, R. D. (1992) A novel di-leucine motif and a tyrosine based motif independently mediate lysosomal targeting and endocytosis of CD3 chains. *Cell*, **69**, 1143.
53. Kirchhausen, T. (1999) Adaptors for clathrin-mediated traffic. *Annu. Rev. Cell Dev. Biol.*, **15**, 705.
54. Mallet, W. G. and Brodsky, F. M. (1996) A membrane-associated protein complex with selective binding to the clathrin coat adaptor AP1. *J. Cell Sci.*, **109**, 3059.
55. Seaman, M. N. J., Sowerby, P. J., and Robinson, M. S. (1996) Cytosolic and membrane-associated proteins involved in the recruitment of AP-1 adaptors onto the *trans*-Golgi network. *J. Biol. Chem.*, **271**, 25446.
56. Zhang, J. Z., Davletov, B. A., Südhof, T. C., and Anderson, R. G. (1994) Synaptotagmin I is a high affinity receptor for clathrin AP-2: implications for membrane recycling. *Cell*, **78**, 751.
57. Haucke, V. and De Camilli, P. (1999) AP-2 recruitment to synaptotagmin stimulated by tyrosine-based endocytic motifs. *Science*, **285**, 1268.
58. Ferguson, S. S., Zhang, J., Barak, L. S., and Caron, M. G. (1998) Role of beta-arrestins in the intracellular trafficking of G protein-coupled receptors. *Adv. Pharmacol.*, **42**, 420.
59. Lin, F.-T., Krueger, K. M., Kendall, H. E., Daaka, Y., Fredericks, Z. L., Pitcher, J. A., *et al.*

(1997) Clathrin-mediated endocytosis of the β-adrenergic receptor is regulated by phosphorylation/dephosphorylation of β-arrestin 1. *J. Biol. Chem.*, **49**, 31051.

60. Lindner, R. and Ungewickell, E. (1992) Clathrin-associated proteins of bovine brain coated vesicles. An analysis of their number and assembly-promoting activity. *J. Biol. Chem.*, **267**, 16567.
61. Zhang, B., Koh, Y. H., Beckstead, R. B., Budnik, V., Ganetzky, B., and Bellen, H. J. (1998) Synaptic vesicle size and number are regulated by a clathrin adaptor protein required for endocytosis. *Neuron*, **21**, 1465.
62. Hao, W., Luo, Z., Zheng, L., Prasad, K., and Lafer, E. M. (1999) AP180 and AP-2 interact directly in a complex that cooperatively assembles clathrin. *J. Biol. Chem.*, **274**, 22785.
63. Dreyling, M. H., Martinez-Climent, J. A., Zheng, M., Mao, J., Rowley, J. D., and Bohlander, S. K. (1996) The t(10;11) (p13;q14) in the U937 cell line results in the fusion of the AF10 gene and CALM, encoding a new member of the AP-3 clathrin assembly protein family. *Proc. Natl. Acad. Sci. USA*, **93**, 4804.
64. Tebar, F., Bohlander, S. K., and Sorkin, A. (1999) Clathrin assembly lymphoid myeloid leukemia (CALM) protein: localization in endocytic-coated pits, interactions with clathrin, and the impact of overexpression on clathrin-mediated traffic. *Mol. Biol. Cell*, **10**, 2687.
65. Fazioli, F., Minichiello, L., Matoskova, B., Wong, W. T., and Di Fiore, P. P. (1993) Eps15, a novel tyrosine kinase substrate, exhibits transforming activity. *Mol. Cell. Biol.*, **13**, 5814.
66. Wong, W. T., Schumacher, C., Salcini, A. E., Romano, A., Castagnino, P., Pelicci, P. G., *et al.* (1995) A protein-binding domain, EH, identified in the receptor tyrosine kinase substrate Eps15 and conserved in evolution. *Proc. Natl. Acad. Sci. USA*, **92**, 9530.
67. Cupers, P., ter Haar, E., Boll, W., and Kirchhausen, T. (1997) Parallel dimers and antiparallel tetramers formed by epidermal growth factor receptor pathway substrate clone 15. *J. Biol. Chem.*, **272**, 33430.
68. Iannolo, G., Salcini, A. E., Gaidarov, I., Goodman, O. B. J., Baulida, J., Carpenter, G., *et al.* (1997) Mapping of the molecular determinants involved in the interaction between eps15 and AP-2. *Cancer Res.*, **57**, 240.
69. Chen, H., Fre, S., Slepnev, V. I., Capua, M. R., Takei, K., Butler, M. H., *et al.* (1998) Epsin is an EH domain-binding protein implicated in clathrin-mediated endocytosis. *Nature*, **394**, 793.
70. Cupers, P., Jadhav, A. P., and Kirchhausen, T. (1998) Assembly of clathrin coats disrupts the association between Eps15 and AP-2 adaptors. *J. Biol. Chem.*, **273**, 1847.
71. Benmerah, A., Bayrou, M., Cerf-Bensussan, N., and Dautry-Varsat, A. (1999) Inhibition of clathrin-coated pit assembly by an Eps15 mutant. *J. Cell Sci.*, **112**, 1303.
72. Chen, H., Slepnev, V. I., Di Fiore, P. P., and De Camilli, P. (1999) The interaction of Epsin and Eps15 with clathrin adaptor AP-2 is inhibited by mitotic phosphorylation and enhanced by stimulation-dependent dephosphorylation in nerve terminals. *J. Biol. Chem.*, **274**, 3257.
73. Wilde, A. and Brodsky, F. M. (1996) *In vivo* phosphorylation of adaptors regulates their interaction with clathrin. *J. Cell Biol.*, **135**, 635.
74. Cao, H., Garcia, F., and McNiven, M. A. (1998) Differential distribution of dynamin isoforms in mammalian cells. *Mol. Biol. Cell*, **9**, 2595.
75. Schmid, S. L., McNiven, M. A., and De Camilli, P. (1998) Dynamin and its partners: a progress report. *Curr. Opin. Cell Biol.*, **10**, 504.
76. Sweitzer, S. M. and Hinshaw, J. E. (1998) Dynamin undergoes a GTP-dependent conformational change causing vesiculation. *Cell*, **93**, 1021.
77. Sever, S., Muhlberg, A. B., and Schmid, S. L. (1999) Impairment of dynamin's GAP domain stimulates receptor-mediated endocytosis. *Nature*, **398**, 481.

78. Ringstad, N., Nemoto, Y., and De Camilli, P. (1997) The SH3p4/Sh3p8/SH3p13 protein family: binding partners for synaptojanin and dynamin via a Grb2-like Src homology 3 domain. *Proc. Natl. Acad. Sci. USA*, **94**, 8569.
79. Greene, L. E. and Eisenberg, E. (1990) Dissociation of clathrin from coated vesicles by the uncoating ATPase. *J. Biol. Chem.*, **265**, 6682.
80. Buxbaum, E. and Woodman, P. G. (1995) Selective action of uncoating ATPase towards clathrin-coated vesicles from brain. *J. Cell Sci.*, **108**, 1295.
81. Ungewickell, E., Ungewickell, H., Holstein, S. E. H., Lindner, R., Prasad, K., Barouch, W., *et al.* (1995) Role of auxilin in uncoating clathrin-coated vesicles. *Nature*, **378**, 632.
82. Martin, T. F. (1998) Phosphoinositide lipids as signaling molecules: common themes for signal transduction, cytoskeletal regulation, and membrane trafficking. *Annu. Rev. Cell Dev. Biol.*, **14**, 231.
83. Fujimoto, L. M., Roth, R., Heuser, J. E., and Schmid, S. L. (2000) Actin assembly plays a variable, but not obligatory role in receptor-mediated endocytosis in mammalian cells. *Traffic*, **1**, 161.
84. Geli, M. I. and Riezman, H. (1998) Endocytic internalization in yeast and animal cells: similar and different. *J. Cell Sci.*, **111**, 1031.
85. Wendland, B., Emr, S. D., and Riezman, H. (1998) Protein traffic in the yeast endocytic and vacuolar protein sorting pathways. *Curr. Opin. Cell Biol.*, **10**, 513.
86. Gruenberg, J., Griffiths, G., and Howell, K. E. (1989) Characterization of the early endosome and putative endocytic carrier vesicles *in vivo* and with an assay of vesicle fusion *in vitro*. *J. Cell Biol.*, **108**, 1301.
87. Santini, F. and Keen, J. H. (1996) Endocytosis of activated receptors and clathrin-coated pit formation: deciphering the chicken or egg relationship. *J. Cell Biol.*, **132**, 1025.
88. Wiley, H. S., Herbst, J. J., Walsh, B. J., Lauffenburger, D. A., Rosenfeld, M. G., and Gill, G. N. (1991) The role of tyrosine kinase activity in endocytosis, compartmentation, and down-regulation of the epidermal growth factor receptor. *J. Biol. Chem.*, **266**, 11083.
89. Wilde, A., Beattie, E. C., Lem, L., Riethof, D. A., Liu, S.-H., Mobley, W. C., *et al.* (1999) EGF receptor signaling stimulates SRC kinase phosphorylation of clathrin, influencing clathrin redistribution and EGF uptake. *Cell*, **96**, 677.
90. Slepnev, V. I., Ochoa, G.-C., Butler, M. H., Grabs, D., and De Camilli, P. (1998) Role of phosphorylation in regulation of the assembly of endocytic coat complexes. *Science*, **281**, 821.
91. McLauchlan, H., Newell, J., Morrice, N., Osborne, A., West, M., and Smythe, E. (1997) A novel role for Rab5-GDI in ligand sequestration into clahtrin-coated pits. *Curr. Biol.*, **8**, 34.
92. Marsh, M. and McMahon, H. T. (1999) The structural era of endocytosis. *Science*, **285**, 215.
93. Wakeham, D. E., Ybe, J. A., Brodsky, F. M., and Hwang, P. K. (2000) Molecular structures of proteins involved in vesicle coat formation. *Traffic*, **1**, 393.
94. Pishvaee, B. and Payne, G. S. (1998) Clathrin coats – threads laid bare. *Cell*, **95**, 443.
95. ter Haar, E., Musacchio, A., Harrison, S. C., and Kirchhausen, T. (1998) Atomic structure of clathrin: A β propeller terminal domain joins an α zigzag linker. *Cell*, **95**, 563.
96. Goodman, O. B., Jr., Krupnick, J. G., Gurevich, V. V., Benovic, J. L., and Keen, J. H. (1997) Arrestin/clathrin interaction. Localization of the arrestin binding locus to the clathrin terminal domain. *J. Biol. Chem.*, **272**, 15017.
97. Ybe, J. A., Brodsky, F. M., Hofmann, K., Lin, K., Liu, S.-H., Chen, L., *et al.* (1999) Clathrin self-assembly is mediated by a tandemly repeated superhelix. *Nature*, **399**, 371.
98. Musacchio, A., Smith, C. J., Roseman, A. M., Harrison, S. C., Kirchhausen, T., and Pearse, B. M. F. (1999) Functional organization of clathrin in coats: Combining electron cryomicroscopy and x-ray crstallography. *Mol. Cell*, **3**, 761.

99. Owen, D. J. and Evans, P. R. (1998) A structural explanation for the recognition of tyrosine-based endocytotic signals. *Science*, **282**, 1327.
100. Trowbridge, I. S., Collawn, J. F., and Hopkins, C. R. (1993) Signal-dependent membrane protein trafficking in the endocytic pathway. *Annu. Rev. Cell Biol.*, **9**, 129.
101. Owen, D. J., Vallis, Y., Noble, M. E., Hunter, J. B., Dafforn, T. R., Evans, P. R., *et al.* (1999) A structural explanation for the binding of multiple ligands by the alpha-adaptin appendage domain. *Cell*, **97**, 805.
102. Traub, L. M., Downs, M. A., Westrich, J. L., and Fremont, D. H. (1999) Crystal structure of the alpha appendage of AP-2 reveals a recruitment platform for clathrin-coat assembly. *Proc. Natl. Acad. Sci. USA*, **96**, 8907.
103. Wang, L. H., Südhof, T. C., and Anderson, R. G. W. (1995) The appendage domain of α-adaptin is a high affinity binding site for dynamin. *J. Biol. Chem.*, **270**, 10079.
104. Wigge, P., Kohler, K., Vallis, Y., Doyle, C. A., Owen, D., Hunt, S. P., *et al.* (1997) Amphiphysin heterodimers: potential role in clathrin-mediated endocytosis. *Mol. Biol. Cell*, **8**, 2003.
105. Benmerah, A., Gagnon, J., Bègue, B., Mégarbané, B., Dautry-Varsat, A., and Cerf-Bansussan, N. (1995) The tyrosine kinase substrate eps15 is constitutively associated with the plasma membrane adaptor AP-2. *J. Cell Biol.*, **131**, 1831.
106. Vallis, Y., Wigge, P., Marks, B., Evans, P. R., and McMahon, H. T. (1999) Importance of the pleckstrin homology domain of dynamin in clathrin-mediated endocytosis. *Curr. Biol.*, **9**, 257.
107. Ferguson, K. M., Lemmon, M. A., Schlessinger, J., and Sigler, P. B. (1994) Crystal structure at 2.2 Å resolution of the pleckstrin homology domain from human dynamin. *Cell*, **79**, 199.
108. Timm, D., Salim, K., Gout, I., Guruprasad, L., Waterfield, M., and Blundell, T. (1994) Crystal structure of the pleckstrin homology domain from dynamin. *Nat. Struct. Biol.*, **1**, 782.
109. Ferguson, K. M., Lemmon, M. A., Schlessinger, J., and Sigler, P. B. (1995) Structure of the high affinity complex of inositol trisphosphate with a phospholipase C pleckstrin homology domain. *Cell*, **83**, 1037.
110. Owen, D. J., Wigge, P., Vallis, Y., Moore, J. D., Evans, P. R., and McMahon, H. T. (1998) Crystal structure of the amphiphysin-2 SH3 domain and its role in the prevention of dynamin ring formation. *EMBO J.*, **17**, 5273.
111. de Beer, T., Carter, R. E., Lobel-Rice, K. E., Sorkin, A., and Overduin, M. (1998) Structure and Asn-Pro-Phe binding pocket of the Eps15 homology domain. *Science*, **281**, 1357.
112. Haffner, C., Takei, K., Chen, H., Ringstad, N., Hudson, A., Butler, M. H., *et al.* (1997) Synaptojanin 1: localization on coated endocytic intermediates in nerve terminals and interaction of its 170 kDa isoform with Eps15. *FEBS Lett.*, **419**, 175.
113. Moore, M. S., Mahaffey, D. T., Brodsky, F. M., and Anderson, R. G. W. (1987) Assembly of clathrin-coated pits onto purified plasma membranes. *Science*, **236**, 558.
114. Schmid, S. L. (1993) Coated-vesicle formation *in vitro*: Conflicting results using different assays. *Trends Cell Biol.*, **3**, 145.
115. Takei, K., Haucke, V., Slepnev, V., Farsad, K., Salazar, M., Chen, H., *et al.* (1998) Generation of coated intermediates of clathrin-mediated endocytosis on protein-free liposomes. *Cell*, **94**, 131.
116. O'Halloran, T. J. and Anderson, R. G. W. (1992) Clathrin heavy chain is required for pinocytosis, the presence of large vacuoles, and development in *Dictyostelium*. *J. Cell Biol.*, **118**, 1371.

117. Nonet, M. L., Holgado, A. M., Brewer, F., Serpe, C. J., Norbeck, B. A., Holleran, J., *et al.* (1999) UNC-11, a *Caenorhabditis elegans* AP180 homologue, regulates the size and protein composition of synaptic vesicles. *Mol. Biol. Cell*, **10**, 2343.
118. Fagan, M. H. and Dewey, T. G. (1984) Statistical mechanical model of the energetics of coated vesicle formation. *Biophys. J.*, **45**, 383.
119. Damke, H., Baba, T., Warnock, D. E., and Schmid, S. L. (1994) Induction of mutant dynamin specifically blocks endocytic coated vesicle formation. *J. Cell Biol.*, **127**, 915.
120. Henley, J. R., Krueger, E. W., Oswald, B. J., and McNiven, M. A. (1998) Dynamin-mediated internalization of caveolae. *J. Cell Biol.*, **141**, 85.
121. Oh, P., McIntosh, D. P., and Schnitzer, J. E. (1998) Dynamin at the neck of caveolae mediates their budding to form transport vesicles by GTP-driven fission from the plasma membrane of endothelium. *J. Cell Biol.*, **141**, 101.
122. Liu, S.-H., Marks, M. S., and Brodsky, F. M. (1998) A dominant negative clathrin mutant differentially affects trafficking of molecules with distinct sorting motifs in the class II MHC pathway. *J. Cell Biol.*, **140**, 1023.
123. Nesterov, A., Carter, R. E., Sorkina, T., Gill, G. N., and Sorkin, A. (1999) Inhibition of the receptor-binding function of clathrin adaptor protein AP-2 by dominant-negative mutant μ2 subunit and its effects on endocytosis. *EMBO J.*, **18**, 2489.
124. Heuser, J. E., Keen, J. H., Amende, L. M., Lippoldt, R. E., and Prasad, K. (1987) Deep-etch visualization of 27S clathrin: a tetrahedral tetramer. *J. Cell Biol.*, **105**, 1999.
125. Brodsky, F. M. (1985) Clathrin structure characterized with monoclonal antibodies. I. Analysis of multiple antigenic sites. *J. Cell Biol.*, **101**, 2047.
126. Chin, D. J., Straubinger, R. M., Acton, S., Näthke, I., and Brodsky, F. M. (1989) 100-kDa polypeptides in peripheral clathrin-coated vesicles are required for receptor-mediated endocytosis. *Proc. Natl. Acad. Sci. USA*, **86**, 9289.
127. Brodsky, F. M. (1999) Clathrin. In *Guidebook to cytoskeletal and motor proteins* (ed. T. Kreis and R. Vale), p. 512. Oxford University Press.
128. Heuser, J. E. and Keen, J. (1988) Deep-etch visualization of proteins involved in clathrin assembly. *J. Cell Biol.*, **107**, 877.
129. Page, L. J. and Robinson, M. S. (1995) Targeting signals and subunit interactions in coated vesicle adaptor complexes. *J. Cell Biol.*, **131**, 619.
130. Brodsky, F. M. (1997) New fashions in vesicle coats. *Trends Cell Biol.*, **7**, 175.

2 | Clathrin-independent endocytosis

ALICE DAUTRY-VARSAT

1. Introduction

The most extensively characterized mechanism of endocytosis is clathrin-mediated (see Chapter 1). Irrefutable evidence has accumulated over the years that clathrin-coated pits form at the plasma membrane, bud, and form clathrin-coated vesicles (CCVs) that transport membrane receptors and their ligands as well as solutes into the cells. For many years the enormous interest in clathrin-coated pits and CCVs overshadowed the possibility that other endocytic pathways may exist. Recently, however, the existence of alternative pathways has become more widely accepted, although the molecular mechanisms involved are still far from being understood.

2. Evidence in favour of clathrin-independent endocytosis: the morphological approach

At the same time as endocytosis via clathrin-coated pits was uncovered, results suggesting the existence of non-clathrin mediated endocytosis were regularly published. Most of these initial studies relied on morphological approaches. The characterization of clathrin-coated structures largely benefited from the characteristic coats seen by electron microscopy (EM). By contrast, the absence of a coat made it difficult to distinguish an endocytic activity from smooth areas of the plasma membrane. Since the uncoating of CCVs occurs within minutes and endocytosis cannot be synchronized, how could one prove that smooth invaginations were about to give rise to vesicles detached from the membrane, or that what appeared as 'smooth' vesicles were not still connected to the plasma membrane or did not derive from CCVs that had just lost their coat? EM, which produces images of fixed samples, was not sufficient to study a very dynamic and asynchronous process such as endocytosis. Despite these difficulties, sequential images were observed in these early times by labelling the cell surface at 4 °C, a temperature that does not allow endocytosis to proceed, and then warming the cells to 37 °C for different times, fixing them, and analysing them by EM on thin sections. Several receptors and ligands were studied using this protocol. The aggregation and endocytosis of the major histocompatibility complex class I (MHC I) induced by antibody recognizing β_2 microglobulin was observed in fibroblasts. Neither the invaginations containing MHC molecules, nor the vesicles in which they were localized, were covered with a characteristic clathrin coat (1).

Cholera and tetanus toxin which bind to membrane glycolipids, in particular to either GM_1 monosialoganglioside or di- and trisialogangliosides respectively, were found to bind and to be internalized in non-coated pits and vesicles in experiments performed on cultured liver cells with gold-labelled toxins (2). This was later confirmed and extended using ^{125}I- and gold-labelled cholera toxin following the internalization of the ligand which progressively moved from non-coated pits into non-coated vesicles and a tubulo-vesicular compartment (3).

In rat adipocytes, it was reported that gold-labelled α_2-macroglobulin was internalized via coated pits, while gold or ferritin-labelled insulin were internalized via non-coated pits. However, in other cell types, insulin was found in coated pits and vesicles and was internalized via the clathrin-mediated pathway (4, 5 and references therein).

In human epidermoid A431 carcinoma cells, the beta-adrenergic catecholamine receptors were found in non-coated invaginations on the cell surface and, upon warming up to 37 °C, non-coated vesicles were labelled (6). Similarly, the transmembrane receptor thrombomodulin expressed in COS cells probed with antibodies or with gold-conjugated thrombin was predominantly found in non-coated pits and rarely in clathrin-coated pits (7). Interestingly, thrombomodulin lacking the cytoplasmic domain internalized thrombin as efficiently as the wild-type receptor and there was a notable absence of gold particles in clathrin-coated pits and vesicles (8). To better characterize morphologically these non-coated vesicles, Hansen and co-workers used the following approach: gold-labelled concanavalin A was bound to cells at 4 °C which were then warmed to 37 °C for a short time (30–60 seconds) and fixed. The cells were then incubated with horse radish peroxidase (HRP) which stained the membranes in topological continuity with the extracellular medium while endocytic vesicles were only stained with gold (9). This revealed the presence of uncoated vesicles, about as abundant as coated vesicles but slightly smaller (95 nm versus 110 nm). However, considering that CCVs are very short-lived (in the range of a minute), it could not be ruled out that these smooth vesicles were derived from vesicles that had lost their coat or were recycling. Furthermore, this approach did not allow the rate of smooth vesicle formation to be estimated or the extent of their contribution to endocytic processes.

In conclusion, although the morphological approach could not definitively demonstrate the existence of clathrin-independent endocytosis, it brought together many data in favour of its existence.

3. Inhibition of clathrin-dependent endocytosis to detect clathrin-independent endocytosis

One major difficulty in studying clathrin-independent endocytosis has been the dearth of specific markers that are strictly internalized by this pathway. Hence, the alternative pathways have been (and still are) mostly studied in perturbed systems. Thus, an indirect way to demonstrate the existence of clathrin-independent endocytosis has been to study the internalization of fluid phase markers or specific ligands under conditions in which CCV formation is inhibited. This has been achieved mainly through pharmacological methods and, more recently, using several dominant negative mutants of endocytosis. The key question is then to establish how specific these methods may be, and if and how the perturbation introduced may affect the process of clathrin-independent endocytosis under study.

3.1 Pharmacological methods

Three main methods have been used to inhibit the coated-pit pathway: potassium depletion, acidification of the cytosol, or incubation in hypertonic media. These methods may sometimes be tricky and somewhat cell type-dependent. Nevertheless, despite some discrepancies, many published studies support the existence of clathrin-independent pathways. Although the molecular mechanism by which these treatments inhibit clathrin-dependent endocytosis is not fully understood, they seem to affect it at different steps.

The first method, potassium depletion in conjunction with hypotonic shock, was initially described to inhibit the endocytosis of low density lipoproteins (LDL) (10). It inhibits the assembly of coated pits as seen by EM, and causes an increase in the diffuse clathrin immunofluorescence staining throughout the cytoplasm. This diffuse fluorescence is due to the precipitation of cytoplasmic clathrin, into unusually small microcages (11). Thus this treatment results in the depletion of coated pits from the cell surface (12, 13).

Potassium depletion turned out to be a useful method. It should be pointed out that, for unknown reasons, its efficiency in preventing coated pit assembly varies depending on the cells used (14) and that it varies somewhat from experiment to experiment. Therefore, it is necessary to probe in every experiment for the efficiency of inhibition of endocytosis via coated pits using a well characterized marker of this pathway. Following internalization of transferrin or its receptor is, in most cases, a good choice because this receptor is expressed in most cells and can be easily studied.

The second method, exposure of cells to hypertonic media, was initially used in polymorphonuclear leukocytes (15) using sodium chloride, sucrose, or lactose to increase the tonicity. Later it was applied to isolated rat hepatocytes (16). Initially, the receptor-mediated uptake of a chemotactic peptide was inhibited while uptake of a fluid phase marker was not. Similarly, in hepatocytes endocytosis of asialo-orosomucoid bound to the asialoglycoprotein receptor was inhibited, while the uptake of fluids was not. In both cases this inhibition was reversed by washing the cells. In Vero cells, by contrast, high salt treatment inhibited uptake of the clathrin pathway marker, transferrin, and also the entry of ricin which can enter cells by both clathrin-dependent and -independent mechanisms. Therefore one should be cautious about using this method to differentiate between different forms of receptor-mediated endocytosis (17). In an effort to analyse the effect of such a treatment,

Heuser and Anderson reported that in fibroblasts incubated in hypertonic medium, normal clathrin lattices underlying the plasma membrane disappeared and were replaced by accumulations of numerous microcages of clathrin. Concomitantly, LDL receptors lost their normal clustered distribution and became dispersed over the cell surface (18). Within minutes of a return to standard medium, normal clathrin lattices began to reappear.

The third method is cytosol acidification (17, 19). It can be achieved by treating cells with acetic acid, or with isotonic KCl and the ionophores valinomycin and nigericin, or with pre-incubating cells with NH_4Cl and then inhibiting the Na^+/H^+ exchanger by adding amiloride. The effect of this treatment on clathrin-coated pits differs somewhat from the two previously described treatments which prevent coated pit assembly. Cytoplasmic acidification inhibits a later step in clathrin-mediated endocytosis since coated pits are still visible by EM but vesicles do not seem to pinch off the membrane. Instead, during acidification, clathrin lattices with increased curvature but persistent attachment with the plasma membrane accumulate, resulting in what was described as a 'paralysis' of coated pits (11).

There are some discrepancies as to the effect of cytosol acidification on fluid phase endocytosis. This treatment inhibited the internalization of fluid phase markers in BHK cells (20, 21). In other cell types, it affected very slightly fluid phase endocytosis while clathrin-dependent endocytosis was inhibited by more than 90% (19). It is likely that the relative importance of the various endocytic pathways differ depending on the cell type and culture conditions (for instance macropinocytosis can be efficiently induced in some cells by growth factors and would allow fluid phase internalization, see below). Also, it should be noted that while an insufficient acidification does not efficiently inhibit clathrin-dependent internalization, excessive acidification inhibits other endocytic pathways (A. Subtil and A. Dautry-Varsat, unpublished).

The potential effects of these pharmacological treatments on clathrin-independent pathways has not been checked in most published studies simply because it was, and still is, difficult to do. To check this, one would need markers of clathrin-independent uptake. The easiest to use would probably be ricin. However, as discussed below, it is clear that more than one clathrin-independent pathway exists, and ricin would not distinguish between these pathways. Also a careful kinetic study of the effect of potassium depletion, as opposed to acidification, on fluid phase uptake in fibroblasts showed that there was an accelerated 'regurgitation' in treated cells (22). The conclusion from this work was that neither internalization of a fluid phase marker, nor that of two lipid markers were affected by a treatment which affected entry via CCVs. Again this illustrated that when studying the initial steps of endocytosis, i.e. the formation of vesicles from the plasma membrane, which takes place within a few minutes, one should be careful to look at the initial rates of uptake and not at overall effects after 15 minutes or more, as is often done.

Potassium depletion, cytosol acidification and, to a lesser extent hyperosmolarity, were also used to analyse the internalization of some receptors. The plant toxin ricin was the first ligand shown to be internalized when coated pit function is inhibited and remains the most studied probe. Thus, potassium depletion (14) as well as cytosol acidification (19) blocked transferrin uptake whereas ricin internalization continued. Ricin is part of a group of bacterial and plant toxins that bind to cells, are endocytosed, and later transported to the cytosol where they inhibit protein synthesis (23). Ricin binds to glycolipids and glycoproteins with terminal galactose. It therefore labels the cell surface extensively and can be taken up into cells by any vesicle that pinches off. Thus ricin can be internalized by both clathrin-dependent and -independent routes (14, 19, 23 and references therein).

Following the initial study on ricin, several receptors were reported to be internalized by a clathrin-independent mechanism based on the lack of, or partial, inhibition by one or more of the pharmacological inhibitors described above; for example the β-adrenergic receptor in A431 cells (6), the IL-2 (IL-2) receptors in lymphocytes (24, 25), the cholecytokinin receptor in CHO cells (26), fibronectin receptors in fibroblasts (27). Also, insulin and epidermal growth factor (EGF) receptors may enter cells in two ways depending or not on clathrin. For instance, in IM9 cultured lymphocytes, insulin uptake remained unchanged after potassium depletion (28) and in CHO cells, insulin-stimulated entry of the receptor into a rapid, coated pit-mediated pathway, was saturable and there was also a non-saturable pathway which was not affected by potassium depletion (29). Similarly, EGF receptors may be internalized by two distinct pathways, a low capacity one with high affinity for occupied EGF receptors, and a high capacity one which is followed for instance by EGF receptors lacking intrinsic kinase activity (30).

In conclusion, pharmacological strategies have proven useful for showing the existence of non-clathrin mediated uptake of fluid phase markers and some specific ligands. As already pointed out, the use of these strategies is difficult because the results are not reproducible and their effect is variable between cell types. Moreover, the treatments are unspecific and may affect other cellular processes that in turn affect endocytosis. Thus, the treatment must be carefully controlled, not only with regard to its effects on the clathrin pathway but also because excessive treatment may also affect other pathways. Using more than one method is certainly recommended. Because these treatments modify the cells in unknown ways and generate conditions whereby, among other things, clathrin-mediated endocytosis is inhibited, the validity of the results obtained may be questioned. Such considerations have slowed our efforts to better define clathrin-independent endocytosis. Nevertheless, it may be a useful way of determining rapidly whether a process may be clathrin-mediated.

More recently dominant negative mutants have been obtained that provide much more specific means to inhibit clathrin-mediated internalization.

3.2 The use of dominant negative mutants of endocytosis

As our knowledge of clathrin-mediated endocytosis improves and as new proteins required for the early steps of this pathway are discovered, new tools are emerging in the form of dominant negative mutants of these proteins. We first present here some aspects concerning the formation of clathrin-coated pits and vesicles and in each case discuss the properties of the tools derived from our knowledge of the proteins involved. Clathrin-dependent endocytosis is discussed in depth in Chapter 1. Here, I review only the information necessary to understand how different mutants may function.

Cargo selection and vesiculation in clathrin-mediated endocytosis, as in many other membrane traffic events, are linked to the formation of a coat. Coated pit formation at the plasma membrane requires clathrin and the adaptor complex AP-2. Coated vesicle formation also depends on, and is regulated by, the activities of a set of cytosolic proteins that are recruited through protein–protein and protein–lipid interactions (see Chapter 1). Over the past years, a number of these proteins have been discovered that interact with AP-2 and appear to be part, at some point, of the coat complex: these include clathrin, amphiphysins, Eps15, epsin, AP180, auxilin, and arrestins. These proteins, in turn, may interact with other proteins, namely dynamin, Ese/intersectin, synaptojanin, endophilin, and syndapin. The precise order in which these proteins act during formation of a CCV remains to be resolved. Many of them have been discovered only recently, and appear to be necessary at some step of the CCV formation, but their function is not completely understood.

Nevertheless, overexpression of some of them or of mutated defective forms have provided fairly specific tools to inhibit clathrin-mediated endocytosis. One of the key questions is how specific each of these tools may be and we will therefore briefly review the properties of each of the proteins involved.

3.2.1 Clathrin

Clathrin coats are involved in two crucial transport steps: endocytosis from the plasma membrane and transport from the *trans*-Golgi network (TGN) to endosomes. They may also have other functions since there are other cellular sites to which clathrin has been localized such as endosomes (31).

In endocytosis, clathrin coats assemble on the cytoplasmic face of the plasma membrane, forming pits that invaginate and pinch off the membrane within ≅ 1 min to form CCVs. Clathrin, the main scaffold protein of the coat forms trimers, the so-called triskelions, that oligomerize both *in vivo* and *in vitro* to form the polygonal clathrin cages. The carboxy terminal third of the clathrin heavy chain alone can trimerize and folds to produce the central portion of the triskelion, the hub fragment (see Chapter 1). These hubs can also polymerize into open-ended lattices instead of a closed polyhedron. The hub fragment can be used as a dominant negative form of clathrin (32). When overexpressed by transfection in mammalian cells, hubs significantly inhibited clathrin-dependent receptor-mediated endocytosis of FITC-labelled transferrin, as observed by fluorescence microscopy. In contrast, the uptake of FITC–dextran, a marker for bulk flow fluid phase endocytosis, was not significantly affected (32).

To function as a dominant negative mutant, the hub fragment had to be overexpressed approx. 15-fold. Not surprisingly, because of the various functions of clathrin, hub overexpression also blocked lysosomal delivery of chimeric molecules containing either a tyrosine-based or a di-leucine-based sorting signal. It also blocked export of MHC II associated HLA-Dm from the TGN (32).

Use of the hub fragment clearly shows that bulk fluid phase endocytosis can proceed when clathrin functions are impaired. However, this mutant is not a specific inhibitor of clathrin-mediated internalization.

3.2.2 Adaptor complexes

Two adaptor complexes, AP-1 and AP-2, were initially found to be associated with CCVs derived from the Golgi and plasma membranes, respectively (see Chapter 1). Recently, two other adaptor complexes AP-3 and AP-4 were discovered through their sequence homology to AP-1 and AP-2. AP-3 and AP-4 have not been characterized in detail and may not necessarily involve clathrin in their functions. They will not be discussed further. In clathrin-coated pits and vesicles, AP-2 is located between the lipid bilayer and clathrin lattice. It provides a link between the two players by interacting with clathrin and with membrane proteins that contain appropriate signals for sorting into clathrin-coated pits.

Membrane proteins internalized in CCVs contain one or more signal sequences in their cytosolic tail that direct them into clathrin-coated pits (see Chapter 7). Two major families of signal have been identified:

(a) Tyrosine-based signals of the form Yxxϕ (where ϕ is a large hydrophobic residue and x can be any amino acid with a preference for an R at position Y + 2).

(b) Di-leucine motifs (see Chapter 7; reviewed in ref. 33).

AP-2 has been shown to bind both of these motifs. It is well established that Yxxϕ motifs interact with the μ2 subunit of AP-2 (see Chapter 1). The binding site of di-leucine motifs on AP-2 is less well established and it is currently controversial whether di-leucine motifs are recognized by the μ2 (34) or the β2 (35) subunit.

Based on the knowledge of AP-2 interaction with Yxxϕ motifs, an epitope-tagged mutant of AP-2 was constructed which incorporates into AP-2 complexes (36). In this mutant, Asp176 and Trp421 of μ2, both located within the Yxxϕ binding pockets, were replaced by alanines. When stably expressed in HeLa cells, under the control of a tetracycline-regulated promoter, this mutant decreased the rate of transferrin internalization fourfold. However, internalization of EGF was not significantly affected, despite the fact that AP-2 complexes from cells expressing the μ2 mutant failed to bind GST-EGF receptor constructs (36). These data suggest that endocytosis of the EGF receptor does not require μ2 even though both the EGF and transferrin receptors are internalized via coated pits in HeLa cells and have Yxxϕ type signals. This μ2 mutant has not been yet further used to study endocytosis of other receptors, and it remains unclear whether the EGF receptor is a peculiar case or not. It has been proposed that alternative mechanisms control internalization of the EGF receptor (37). One possibility would be that it involves arrestins (see below) or other proteins, such as proteins containing SH2 or PTB domains, known to bind to the activated EGF receptor.

In conclusion, this AP-2 mutant may turn out to very specifically inhibit clathrin-mediated endocytosis of receptors interacting with the μ2 subunit of AP-2 and not using other adaptors, such as arrestins, but more data are needed to draw this conclusion.

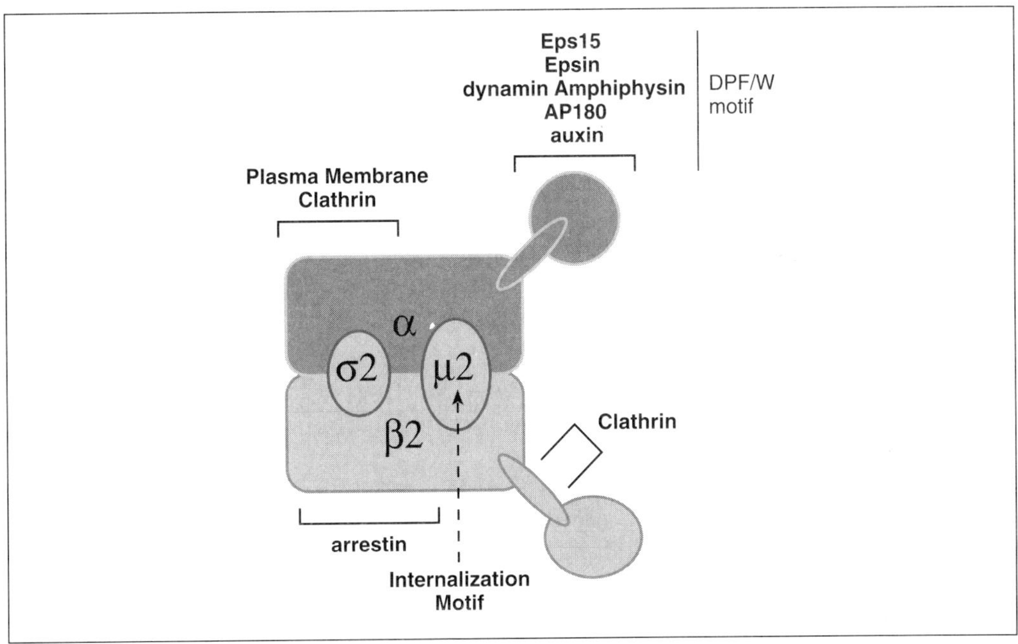

Fig. 1 AP-2 and its partners. The adaptor protein AP-2 interacts with several proteins involved in the formation of clathrin-coated pits making it a key player in this process. The AP-2 tetramer has four subunits, α, β2, μ2, and σ2, which form a core and two appendages or ears. Receptors containing a Yxxϕ signal interact with the μ subunit. Clathrin binds to the appendage domain of the β2 chain. The α adaptin ear binds proteins containing DPF/W motifs: Eps15, epsin, amphiphysin, auxilin, AP180. Other accessory proteins, arrestins, bind both the β subunit of AP-2 and G-coupled receptors. Epsin and mutated forms of Eps15 and amphipysin inhibit clathrin-mediated endocytosis when overexpressed. Mutated forms of arrestin inhibit the internalization of some G-coupled receptors.

3.2.3 Proteins binding to the AP-2 complex: mutants of Eps15, epsin, amphiphysin, arrestin

Several proteins that bind to the α-adaptin appendage domain of AP-2 have recently been identified. These include Eps15, epsin, amphipysins, AP180, and auxilin (recently reviewed in refs 38, 39). Remarkably, the crystal structure of the appendage domain reveals a single binding site for these ligands where a group of charged amino acids is centred on Trp840 (40). The regions of Eps15, epsin, amphipysins, AP180, and auxilin that bind the appendage domain all contain DPF/W motifs. Their affinity appears to be correlated with the number of DPF/W motifs, in that Eps15 and epsin, which contain many copies of this motif, bind more tightly than the other proteins that contain fewer copies. Interestingly, overexpression of the α-ear in COS cells resulted in a significant inhibition of transferrin uptake. Mutant forms of the α-ear unable to bind either Eps15 or amphiphysin did not affect transferrin uptake (40).

i. Eps15 and its mutants

Eps15 (EGF receptor pathway substrate 15) was initially identified as a substrate for ligand-activated EGF receptor tyrosine kinase (41). It was subsequently discovered that Eps15 is constitutively associated with AP-2 (42–44). It is organized in three structural domains: the N-terminal domain is composed of three imperfect repeats of 70 amino acids homologous to each other and to domains found in proteins from different species (called EH domains for Eps15-homology); the central coiled-coil domain is involved in Eps15 oligomerization to form tetramers (45); the C-terminal proline-rich domain contains all the AP-2 binding sites (43, 46). Both EH domains and AP-2 binding sites play an important role in Eps15 targeting to coated pits (47). EM images have shown that Eps15 is composed of a globular head corresponding to the EH domains, joined to a stalk, the coiled-coil domain, linked via a kink to the C-terminal AP-2 binding domain (48). EH domains are protein recognizing modules that mostly bind sequences containing NPF (49, 50). In yeast EH domains containing proteins are required for endocytosis (51, 52).

When the C-terminal AP-2 binding domain of Eps15 fused to the green fluorescent protein (GFP) was overexpressed by transient transfection of HeLa cells, transferrin endocytosis was strongly inhibited. In a cell-free assay using perforated cells to follow the initial steps of CCV formation at the plasma membrane, constitutive endocytosis of transferrin and ligand-induced endocytosis of EGF receptor were inhibited (53). This inhibition could be ascribed to a competition between the mutant and endogenous wild-type Eps15 for AP-2 binding. Another dominant negative mutant was obtained by deleting the EH domains of Eps15. This latter mutant inhibited transferrin endocytosis even more efficiently than the preceding one and induced the loss of the characteristic AP-2 and clathrin punctate immunofluorescence staining implying the absence of coated pits (54). However, it did not appear to markedly affect fluid phase uptake. These results are supported by those obtained by microinjecting antibodies against Eps15 and the related Eps15r (55) which inhibited internalization of transferrin and EGF (56). Because Eps15 interacts with AP-2, these mutants appear to be rather specific of clathrin-coated pit endocytosis. Interestingly, the clathrin-independent internalization pathway of interleukin 2 (IL-2) receptors is not affected by these mutants while it is inhibited by dynamin mutants (Lamaze *et al.*, submitted).

ii. Epsins

Epsins 1 and 2 (Eps15 interactors) were recently identified as interacting partners for Eps15 (57, 58). They have a widespread tissue distribution and are enriched in brain. Their C-terminal regions contain three repeats of the EH domain binding consensus NPF and interact with Eps15. Their central regions containing DPW motifs and interacts with AP-2. Epsins have a clathrin binding consensus motif, bind the N-terminal domain of clathrin, and are partially co-localized with clathrin (57, 58). The N-terminal domain of epsins constitute a phylogenetically conserved module named ENTH (epsin N-terminal homology domain). Yeast homologues of epsin, Ent1p and Ent2p (see Chapter 10), are involved in endocytosis and actin function and their NPF motif binds the EH domains of Pan1, a yeast homologue of Eps15 (59). Both epsins appear to have binding sites in the cell cortex in addition to components of the clathrin coat (57, 58).

When greatly overexpressed by transfection, both epsins mislocalized clathrin coat components and inhibited clathrin-mediated endocytosis of transferrin (57, 58).

To illustrate the possible complex effects of some of these mutants, it is interesting to note that epsin 1, not only is part of coated pits, but also that it shuttles between the nucleus and the cytosol and that its Eps15 interacting domain interacts with a transcription factor (60). Similarly, Eps15 has a nuclear export signal, indicating that it may undergo nucleocytosolic shuttling (Benmerah *et al.*, in preparation).

iii. Amphiphysin

Two major forms of amphiphysin have been identified: amphiphysin 1 is expressed primarily in the brain and amphiphysin 2 is ubiquitously expressed and has a wider diversity of splice variants (61). Both are members of a protein family conserved from yeast to human. The N-terminal domain of amphiphysin forms a coiled-coil domain and also contains a lipid binding site accounting for the property of amphiphysin to bind liposomes and evaginate them in narrow tubules (62). The C-terminal region contains an SH3 domain which interacts with the proline-rich domain of dynamin in a phosphorylation-dependent manner (63). It co-localizes with dynamin in nerve terminals and assembles with this protein into ring-like structures around membrane tubules. The central region of amphiphysin 1 and 2 binds clathrin and the ear of AP-2, providing a possible mechanism through which amphiphysin can be recruited to clathrin coats where it may in turn participate in the recruitment of dynamin (64).

Cells transfected with a myc-tagged amphiphysin 1 SH3 domain showed an 80% inhibition of transferrin and EGF uptake, while other SH3 domains had no effect. The blockade was rescued by co-expression of dynamin, showing dynamin to be the cellular target of amphiphysin. Fluid phase endocytosis of dextran was not affected (65, 66). In agreement with this, microinjection of the amphiphysin SH3 domain inhibited synaptic vesicle endocytosis at the stage of invaginated coated pits (64) and inhibited CCV formation in an *in vitro* assay (67). A fragment of amphiphysin 1 that binds clathrin and AP-2 inhibited transferrin internalization when overexpressed (63). When either the clathrin or the AP-2 binding site of this fragment was mutated, it still inhibited transferrin internalization (68).

Very recently, a novel isoform of amphiphysin 2, amphiphysin 2m, was found to associate with phagosomes in macrophages. Overexpression of a mutant in which the dynamin binding site was ablated inhibited phagocytosis of particles by macrophages at the stage of membrane extensions around the bound particles (69).

iv. Arrestins

Another type of interaction with AP-2 occurs through its β subunit (Fig. 1). Non-visual arrestins (β-arrestin/arrestin 3) bind clathrin and AP-2 and have been shown to mediate the internalization of some receptors that couple to heterotrimeric GTP-binding proteins (G protein-coupled receptors; GPCRs) (70–72). β-arrestin binds to activated β_2-adrenergic receptors (β_2AR) and mediates its internalization (73). The clathrin-binding site of β-arrestin (LIEFE) can be deleted without affecting its function. However, expressing only the binding site for the β2 subunit of AP-2 (amino acids 378–410) inhibits ligand-stimulated internalization of the β_2AR (73). Also, arrestins can bind PPIs through amino acids 223–285. Mutations that inhibit this binding cause inefficient recruitment of receptors to coated pits and consequently inhibit their internalization (74). In conclusion, arrestin mutants are now available which inhibit clathrin-mediated endocytosis of activated GPCRs.

3.2.4 Mutants of dynamin and its partners

The first dominant negative mutant of clathrin-dependent endocytosis reported, and the most extensively used one, is a mutant of dynamin. Dynamin is a key player in the process by which an invaginated pit becomes a vesicle detached from the membrane (see Chapter 1). Dynamin was initially isolated from mammalian brain due to its ability to bind microtubules (75). In *Drosophila*, it is the product of the *shibire* gene (76, 77). Flies with a temperature-sensitive mutation in this gene are paralysed at the non permissive temperature. Their synapses are depleted of synaptic vesicles and invaginated coated and non-coated pits accumulate at the plasma membrane (78). These pits exhibit 'collars' that were later suggested to represent dynamin polymers.

Three different dynamin genes have been identified in mammalian cells; the neuronal-specific dynamin 1, dynamin 2 which is ubiquitously expressed, and dynamin 3 which is expressed in testes and to a lesser extent in neurons and lung. Each of these produce several splice variants (79). A number of reviews have been recently published on these proteins (80–86). Dynamins contain four conserved domains: an N-terminal GTP-binding domain located within the 300 first amino acids, a pleckstrin homology (PH) domain of 100 amino acids, a coiled-coil domain, and a C-terminal proline-rich domain. Dynamin binds to phosphoinositides through its PH domain, mediating its interactions with membranes. The coiled-coil domain has been characterized as a GTPase-effector domain (87), and the proline-rich domain mediates dynamin's interactions with cytoskeletal and SH3-containing proteins (reviewed in ref. 85). There is substantial evidence supporting the hypothesis that dynamin is a 'pinchase'; a mechanoenzyme capable of constricting and severing membrane tubules into discrete vesicles. Dynamin polymerizes *in vitro* to form helical ring structures (88), was observed forming stacked collars in isolated synaptosome fractions treated with the non-hydrolysable analogue of GTP, GTPγS (89), and can tubulate and vesiculate synthetic liposomes (90, 91). The precise mechanism of dynamin function is unclear. One proposal is that dynamin is a mechanochemical enzyme that generates force. An alternative model is that it could play a regulatory role, acting as a molecular switch like other GTPases, by monitoring the neck dimensions of a membrane bud and serving as a geometrical sensor that regulates endocytosis (87). Recently, it was reported that the protein endophilin I modifies lipids in the membrane bilayer, thereby promoting membrane curvature and subsequently vesiculation (92). Endophilin I binds to dynamin and it has been proposed that the combined functions of these proteins could be involved in membrane severing.

When dynamin was shown to be the mammalian homologue of the *shibire* protein of *Drosophila*, studies in mammalian cells showed its participation in clathrin-mediated endocytosis (93–95). At that time, it was reported that dynamin acts exclusively in the scission of CCVs from the plasma membrane (96). A mutant of dynamin 1 defective in GTP binding and hydrolysis was obtained by mutating the neuronal dynamin 1

(K44A mutant) and was stably expressed in HeLa cells under the control of a tetracycline inducible promoter (95). More recently, mutants in the PH domain of dynamin were obtained which when overexpressed inhibited transferrin uptake: these were either a deletion of the PH domain or mutations in the PH domain so that it becomes defective in phosphoinositide binding and have not yet been largely used (80, 97–99).

Since it was published, the K44A dynamin 1 mutant expressed in HeLa cells under an inducible promoter has been widely used to show that a number of receptors are internalized via coated pits based on the fact that their internalization was inhibited when this dominant negative mutation was expressed. If this conclusion is based solely on the use of the dynamin mutant, it must be taken with caution, as it may turn out to be wrong in some instances. It has recently turned out that dynamin mediates the liberation of several different forms of nascent vesicle. It acts to sever caveolae, small flask-shaped invaginations, from the plasma membrane (see below) (100, 101). It is also involved in the clathrin- and caveolae-independent internalization of IL-2 receptors (Lamaze *et al.*, submitted). Dynamin may also be required for phagocytosis in macrophages (102), though it is not involved in the phagocytosis of the bacteria *Chlamydia* in epithelial cells (103). Studies using the dynamin 1 mutant in non-neuronal cells have suggested that dynamin is not involved in fluid phase uptake HRP (96). However, cells from *shibire*^{ts} flies do not take up fluid phase markers at the restrictive temperature (104), and mammalian cells microinjected with anti-dynamin antibodies or overexpressing a mutant of the ubiquitous dynamin 2 also show an inhibition of fluid phase endocytosis (100). The apparent contradiction between these results may simply due to the fact that the neuronal dynamin 1 mutant may not function as efficiently as the dynamin 2 mutant in non-neuronal cells. Nevertheless, dynamin might be required for the formation of four distinct endocytic carrier vesicles, including phagosomes, at the plasma membrane.

In addition to mediating these early stages of endocytic vesicle formation, dynamin has also been implicated in later stage of endocytosis. Internalization of ricin was not inhibited in HeLa cells expressing the dynamin 1 mutant (105). Ricin is normally transported to the secretory pathway from late endosomes, and its transport from late endosomes to Golgi was inhibited in the mutant cells, greatly reducing the toxicity of ricin (105). Recently, it has been observed by EM studies that endogenous dynamin is localized on late endosomes containing the cation-independent mannose-6-phosphate receptor (MPR). In cells expressing the dynamin K44A mutant, this receptor was redistributed. These results suggest that dynamin also plays an important role in the recycling from endosomes to the TGN, and is required for the scission of vesicles budding from endosome tubules (106). Consistent with this, a dynamin-related protein has been reported to function in trafficking along the endo-lysosomal pathway in *Dictyostelium discoideum* (107).

Dynamin also appears to participate in the formation and fission of coated vesicles from secretory compartments (85, 108). Interactions with the actin cytoskeleton have also been described. The neuronal syndapin 1 and ubiquitous syndapin 2 are proteins that bind dynamin 1 and also interact with proteins involved in actin dynamics.

A syndapin SH3 domain partially inhibited clathrin-mediated endocytosis of transferrin, and overexpression of the full-length protein had a strong effect on cortical actin organization and induced filopodia (109). Profilin and cortactin may also form complexes with dynamin. Finally, results obtained in yeast, mammalian cells, and the nematode, implicate dynamin in the maintenance of mitochondrial morphology (reviewed in ref. 85) and a functional role for dynamin in mitogenic signal transduction by certain GPCRs can be dissociated from its role in their endocytosis (110).

In conclusion, these multiple functions of dynamin are consistent with the existence of multiple spliced forms that localize to different cellular compartments (79). It is now clear that dynamins can function at multiple steps along the secretory and endocytic pathways. Therefore, if endocytosis of ligands or receptors is inhibited in cells overexpressing a dynamin mutant, one cannot conclude that they are internalized via CCVs. Moreover, if signalling by a given receptor is inhibited it cannot be directly concluded that endocytosis of the receptor is necessary for signal transduction. The effect of the dynamin mutant may be very indirect. Since a large body of data has been obtained using these dynamin mutants, one must be very cautious about the general conclusions at this stage, unless they have been confirmed by independent methods.

Few dynamin partners have been characterized from which mutants have been derived. The recently identified Ese1 and 2 proteins, also referred to as intersectins, interact with dynamin. They contain multiple EH and SH3 domains, and are homologous to *Xenopus* intersectin and *Drosophila* dynamin-associated protein DAP160–1 (111–113). They are associated with Eps15 and bind epsins and dynamin through their EH and SH3 domains, respectively. When overexpressed in COS cells, Ese1 strongly inhibited transferrin endocytosis, maybe through sequestration of dynamin or other Ese1 partners (111). In an *in vitro* assay, the intersectin SH3 domain inhibited the formation of constricted coated pits (67).

In conclusion, since the first AP-2 interacting proteins were found, more and more proteins are discovered that may interact with it, directly or indirectly. From our knowledge of these proteins, a picture is emerging in which the formation of coated pits and vesicles is assisted and controlled by a number of accessory factors that interact with each other to form large macromolecular complexes. They have provided, and probably will keep providing, a number of new methods to inhibit clathrin-dependent endocytosis, some of which have just been described and have not yet been investigated in a detailed manner.

4. Multiple clathrin-independent pathways?

A number of other endocytic pathways do not involve clathrin. Most are poorly understood because of the lack of markers, and because in unperturbed systems, they are studied with a high background of clathrin-dependent endocytosis. They all function in a temperature- and energy-dependent fashion. Phagocytosis, a clathrin-independent, actin-dependent process that allows the ingestion of large particles and micro-organisms, and endocytosis in yeast will not be considered here (see Chapter 3 and 10, respectively).

4.1 Macropinocytosis

Macropinocytosis is a non-clathrin mediated endocytic pathway (see Chapter 4 and ref. 114). It constitutively occurs in macrophages and many tumour cells and can be induced, for instance with growth factors, in other cells. Macropinosomes form from membrane ruffles that close to form vesicles that can reach sizes of 5 μm in diameter. This route may account for the ingestion of large amounts of plasma membrane as well as large volumes of fluid. It resembles phagocytosis in that it involves large membrane extensions that fuse to form vesicles. Macropinocytosis can be distinguished from other pinocytic pathways by its susceptibility to agents such as cytochalasin D that depolymerize actin. Also, dimethyl amiloride is an inhibitor of macropinocytosis and not of other known pinocytic pathways. Since macropinocytosis can be induced and can take up large amounts of membrane or fluid, even limited induction would allow significant pinocytosis. Thus, after protein kinase C stimulation in A431 cells, ricin uptake increased due to macropinocytosis (115). Macropinocytosis might account, for instance, for some of the induced pinocytosis observed when cells are treated with inhibitors of clathrin-dependent endocytosis.

4.2 Caveolae

One clathrin-independent route for endocytosis involves caveolae (small caves), which are specialized microdomains of the plasma membrane defined by their unique morphology as viewed by EM (reviewed in refs 116–118; Fig. 2). They were initially observed over forty years ago, are present in many, but not all, cell types, and are very abundant in some cells, such as capillary endothelial cells, smooth muscle, and adipocytes. Caveolae are small flasked-shaped membrane invaginations that can be distinguished from coated pits by their size (50–80 nm diameter, compared to 100–110 nm for coated pits), their morphology and their striated coat consisting of evenly spaced filaments that are arranged in concentric whorls to form a 'fingerprint-like' structure. The study of caveolae was limited to morphological descriptions until a specific marker was uncovered, caveolin. The striated coat may represent caveolin complexes visible using scanning EM. Caveolae have a characteristic high content of this 21–24 kDa integral membrane protein. The caveolin family presently comprises three distinct gene products: caveolin 1 (VIP21/caveolin) and caveolin 2 are present in many cell types, but not in lymphocytes or neuroblastoma cells (118, 119), and co-associate in many cells. Caveolin 3 is present in skeletal and cardiac muscle and astrocytes (reviewed in ref. 120). Other caveolae-associated proteins, flotillins/cavatellins, which form a complex with caveolin family members when co-expressed, have been recently described and may participate in the formation of caveolae or caveolae-like vesicles (121). The precise function of caveolin is still not understood; on one hand it is a cholesterol-binding protein (122), and on the other hand, it has been implicated in

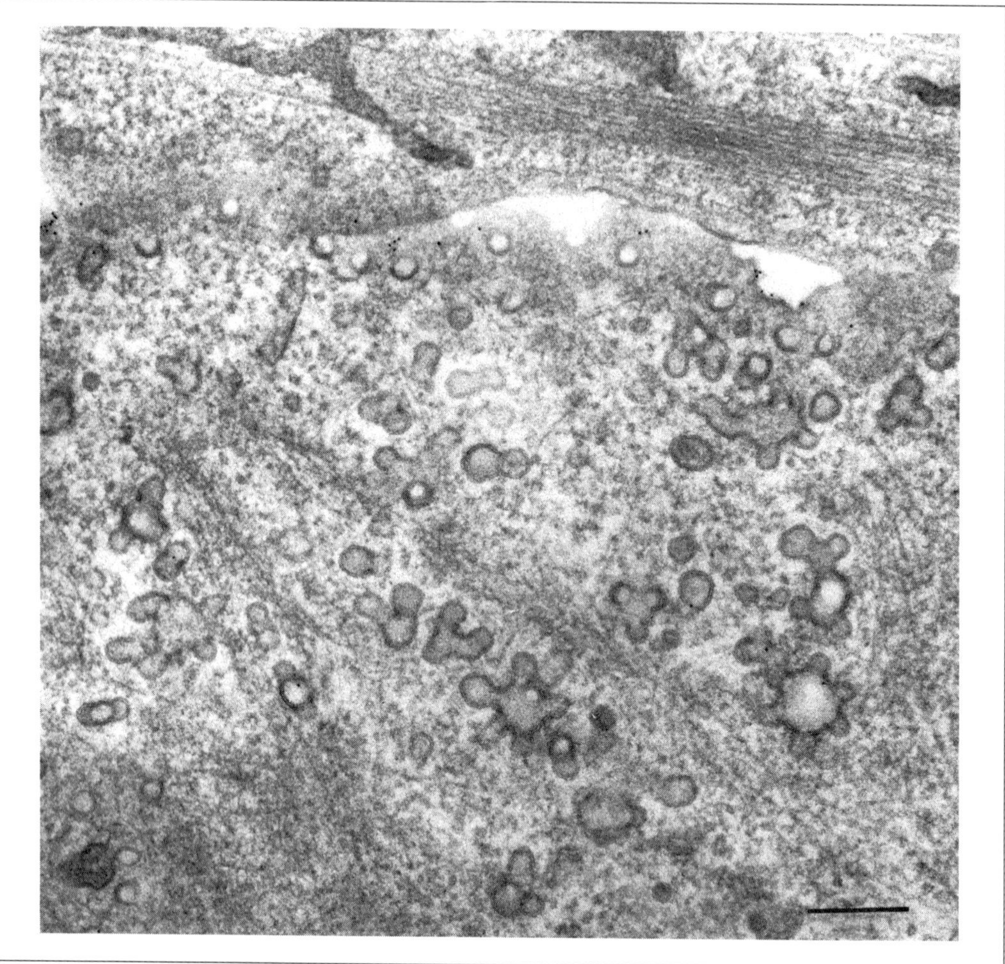

Fig. 2 Caveolae. Image from thin plastic section of mink lung (Mv-1-Lu) epithelial cells showing the distribution of caveolae associated with the plasma membrane and sub-plasma membrane regions of the cell. The cells were incubated with 5 nm BSA-gold particles for 10 min at 37 °C prior to fixation. Gold particles can be seen in numerous typical flask-shaped plasma membrane-associated caveolae and in vesicular profiles. This image is taken from a grazing section cut nearly parallel to the cell surface. Consequently, many of the vesicular profiles and caveolae clusters may still be connected to the plasma membrane. Bar = 200 nm. Provided by Annegret Pelchen-Matthews (MRC-LMCB University College London).

several signal transduction events (117, 120). A good correlation exists between the tissue distribution of caveolins and the presence of caveolae. Consistent with this, heterologous expression of caveolin 1 in cells devoid of caveolae induced their formation (123, 124). While caveolins are present in cell surface caveolae, caveolin 1 is also associated with the Golgi complex and cycles to the Golgi and back to the surface via the TGN (125).

Caveolae have a distinct lipid composition: they are highly enriched for glycosphingolipids and cholesterol. Consequently, they partition into a low-density Triton X-100-insoluble cell fraction and depend on cholesterol for their integrity (117, 120, 126).

The existence of lateral assemblies of lipids in membranes, called 'rafts', was proposed a few years ago. Since then, these structures have been a subject of great interest (127, 128). They form by the association of sphingolipids packed together with cholesterol and constitute microdomains within the membrane bilayer. Rafts are viewed as platforms in which some molecules may be concentrated and are maintained by lipid–lipid interactions. Several reviews on rafts have been published recently (117, 120, 126, 129–131). Experimental evidence is accumulating to support the existence of rafts. Thus, using non-ionic detergent extraction of cells at low temperature, some lipids and proteins were found in a detergent insoluble fraction that can be isolated on a sucrose gradient. This fraction, called DIG (detergent insoluble glycolipid-rich domain), or DIMs (detergent insoluble membranes), DRMs (detergent-resistant membranes), is enriched in glycosphingolipids, sphingomyelin, and cholesterol. Rafts appear to be a general organizing principle of mammalian cell membranes and clearly function in concentrating signalling molecules (117, 120).

Like DIGs, caveolae are insoluble in certain non-ionic detergent and, although caveolae are not the same as DIGs, caveolae appear to be specific forms of DIGS or rafts which may be formed through the coalescence of smaller raft domains (discussed in refs 117, 126, 130, 132). Caveolin 1 can polymerize into large protein complexes (133) and tightly binds cholesterol. In this way, it may function as a raft organizer.

It has long been proposed that in endothelial cells, caveolae are involved in transcytosis of solutes across the endothelium (134–137). In fibroblasts and hepatoma cell lines, several reports suggested that caveolae may play a role in uptake of glycolipids and glycosylphosphatidylinositol (GPI) anchored proteins (2, 3, 138). Although caveolae were initially implicated in the uptake of small solutes such as folate in a process named potocytosis (139), their role in this process has been questioned (140, 141).

There has been some controversy over the years as to whether caveolae-mediated internalization occurs via true vesicles. Recently, several lines of evidence have suggested that caveolae do undergo budding, docking, and fusion events, as do other trafficking vesicles, and participate in membrane trafficking. An assay in fibroblasts, based on clustering of a GPI-anchored protein by antibodies and determining the relative amount of protein remaining on the surface, showed that caveolae are dynamic structures which can be internalized into the cell in a time- and temperature-dependent way (138). Purification provided evidence that endothelial cell caveolae have the molecular machinery necessary for fusion of budded vesicles such as NSF and a v-SNARE (142, 143). Furthermore, a cell-free system that reconstituted fission of caveolae from plasma membranes was developed. In this assay, fission required cytosol, GTP hydrolysis, and dynamin, and the budded caveolae contained the molecular machinery for endocytosis and transcytosis (101, 136). GTP, but not GTPγS, stimulated caveolar budding and endocytosis of the cholera toxin B chain, bound to its receptor, the ganglioside GM1, a marker of caveolae, to endosomes. A cell-free assay measuring the detachment of caveolae containing cholera toxin and established with murine fibroblasts confirmed these results (144).

Some receptors appear to be internalized via caveolae as well as clathrin-coated pits. The insulin receptor was reported to utilize caveolae in endothelium in addition to clathrin-coated pits (135) and the cholecystokinin receptor was also internalized by both pathways (26). A portion of the growth hormone receptor was aggregated in caveolae upon ligand stimulation as observed by EM and was internalized in caveolin containing vesicles in CHO cells expressing the receptor. Transient transfection of caveolin and growth hormone receptor cDNAs increased hormone internalization (145). The autocrine motility factor receptor was localized within caveolae at the surface of murine fibroblasts as seen by EM, suggesting that it may be internalized via caveolae, and after 30 minutes internalization, it was present in the ER (146).

Some viruses and toxins represent interesting examples of caveolae uptake. It has been shown that the infectious entry process of SV40 involves caveolae-mediated uptake of virus particles (147–150). By EM SV40 appeared to be concentrated in caveolin 1-labelled invaginations while it was not found in clathrin-coated pits (148). In cells treated with drugs that disrupted caveolae but did not affect transferrin internalization, the cholesterol-binding drug nystatin and the phorbol ester PMA, SV40 particles pre-adsorbed on the cell surface were not taken up, and the drug effect was reversible (147). Several results suggest that MHC I is part of the receptor for SV40 (147, 150). Antibodies to MHC I induced their clustering in caveolin 1-positive invaginations suggesting that MHC I is part of a group of surface proteins, including GPI-anchored ones, which upon cross-linking associate with caveolae (141, 151) and can even participate in caveolae induction on the apical plasma membrane of MDCK cells (152). A model proposed for SV40 internalization is that the virus, presumably bound to MHC I, recruits caveolin around the particle and participates in generating a caveola-like pit (149). It is noteworthy that SV40 entry is notaby slow: it does not enter until about 1.5 hours after it is pre-adsorbed on the cell surface (147).

The receptor for cholera toxin, the ganglioside GM1, was shown, in a study of the ultrastructural distribution of gold-labelled cholera toxin B, to be enriched fourfold in caveolae. Consequently, it was inferred that caveolae may be the major vehicle for toxin internalization (reviewed in refs 149, 153). Recently, it was reported that both caveolae and caveolae-like glycolipid rafts act as vehicles for cholera toxin internalization and activation (154).

As already mentioned, dynamin appears to mediate scission of caveolae from the plasma membrane as shown in cells expressing the dynamin K44A mutant (100, 101). Also, when specific anti-dynamin antibodies were injected into cultured mammalian cells, non-coated pits resembling plasmalemmal caveolae were accumulated and the cellular uptake of cholera toxin B was inhibited (155). Specific inhibitors of caveolae uptake are not yet well characterized. However, it has been reported that filipin, a drug that binds sterol, under specific conditions, inhibits caveolae-mediated but not clathrin-coated pit mediated processes (154).

4.3 Other pathways

Several lines of evidence coming from the study of different endocytic markers suggest the existence of additional pathways. Thus, interesting results were obtained studying the plant toxin ricin that binds both glycoproteins and glycolipids with terminal galactose. Ricin labels the cell surface and may be taken up by various structures that pinch off (reviewed in ref. 23). It is transported through the endosomal system to lysosomes and also to the Golgi apparatus from where some of it enters the ER by retrograde transport. The toxin is then translocated to the cytosol where it exerts its final action, i.e. the inactivation of the 60S ribosome subunit, and thereby kills the cell (23). Early studies, based on pharmacological inhibitors of clathrin-mediated uptake, revealed that ricin is taken up by clathrin-dependent as well as clathrin-independent mechanisms (153). This was confirmed by the fact that in cells overexpressing the clathrin hub fragment discussed above, ricin was still taken up (23). Remarkably, in HeLa cells overexpressing the K44A dynamin 1 mutant, ricin internalization still took place. Therefore, clathrin-independent internalization of ricin can clearly be different from uptake which might occur from caveolae, since the latter is inhibited when this dynamin mutant is overexpressed (100, 101, 155). In agreement with this, in polarized MDCK cells there is a regulated uptake of ricin from the apical side while caveolae are found on the basolateral side but are absent from the apical membrane (156, 157). Another argument in favour of a clathrin- and caveolae-independent pathway for ricin comes from the use of β methyl-cyclodextrin, which causes an acute cholesterol depletion in treated cells. When β methyl-cyclodextrin was used, caveolae were not detected, and the formation of CCVs and transferrin internalization were impaired (158, 159). However, ricin uptake still took place. This apical clathrin-independent ricin uptake could function in the presence of the non-hydrolysable GTPγS, in contrast to other internalization processes described so far (23).

Similarly, the uptake of a GPI-linked version of the diphteria toxin was analysed in HeLa cells expressing the dominant negative dynamin mutation. The GPI-diptheria toxin uptake was not affected by treatments which affect both clatrin-coated pit and caveolae-mediated internalization: overexpression of the dynamin mutation, treatment with β methyl-cyclodextrin or by nystatin (160). The internalization of another GPI-anchored protein, the urokinase plasminogen activator receptor, was studied in polarized and non-polarized MDCK cells and appeared to follow different routes. In non-polarized cells, in the presence of cross-linking antibodies, it clustered in caveolae. In addition, it entered the apical side of MDCK cells which lack caveolae (161).

A virus, murine leukaemia virus also entered cells after binding to a cell surface receptor in a way that did not require dynamin (162).

Recently, the uptake of cholera toxin was studied in cells that lack caveolae or which were treated with filipin, a sterol-binding drug that did not inhibit clathrin-coated pit internalization under the conditions used. This is in contradiction with the fact that cells treated with β methyl-cyclodextrin are inhibited for CCV uptake, as just mentioned; this apparent contradiction may be related to the extent of cholesterol depletion. The results from this study indicate that the internalization and activation of cholera toxin B was mediated through cholesterol- and glycolipid-rich microdomains at the plasma membrane rather than through a specific morphological structure (154).

Physiologically important receptors also enter cells by a clathrin- and caveolae-independent pathway. Some GPCRs represent an interesting case. Desensitization, which is important for turning off receptor-mediated signal transduction, takes place by two processes: receptor uncoupling from G protein and receptor internalization. There is increasing evidence that certain of these signalling receptors can be internalized by alternate mechanisms, in addition to agonist-induced internalization by coated pits. The angiotensin II type 1A receptor (AT1AR) does not require dynamin for its internalization and is independent of the function of arrestins (163). Arrestins are critical components for cellular trafficking of other GPCRs such as the β2-adrenergic receptors, specifically targeting them for dynamin-dependent endocytosis via CCVs (see above). Another case is that of the m2 muscarinic acetylcholine receptor (m2 mAchR), a member of the mAchR family that includes five subtypes termed m1–m5. It was reported that the m2 mAchR internalized predominantly by a dynamin- and arrestin-independent pathway, even when arrestins were overexpressed (164). Furthermore, while β2-adrenergic receptors are internalized via coated pits in some cell types, a previous study suggested that they may enter in a different manner in other cells (6). Other members of the same family, the m1, m3, and m4 mAchR, were internalized in an arrestin-independent but dynamin-dependent pathway (165). The fact that it is arrestin-independent does not prove that it cannot interact with clathrin-coated pits, and therefore, it is not yet established whether the latter receptors follow a clathrin-dependent or -independent way.

The D1 and D2 dopamine receptors are also taken up by distinct dynamin-dependent and -independent mechanisms, respectively (166). Both operate with similar kinetics and are rapidly internalized ($t_{1/2} \cong 5$ min). In the presence of the physiological agonist dopamine, both D1 and D2 were observed in numerous endocytic vesicles in neuroblastoma and non-neuronal cells. The dynamin 1 mutant inhibited dopamine-induced internalization of the D1 receptor which co-localized with clathrin, while it did not affect internalization of the D2 receptor. At early internalization times, both receptors were found in distinct intracellular vesicles (166).

The dynamin-independence of the pathway followed by these GPCRs strongly suggest that they are not internalized via caveolae. Their pathway has similar characteristics, at far as is now known, to the clathrin-independent one of ricin, same clathrin- and dynamin-independence, except that ricin uptake is a much less efficient process (less than 10% internalized after 10 min) (115) than uptake of receptors such as dopamine receptors (60% internalized in less than 10 min) (166).

The IL-2 receptor is another interesting case of a signalling growth factor receptor. It is composed of three chains, α, β, and γ, and is normally expressed on activated lymphocytes. The IL-2 receptor is rapidly internalized by a clathrin-independent pathway in these cells, that are devoid of caveolae. This was shown using pharmacological approaches (24), as well as specific inhibition of clathrin-mediated entry, using overexpression of Eps15 mutants. In agreement with their clathrin-independent entry, the IL-2 receptor β and γ chains have several weak internalization signals that do not resemble the typical clathrin-coated pit localization signal (167, 168). Remarkably, based on the involvement of dynamin, the mechanism of uptake of the IL-2 receptor appears to be different from that of ricin and of the GPCRs described above. Indeed, its internalization is inhibited by overexpression of the dynamin K44A mutant, but not of Eps15 (Lamaze *et al.*, submitted).

What could be the characteristics of these clathrin- and caveolae-independent pathways? It is generally accepted in the field of intracellular trafficking that sorting depends on the presence of some protein coat. Except for clathrin-coated pits and caveolae, no other coated structures have been observed on the plasma membrane. It is possible that some other coats exist which could not be detected by the methods used. Another interesting alternative is that membrane microdomains, defined by their composition in lipids, might participate in sorting some membrane components that would interact, even weakly, with these lipids. Considering that transferrin receptors are concentrated only about fivefold in clathrin-coated pits (169, 170) and that this interaction is sufficient to cause their rapid internalization in these structures, a small concentration factor in some membrane domains would represent a sufficient sorting mechanism. As discussed above, there is growing evidence that such domains, i.e. localized regions with non-random lipid compositions, exist in biological membranes. Some membrane proteins may interact preferentially with a subclass of lipids. Examples of lipid-based sorting in the endocytic pathway are already known (reviewed in ref. 171). Cross-linked GPI-anchored proteins associate preferentially with cholesterol/sphingolipid-enriched raft domains, even more so following aggregation or cross-linking by a ligand or antibody (141, 172). The sorting and trafficking of lipid analogues in sorting endosomes has been reported to be regulated by lipid acyl chain properties and the resultant partitioning preferences (173). Concerning late endosomes, preferential retention of certain lipids has been reported (174) and a negatively charged lipid, LBPA, has been shown to concentrate in late endosome internal membranes. These LBPA-enriched membranes were critical for the sorting of transmembrane proteins such as MPR from late endosomes to the TGN (175). Similar examples are also found in other intracellular compartments where sorting takes place (171, 176).

How does a membrane invaginate to form a pit and a vesicle? Little is known, but it appears that lipid properties could be responsible for membrane invagination. Thus, a small increase of lipid concentration in the plasma membrane inner layer enhanced internalization (177). Furthermore, it was recently reported that endophilin 1 exhibits lysophosphatidic acid acyl transferase activity, and thereby may induce negative membrane curvature by converting an inverted-cone-shaped lipid to a cone-shaped one in the cytoplasmic leaflet of the bilayer. Based on these data, it was proposed that endophilin 1, acting together with dynamin, mediates synaptic vesicle invagination from the plasma membrane and fission (92).

4.4 Later steps along the endocytic pathway

Very few data are available as to the intracellular destination of components taken up by clathrin-independent pathways. This might differ depending on their entry route. Also, it is difficult to address this question for the following reasons. In many cases, the molecules studied enter cells both in clathrin-dependent and -independent ways and it may not be simple to distinguish which pathway leads where. Also, as these studies are usually performed on perturbed cells to inhibit the initial steps of clathrin-mediated internalization, later steps of sorting may also be affected.

In some instances, internalized components are found in early/recycling endosomes from where they can reach their final destination, as is the case for clathrin-mediated endocytosis. The cholera toxin endocytic pathway was one of the first clathrin-independent ones to be followed by EM in cultured liver cells and murine fibroblasts (2, 3). After internalization, it was found in a tubulovesicular compartment and multivesicular bodies where it co-localized with markers known to enter via coated pits. This suggested a common intracellular pathway for ligands that enter a cell via coated and non-coated invaginations on the plasma membrane. Gold-labelled concanavalin A, initially bound to the plasma membrane, was first found in non-coated vesicles and then in endosomes (9). Similarly, ricin is transported to endosomes from where it gets to lysosomes and also to the Golgi apparatus and then the ER by retrograde transport (23). The GPI-diphteria toxin, internalized independently of clathrin-coated pits and caveolae, was shown to be transported to vesicles containing EEA1, a marker for early endosomes (160).

The IL-2 receptors also reach early endosomes where they co-localize with transferrin (178). However, considering the small number of receptors and the size and morphology of lymphocytes that have a small, packed cytoplasm, it is not possible to rule out that some receptors might reach another intracellular destination. Nevertheless, the cytosolic tail of the IL-2 receptor β chain carries a motif of eight amino acids that is necessary for its intracellular sorting. This signal is sufficient to sort a normally recycling receptor towards late endosomes/lysosomes. It does not function as an internalization signal and is different from the signals involved in clathrin-dependent endocytosis (179, 180).

In other cases, clathrin-dependent and -independent delivery to intracellular compartments may differ. The D1 and D2 G protein-coupled dopamine receptors are internalized by distinct pathways as defined by their dependence on dynamin. The receptors are then specifically delivered to different primary endocytic vesicles when co-expressed in neuroblastoma cells as well as in human fibroblastic kidney cells (166). After this segregation in different early endosomes, both receptors recycle to the plasma membrane.

SV40 particles, after internalization via caveolin-enriched pits, accumulate in the ER. How they get there is still not known. They could either fuse directly with the ER, or alternatively be transported to early endosomes and from there to the Golgi and the ER (see refs 149, 150).

4.5 Potential functions

Why should several pathways exist? The answers to that question are mostly speculative but several comments can be made. First, it is very striking that the kinetics of uptake by the different pathways vary considerably. The half-time of uptake can be as short as a few minutes for some receptors, even a few seconds in specific cells after stimulated exocytosis, and can be very slow, for instance when caveolae are induced on the apical side of MDCK cells (152), or for the uptake of SV40 particles (half-time of 1.5 hours). The existence of several pathways may be one way to internalize different membrane components at different rates. In some cell types, one pathway may be more developed to answer a particular need. For instance, rapid endocytosis is needed for membrane retrieval after stimulated exocytosis in pituitary cells (181) and in adrenal chromaffin cells (182). Secondly, it is possible that some of the intracellular routes after uptake may differ according to the entry pathway. For instance, one could speculate that in certain cases, uptake via caveolae may facilitate transport to the ER, and that this may be important for some ligands and receptors as in the case of SV40 which is tranported to the ER. Finally, another potential function of clathrin-independent endocytosis may be related to signal transduction. Membrane rafts or caveolae, because of their lipid composition, may have a key role in concentrating molecules involved in signal transduction (120, 132, 183). Their ability to get internalized may provide an efficient way to modulate the activity of receptors and signal transducing complexes that localize to these specialized domains.

5. Conclusions

As we have discussed above, clathrin-independent endocytic traffic in mammalian cells is complex and poorly understood. The initial steps taking place at the plasma membrane are beginning to be described in a few cases but those taking place afterwards are not. The molecular characteristics that govern the uptake of membrane receptors via these pathways are unknown. While many key molecules controlling membrane traffic via coated pits have been cloned, purified, mutated, and well studied, we have no clue as to what they are for clathrin-independent pathways. No specific inhibitors are available, and there is a serious need for specific markers. One of the major tasks for the future is to find such markers, to uncover molecules controlling this traffic, to understand how it is regulated in living cells, and what specific functions it fulfils. Very few studies concerned this topic until recently, and, as some tools have recently become available, there should be a lot of progress in the near future.

Acknowledgements

I am very grateful to A. Benmerah, C. Lamaze, and A. Subtil for numerous discussions and for their constructive comments on this manuscript.

References

1. Huet, C., Ash, J. F., and Singer, S. J. (1980) The antibody-induced clustering and endocytosis of HLA antigens on cultured human fibroblasts. *Cell*, **21**, 429.
2. Montesano, R., Roth, J., Robert, A., and Orci, L. (1982) Non-coated membrane invaginations are involved in binding and internalization of cholera and tetanus toxins. *Nature*, **296**, 651.
3. Tran, D., Carpentier, J.-L., Sawano, F., Gorden, P., and Orci, L. (1987) Ligands internalized through coated or noncoated invaginations follow a common intracellular pathway. *Proc. Natl. Acad. Sci. USA*, **84**, 7957.
4. Smith, R. M. and Jarett, L. (1988) Receptor-mediated endocytosis and intracellular processing of insulin: ultrastructural and biochemical evidence for cell-specific heterogeneity and distinction from nonhormonal ligands. *Lab. Invest.*, **58**, 613.
5. Hamer, I., Haft, C. R., Paccaud, J. P., Maeder, C., Taylor, S., and Carpentier, J. L. (1997) Dual role of a dileucine motif in insulin receptor endocytosis. *J. Biol. Chem.*, **272**, 21685.
6. Raposo, G., Dunia, I., Delavier-Klutchko, C., Kaveri, S., Strosberg, A. D., and Benedetti, E. L. (1989) Internalization of beta-adrenergic receptor in A431 cells involves non-coated vesicles. *Eur. J. Cell Biol.*, **50**, 340.
7. Conway, E. M., Boffa, M. C., Nowakowski, B., and Steiner-Mosonyi, M. (1992) An ultrastructural study of thrombomodulin endocytosis: internalization occurs via clathrin-coated and non-coated pits. *J. Cell. Physiol.*, **151**, 604.
8. Conway, E. M., Nowakowski, B., and Steiner-Mosonyi, M. (1994) Thrombomodulin lacking the cytoplasmic domain efficiently internalizes thrombin via nonclathrin-coated, pit-mediated endocytosis. *J. Cell. Physiol.*, **158**, 285.
9. Hansen, S. H., Sandvig, K., and van Deurs, B. (1991) The preendosomal compartment comprises distinct coated and noncoated endocytic vesicle populations. *J. Cell Biol.*, **113**, 731.
10. Larkin, J. M., Brown, M. S., Goldstein, J. L., and Anderson, R. G. W. (1983) Depletion of intracellular potassium arrests coated pit formation and receptor-mediated endocytosis in fibroblasts. *Cell*, **33**, 273.
11. Heuser, J. (1989) Effects of cytoplasmic acidification on clathrin lattice morphology. *J. Cell Biol.*, **108**, 401.
12. Larkin, J. M., Donzell, W. C., and Anderson, R. G. W. (1985) Modulation of intracellular potassium and ATP: effects on coated pit function in fibroblasts and hepatocytes. *J. Cell. Physiol.*, **124**, 372.
13. Larkin, J. M., Donzell, W. C., and Anderson, R. G. W. (1986) Potassium-dependent assembly of coated pits: new coated pits form as planar clathrin lattices. *J. Cell Biol.*, **103**, 2619.
14. Moya, M., Dautry-Varsat, A., Goud, B., Louvard, D., and Boquet, P. (1985) Inhibition of coated pit formation in HEP2 cells blocks the cytotoxicity of diphteria toxin but not that of ricin toxin. *J. Cell Biol.*, **101**, 548.
15. Daukas, G. and Zigmond, S. H. (1985) Inhibition of receptor-mediated but not fluid-phase endocytosis in polymorphonuclear leukocytes. *J. Cell Biol.*, **101**, 1673.
16. Oka, J. A. and Weigel, P. H. (1988) Effects of hyperosmolarity on ligand processing and receptor recycling in the hepatic galactosyl receptor system. *J. Cell. Biochem.*, **36**, 169.
17. Sandvig, K., Olsnes, S., Petersen, O. W., and Van Deurs, B. (1989) Control of coated-pit function by cytoplasmic pH. *Methods Cell Biol.*, **32**, 365.
18. Heuser, J. E. and Anderson, R. G. W. (1989) Hypertonic media inhibit receptor-mediated endocytosis by blocking clathrin-coated pit formation. *J. Cell Biol.*, **108**, 389.

19. Sandvig, K., Olsnes, S., Petersen, O. W., and van Deurs, B. (1987) Acidification of the cytosol inhibits endocytosis from coated pits. *J. Cell Biol.*, **105**, 679.
20. Davoust, J., Gruengerg, J., and Howell, K. (1987) Two threshold values of low pH block endocytosis at different stages. *EMBO J.*, **6**, 3601.
21. Cosson, P., de Curtis, I., Pouyssegur, J., Griffiths, G., and Davoust, J. (1989) Low cytoplasmic pH inhibits endocytosis and transport from the trans-Golgi network to the cell surface. *J. Cell Biol.*, **108**, 377.
22. Cupers, P., Veithen, A., Kiss, A., Baudhuin, P., and Courtoy, P. J. (1994) Clathrin polymerization is not required for bulk-phase endocytosis in rat fetal fibroblasts. *J. Cell Biol.*, **127**, 725.
23. Sandvig, K. and van Deurs, B. (1999) Endocytosis and intracellular transport of ricin: recent discoveries. *FEBS Lett.*, **452**, 67.
24. Subtil, A., Hémar, A., and Dautry-Varsat, A. (1994) Rapid endocytosis of interleukin 2 receptors when clathrin-coated pit endocytosis is inhibited. *J. Cell Sci.*, **107**, 3461.
25. Subtil, A. and Dautry-Varsat, A. (1997) Microtubule depolymerization inhibits clathrin-coated-pit internalization in non-adherent cell lines while interleukin 2 endocytosis is not affected. *J. Cell Sci.*, **110**, 2441.
26. Roettger, B. F., Rentsch, R. U., Pinon, D., Holicky, E., Hadac, E., Larlin, J. M., *et al.* (1995) Dual pathways of internalization of the cholecystokinin receptor. *J. Cell Biol.*, **128**, 1029.
27. Altankov, G. and Grinnell, F. (1995) Fibronectin receptor internalization and AP-2 complex reorganization in potassium-depleted fibroblasts. *Exp. Cell Res.*, **216**, 299.
28. Ilondo, M. M., Courtoy, P. J., Geiger, D., Carpentier, J. L., Rousseau, G. G., and De Meyts, P. (1986) Intracellular potassium depletion in IM-9 lymphocytes suppresses the slowly dissociating component of human growth hormone binding and the down-regulation of its receptors but does not affect insulin receptors. *Proc. Natl. Acad. Sci. USA*, **83**, 6460.
29. Backer, J. M., Shoelson, S. E., Haring, E., and White, M. F. (1991) Insulin receptors internalize by a rapid, saturable pathway requiring receptor autophosphorylation and an intact juxtamembrane region. *J. Cell Biol.*, **115**, 1535.
30. Lund, K. A., Opresko, L. K., Starbuck, C., Walsh, B. J., and Wiley, H. S. (1990) Quantitative analysis of the endocytic system involved in hormone-induced receptor internalization. *J. Biol. Chem.*, **265**, 15713.
31. Stoorvogel, W., Oorschot, V., and Geuze, H. J. (1996) A novel class of clathrin-coated vesicles budding from endosomes. *J. Cell Biol.*, **132**, 21.
32. Liu, S.-H., Marks, M. S., and Brodsky, F. M. (1998) A dominant-negative clathrin mutant differentially affects trafficking of molecules with distinct sorting motifs in the class II major histocompatibility complex (MHC) pathway. *J. Cell Biol.*, **140**, 1023.
33. Kirchhausen, T., Bonifacino, J. S., and Riezman, H. (1997) Linking cargo to vesicle formation: role in intracellular transport and protein sorting. *Curr. Opin. Cell Biol.*, **9**, 488.
34. Hofmann, M. W., Honing, S., Rodionov, D., Dobberstein, B., von Figura, K., and Bakke, O. (1999) The leucine-based sorting motifs in the cytoplasmic domain of the invariant chain are recognized by the clathrin adaptors AP1 and AP2 and their medium chains. *J. Biol. Chem.*, **274**, 36153.
35. Rapoport, I., Chen, Y. C., Cupers, P., Shoelson, S. E., and Kirchhausen, T. (1998) Dileucine-based sorting signals bind to the beta chain of AP-1 at a site distinct and regulated differently from the tyrosine-based motif-binding site. *EMBO J.*, **17**, 2148.
36. Nesterov, A., Carter, R. E., Sorkina, T., Gill, G. N., and Sorkin, A. (1999) Inhibition of the receptor-binding function of clathrin adaptor protein AP-2 by dominant-negative mutant μ2 subunit and its effects on endocytosis. *EMBO J.*, **18**, 2489.

37. Lamaze, C., Baba, T., Redelmeier, T. E., and Schmid, S. L. (1993) Recruitment of epidermal growth factor and transferrin receptors *in vitro*: differing biochemical requirements. *Mol. Biol. Cell*, **4**, 715.
38. Marsh, M. and McMahon, H. T. (1999) The structural era of endocytosis. *Science*, **285**, 215.
39. Jarousse, N. and Kelly, R. B. (2000) Selective inhibition of adaptor complex-mediated vesiculation. *Traffic*, **1**, 378.
40. Owen, D. J., Vallis, Y., Noble, M. E., Hunter, J. B., Dafforn, T. R., Evans, P. R., *et al.* (1999) A structural explanation for the binding of multiple ligands by the alpha-adaptin appendage domain. *Cell*, **97**, 805.
41. Fazioli, F., Minichiello, L., Matoskova, B., Wong, W. T., and Di Fiore, P. P. (1993) Eps15, a novel tyrosine kinase substrate, exhibits transforming activity. *Mol. Cell. Biol.*, **13**, 5814.
42. Benmerah, A., Gagnon, J., Bègue, B., Mégarbané, B., Dautry-Varsat, A., and Cerf-Bensussan, N. (1995) The tyrosine kinase substrate EPS15 is constitutively associated with the plasma membrane adaptor AP-2. *J. Cell Biol.*, **131**, 1831.
43. Benmerah, A., Bègue, B., Dautry-Varsat, A., and Cerf-Bensussan, N. (1996) The ear of α-adaptin interacts with the COOH terminal domain of the Eps15 protein. *J. Biol. Chem.*, **271**, 12111.
44. Tebar, F., Sorkina, A., Sorkin, A., and Kirchhausen, T. (1997) Eps15 is a component of clathrin-coated pits and vesicles and is located at the rim of coated pits. *J. Biol. Chem.*, **272**, 15413.
45. Tebar, F., Confalonieri, S., Carter, R. E., Di Fiore, P. P., and Sorkin, A. (1997) Eps15 is constitutively oligomerized due to homophilic interaction of its coiled-coil region. *J. Biol. Chem.*, **272**, 15413.
46. Iannolo, G., Salcini, A. E., Gaidarov, I., Goodman, O. B. J., Baulida, J., Carpenter, G., *et al.* (1997) Mapping of the molecular determinants involved in the interaction between eps15 and AP-2. *Cancer Res.*, **57**, 240.
47. Benmerah, A., Poupon, V., Bensussan-Cerf, N., and Dautry-Varsat, A. (2000) Mapping of Eps 15 domains involved in its targeting to clathrin-coated pits. *J. Biol. Chem.*, **275**, 3288.
48. Cupers, P., ter Haar, E., Boll, W., and Kirchhausen, T. (1997) Parallel dimers and anti-parallel tetramers formed by epidermal growth factor receptor pathway substrate clone 15. *J. Biol. Chem.*, **272**, 33430.
49. Di Fiore, P. P., Pelicci, P. G., and Sorkin, A. (1997) EH: a novel protein-protein interaction domain potentially involved in intracellular sorting. *Trends Biochem. Sci.*, **22**, 411.
50. Salcini, A. E., Confalonieri, S., Doria, M., Santolini, E., Tassi, E., Minenkova, O., *et al.* (1997) Binding specificity and *in vivo* targets of the EH domain, a novel protein-protein interaction module. *Genes Dev.*, **11**, 2239.
51. Benedetti, H., Raths, S., Crausaz, F., and Riezman, H. (1994) The END3 gene encodes a protein that is required for the internalization step of endocytosis and for actin cytoskeleton organization in yeast. *Mol. Biol. Cell*, **5**, 1023.
52. Wendland, B., McCaffery, J. M., Xiao, Q., and Emr, S. D. (1996) A novel fluorescence-activated cell sorter-based screen for yeast endocytosis mutants identifies a yeast homologue of mammalian eps15. *J. Cell Biol.*, **135**, 1485.
53. Benmerah, A., Lamaze, C., Bègue, B., Schmid, S. L., Dautry-Varsat, A., and Cerf-Bensussan, N. (1998) AP-2/Eps15 interaction is required for receptor-mediated endocytosis. *J. Cell Biol.*, **140**, 1055.
54. Benmerah, A., Lamaze, C., Bayrou, M., Cerf-Bensussan, N., and Dautry-Varsat, A. (1999) Inhibition of clathrin-coated pit assemby by an Eps15 mutant. *J. Cell Sci.*, **112**, 1303.
55. Schumacher, C., Knudsen, B. S., Ohuchi, T., Di Fiore, P. P., Glassman, R. H., and

Hanafusa, H. (1995) The SH3 domain of Crk binds specifically to a conserved proline-rich motif in Eps15 and Eps15R. *J. Biol. Chem.*, **270**, 15341.
56. Carbone, R., Fre, S., Iannolo, G., Belleudi, F., Mancini, P., Pelicci, P. G., *et al.* (1997) Eps15 and eps15R are essential components of the endocytic pathway. *Cancer Res.*, **57**, 5498.
57. Chen, H., Fre, S., Slepnev, V. I., Capua, M. R., Takei, K., Butler, M. H., *et al.* (1998) Epsin is an EH-domain-binding protein implicated in clathrin-mediated endocytosis. *Nature*, **394**, 793.
58. Rosenthal, J. A., Chen, H., Slepnev, V. I., Pellegrini, L., Salcini, A. E., Di Fiore, P. P., *et al.* (1999) The Epsins define a family of proteins that interact with components of the clathrin coat and contain a new protein module, *J. Biol. Chem.*, **274**, 33959.
59. Wendland, B., Steece, K. E., and Emr, S. D. (1999) Yeast epsins contain an essential N-terminal ENTH domain, bind clathrin and are required for endocytosis. *EMBO J.*, **18**, 4383.
60. Hyman, J., Chen, H., Di Fiore, P. P., De Camilli, P., and Brunger, A. T. (2000) Epsin 1 undergoes nucleocytosolic shuttling and its Eps15 interactor N-terminal domain, structurally similar to Armadillo and HEAT repeats, interacts with the transcripton factor promyelocytic leukemia Zn finger protein. *J. Cell Biol.*, **149**, 537.
61. Wigge, P. and McMahon, H. T. (1998) The amphiphysin family of proteins and their role in endocytosis at the synapse. *Trends Neurosci.*, **21**, 339.
62. Takei, K., Slepnev, V. I., Haucke, V., and De Camilli, P. (1999) Functional partnership between amphiphysin and dynamin in clathrin-mediated endocytosis. *Nat. Cell Biol.*, **1**, 33.
63. Slepnev, V. I., Ochoa, G. C., Butler, M. H., Grabs, D., and Camilli, P. D. (1998) Role of phosphorylation in regulation of the assembly of endocytic coat complexes. *Science*, **281**, 821.
64. Shupliakov, O., Low, P., Grabs, D., Gad, H., Chen, H., David, C., *et al.* (1997) Synaptic vesicle endocytosis impaired by disruption of dynamin-SH3 domain interactions. *Science*, **276**, 259.
65. Wigge, P., Vallis, Y., and McMahon, H. T. (1997) Inhibition of receptor-mediated endocytosis by the amphiphysin SH3 domain. *Curr. Biol.*, **7**, 554.
66. Owen, D. J., Wigge, P., Vallis, Y., Moore, J. D., Evans, P. R., and McMahon, H. T. (1998) Crystal structure of the amphiphysin-2 SH3 domain and its role in the prevention of dynamin ring formation. *EMBO J.*, **17**, 5273.
67. Simpson, F., Hussain, N. K., Qualmann, B., Kelly, R. B., Kay, B. K., McPherson, P. S., *et al.* (1999) SH3-domain-containing proteins function at distinct steps in clathrin-coated vesicle formation. *Nat. Cell Biol.*, **1**, 119.
68. Slepnev, V. I., Ochoa, G. C., Butler, M. H., and De Camilli, P. (2000) Tandem arrangement of the clathrin and AP-2 binding domains in amphiphysin 1 and disruption of clathrin coat function by amphiphysin fragments comprising these sites. *J. Biol. Chem.*, **275**, 17583.
69. Gold, E. S., Morrissette, N. S., Underhill, D. M., Guo, J., Bassetti, M., and Aderem, A. (2000) Amphiphysin IIm, a novel amphiphysin II isoform, is required for macrophage phagocytosis. *Immunity*, **12**, 285.
70. Ferguson, S. S. and Caron, M. G. (1998) G protein-coupled receptor adaptation mechanisms. *Cell Dev. Biol.*, **9**, 119.
71. Goodman, O. B. J., Krupnick, J. G., Gurevich, V. V., Benovic, J. L., and Keen, J. H. (1997) Arrestin/clathrin interaction. Localization of the arrestin binding locus to the clathrin terminal domain. *J. Biol. Chem.*, **272**, 15017.
72. Krupnick, J. G., Goodman, O. B. J., Keen, J. H., and Benovic, J. L. (1997) Arrestin/clathrin interaction. Localization of the clathrin binding domain of nonvisual arrestins to the carboxy terminus. *J. Biol. Chem.*, **272**, 15011.

73. Laporte, S. A., Oakley, R. H., and Zhang, J. (1999) The beta2-adrenergic receptor/beta arrestin complex recruits the clathrin adaptor AP-2 during endocytosis. *Proc. Natl. Acad. Sci. USA*, **96**, 3712.
74. Gaidarov, I., Krupnick, J. G., Falck, J. R., Benovic, J. L., and Keen, J. H. (1999) Arrestin function in G protein-coupled receptor endocytosis requires phosphoinositide binding. *EMBO J.*, **18**, 871.
75. Shpetner, H. S. and Vallee, R. B. (1989) Identification of dynamin, a novel mechanochemical enzyme that mediates interactions between microtubules. *Cell*, **59**, 421.
76. van der Bliek, A. M. and Meyerowitz, E. M. (1991) Dynamin-like protein encoded by the *Drosophila Sibire* gene associated with vesicular traffic. *Nature*, **351**, 411.
77. Chen, M. S., Obar, R. A., Schroeder, C. C., Austin, T. W., Poodry, C. A., Wadsworth, S. C., *et al.* (1991) Multiple forms of dynamin are encoded by *Shibire*, a *Drosophila* gene involved in endocytosis. *Nature*, **351**, 583.
78. Kosaka, T. and Ikeda, K. (1983) Possible temperature-dependent blockage of synaptic vesicle recycling induced by a single gene mutation in *Drosophila*. *J. Neurobiol.*, **14**, 207.
79. Cao, H., Garcia, F., and McNiven, M. (1998) Differential distribution of dynamin isoforms in mammalian cells. *Mol. Biol. Cell*, **9**, 2595.
80. Bottomley, M. J., Surdo, P. L., and Driscoll, P. C. (1999) Endocytosis: How dynamin sets vesicles PHree! *Curr. Biol.*, **9**, R301.
81. Yang, W. and Cerione, R. (1999) Endocytosis: is dynamin a blue collar or white collar worker? *Curr. Biol.*, **9**, R511.
82. Warnock, D. E. and Schmid, S. L. (1996) Dynamin GTPase, a force-generating molecular switch. *BioEssays*, **18**, 885.
83. Schmid, S. L., McNiven, M. A., and De Camilli, P. (1998) Dynamin and its partners: a progress report. *Curr. Opin. Cell Biol.*, **10**, 504.
84. McNiven, M. A. (1998) Dynamin: a molecular motor with pinchase action. *Cell*, **94**, 151.
85. McNiven, M. A., Cao, H., Pitts, K. R., and Yoon, Y. (2000) The dynamin family of mechanoenzymes: pinching in new places. *Trends Biochem. Sci.*, **25**, 115.
86. van der Bliek, A. M. (1999) Functional diversity in the dynamin family. *Trends Cell Biol.*, **9**, 96.
87. Sever, S., Muhlberg, A. B., and Schmid, S. L. (1999) Impairment of dynamin's GAP domain stimulates receptor-mediated endocytosis. *Nature*, **398**, 481.
88. Hinshaw, J. E. and Schmid, S. L. (1995) Dynamin self-assembles into rings suggesting a mechanism for coated vesicle budding. *Nature*, **374**, 190.
89. Takei, K., McPherson, P. S., Schmid, S. L., and De Camilli, P. (1995) Tubular membrane invaginations coated by dynamin rings are induced by GTP-γS in nerve terminals. *Nature*, **374**, 186.
90. Takei, K., Haucke, V., Splepnev, V., Farsad, K., Salazar, M., Chen, H., *et al.* (1998) Generation of coated intermediates of clathrin-mediated endocytosis on protein-free liposomes. *Cell*, **94**, 131.
91. Sweitzer, S. and Hinshaw, J. (1998) Dynamin undergoes a GTP-dependent conformational change causing vesiculation. *Cell*, **93**, 1021.
92. Schmidt, A., Wolde, M., Thiele, C., Fest, W., Kratzin, H., Podtelejnikov, A., *et al.* (1999) Endophilin I mediates synaptic vesicle formation by transfer of arachidonate to lysophosphatidic acid. *Nature*, **401**, 133.
93. van der Bliek, A. M., Redelmeier, T. E., Damke, H., Tisdale, E. J., Meyerowitz, E. M., and Schmid, S. L. (1993) Mutations in human dynamin block an intermediate stage in coated vesicle formation. *J. Cell Biol.*, **122**, 553.

94. Herskovits, J. S., Burgess, C. C., Obar, R. A., and Vallee, R. B. (1993) Effects of mutant rat dynamin on endocytosis. *J. Cell Biol.*, **122**, 565.
95. Damke, H., Baba, T., Warnock, D. E., and Schmid, S. L. (1994) Induction of mutant dynamin specifically blocks endocytic coated vesicle formation. *J. Cell Biol.*, **127**, 915.
96. Damke, H., Baba, T., Van der Bliek, A. M., and Schmid, S. L. (1995) Clathrin-independent pinocytosis is induced in cells overexpressing a temperature-sensitive mutant of dynamin. *J. Cell Biol.*, **131**, 69.
97. Vallis, Y., Wigge, P., Marks, B., Evans, P. R., and McMahon, H. T. (1999) Importance of the pleckstrin homology domain of dynamin in clathrin-mediated endocytosis. *Curr. Biol.*, **9**, 257.
98. Lee, A., Frank, D. W., Marks, M. S., and Lemmon, M. A. (1999) Dominant-negative inhibition of receptor-mediated endocytosis by a dynamin-1 mutant with a defective pleckstrin homology domain. *Curr. Biol.*, **9**, 261.
99. Achiriloaie, M., Barylko, B., and Albanesi, J. P. (1999) Essential role of the dynamin pleckstrin homology domain in receptor-mediated endocytosis. *Mol. Cell. Biol.*, **19**, 1410.
100. Henley, J. R., Krueger, E. W. A., Oswald, B. J., and McNiven, M. A. (1998) Dynamin-mediated internalization of caveolae. *J. Cell Biol.*, **141**, 85.
101. Oh, P., McIntosh, D. P., and Schnitzer, J. E. (1998) Dynamin at the neck of caveole mediates their budding to form transport vesicles by GTP-driven fission from the plasma membrane of endothelium. *J. Cell Biol.*, **141**, 101.
102. Gold, E. S., Underhill, D. M., Morrissette, N. S., Guo, J., McNiven, M. A., and Aderem, A. (1999) Dynamin 2 is required for phagocytosis in macrophages. *J. Exp. Med.*, **190**, 1849.
103. Boleti, H., Benmerah, A., Ojcius, D. M., Cerf-Bensussan, N., and Dautry-Varsat, A. (1999) Chlamydia infection of epithelial cells expressing dynamin and Eps15 mutants: clathrin-independent entry into cells and dynamin-dependent productive growth. *J. Cell Sci.*, **112**, 1487.
104. Kosaka, T. and Ikeda, K. (1983) Reversible blockage of membrane retrieval and endocytosis in the garland cell of the temperature-sensitive mutant of *Drosophilia melanogaster*, *shibire*^{ts1}. *J. Cell Biol.*, **97**, 499.
105. Llorente, A., Rapak, A., Schmid, S. L., van Deurs, B., and Sandvig, K. (1998) Expression of mutant dynamin inhibits toxicity and transport of endocytosed ricin to the Golgi apparatus. *J. Cell Biol.*, **140**, 553.
106. Nicoziani, P., Vilhardt, F., Llorente, A., Hilout, L., Courtoy, P. J., Sandvig, K., *et al.* (2000) Role for dynamin in late endosome dynamics and trafficking of the cation-independent mannose 6-phosphate receptor. *Mol. Biol. Cell*, **11**, 481.
107. Wienke, D. C., Knetsch, M. L., Neuhaus, E. M., Reedy, M. C., and Manstein, D. J. (1999) Disruption of a dynamin homologue affects endocytosis, organelle morphology, and cytokinesis in *Dictyostelium discoideum*. *Mol. Biol. Cell*, **10**, 225.
108. Jones, S. M., Howell, K. E., Henley, J. R., Cao, H., and McNiven, M. A. (1998) Role of dynamin in the formation of transport vesicles from the Trans-Golgi network. *Science*, **279**, 573.
109. Qualmann, Q. and Kelly, R. B. (2000) Syndapin isoforms participate in receptor mediated endocytosis and actin organisation. *J. Cell Biol.*, **148**, 1047.
110. Whistler, J. L. and von Zastrow, M. (1999) Dissociation of functional roles of dynamin in receptor-mediated endocytosis and mitogenic signal transduction. *J. Biol. Chem.*, **274**, 24575.
111. Sengar, A. S., Wang, W., Bishay, J., Cohen, S., and Egan, S. E. (1999) The EH and SH3 domain Ese proteins regulate endocytosis by linking to dynamin and Eps15. *EMBO J.*, **18**, 1159.

112. Yamabhai, M., Hoffman, N. G., Hardison, N. L., McPherson, P. S., Castagnoli, L., Cesareni, G., *et al.* (1998) Intersectin, a novel adaptor protein with two Eps15 homology and five Src homology 3 domains. *J. Biol. Chem.*, **273**, 31401.
113. Guipponi, M., Scott, H. S., Chen, H., Schebesta, A., Rossier, C., and Antonarakis, S. E. (1998) Two isoforms of a human intersectin (ITSN) protein are produced by brain-specific alternative splicing in a stop codon. *Genomics*, **53**, 369.
114. Watts, C. (1997) Capture and processing of exogenous antigens for presentation on MHC molecules. *Annu. Rev. Immunol.*, **15**, 821.
115. Sandvig, K. and van Deurs, B. (1990) Selective modulation of the endocytic uptake of ricin and fluid phase markers without alteration in transferrin endocytosis. *J. Biol. Chem.*, **265**, 6382.
116. Anderson, R. G. (1998) The caveolae membrane system. *Annu. Rev. Biochem.*, **67**, 199.
117. Kurzchalia, T. V. and Parton, R. G. (1999) Membrane microdomains and caveolae. *Curr. Opin. Cell Biol.*, **11**, 424.
118. Parton, R. G. (1996) Caveolae and caveolins. *Curr. Opin. Cell Biol.*, **8**, 542.
119. Fra, A. M., Williamson, E., Simons, K., and Parton, R. G. (1994) Detergent-insoluble glycolipid microdomains in lymphocytes in the absence of caveolae. *J. Biol. Chem.*, **269**, 30745.
120. Ikonen, E. and Parton, R. G. (2000) Caveolins and cellular cholesterol balance. *Traffic*, **1**, 212.
121. Volonté, D., Galbiati, F., Li, S., Nishiyama, K., Okamoto, T., and Lisanti, M. P. (1999) Flotillins/cavatellins are differentially expressed in cells and tissues and form a hetero-oligomeric complex with caveolins *in vivo*. Characterization and epitope-mapping of a novel flotillin-1 monoclonal antibody probe. *J. Biol. Chem.*, **274**, 12702.
122. Murata, M., Peranen, J., Schreiner, R., Wieland, F., Kurzchalia, T. V., and Simons, K. (1995) VIP21/caveolin is a cholesterol-binding protein. *Proc. Natl. Acad. Sci. USA*, **92**, 10339.
123. Fra, A. M., Williamson, E., Simons, K., and Parton, R. G. (1995) *De novo* formation of caveolae in lymphocytes by expression of VIP21-caveolin. *Proc. Natl. Acad. Sci. USA*, **92**, 8655.
124. Lipardi, C., Mora, R., Colomer, V., Paladino, S., Nitsch, L., Rodriguez-Boulan, E., *et al.* (1998) Caveolin transfection results in caveolae formation but not apical sorting of glycosylphosphatidylinositol (GPI)-anchored proteins in epithelial cells. *J. Cell Biol.*, **140**, 617.
125. Kurzchalia, T. V., Dupree, P., Parton, R. G., Kellner, R., Virta, H., Lehnert, M., *et al.* (1992) VIP21, a 21-kD membrane protein is an integral component of trans-Golgi-network-derived transport vesicles. *J. Cell Biol.*, **118**, 1003.
126. Brown, D. A. and London, E. (2000) Structure and function of sphingolipid- and cholesterol-rich membrane rafts. *J. Biol. Chem.*, **275**, 17221.
127. Simons, K. and Ikonen, E. (1997) Functional rafts in cell membranes. *Nature*, **387**, 569.
128. Harder, T. and Simons, K. (1997) Caveolae, DIGs, and the dynamics of sphingolipid-cholesterol microdomains. *Curr. Opin. Cell Biol.*, **9**, 534.
129. Brown, D. A. and London, E. (1998) Structure and origin of ordered lipid domains in biological membranes. *J. Membr. Biol.*, **164**, 103.
130. Hooper, N. M. (1999) Detergent-insoluble glycosphingolipid/cholesterol-rich membrane domains, lipid rafts and caveolae. *Mol. Membr. Biol.*, **16**, 145.
131. Rietveld, A. and Simons, K. (1998) The differential miscibility of lipids as the basis for the formation of functional membrane rafts. *Biochim. Biophys. Acta*, **1376**, 467.

132. Brown, D. A. and London, E. (1998) Functions of lipid rafts in biological membranes. *Annu. Rev. Cell Biol.*, **14**, 111.
133. Monier, S., Parton, R. G., Vogel, F., Behlke, J., Henske, A., and Kurzchalia, T. V. (1995) VIP21-caveolin, a membrane protein constituent of the caveolar coat, oligomerizes *in vivo* and *in vitro*. *Mol. Biol. Cell*, **6**, 911.
134. Simionescu, M. and Simionescu, N. (1991) Endothelial transport of macromolecules; transcytosis and endocytosis. *Cell Biol. Rev.*, **25**, 1.
135. Schnitzer, J. E., Oh, P., Pinney, E., and Allard, J. (1994) Filipin-sensitive caveolae mediated transport in endothelium: reduced transcytosis, scavenger endocytosis and capillary permeability of select macromolecules. *J. Cell Biol.*, **127**, 1217.
136. Schnitzer, J. E., Oh, P., and McIntosh, D. P. (1996) Role of GTP hydrolysis in fission of caveolae directly from plasma membranes. *Science*, **274**, 239.
137. Predescu, D., Predescu, S., McQuistan, T., and Palade, G. E. (1998) Transcytosis of alpha1-acidic glycoprotein in the continuous microvascular endothelium. *Proc. Natl. Acad. Sci. USA*, **95**, 6175.
138. Parton, R. G., Joggerst, B., and Simons, K. (1994) Regulated internalization of caveolae. *J. Cell Biol.*, **127**, 1199.
139. Anderson, R. G. W., Kamen, B. A., Rothberg, K. G., and Lacey, S. W. (1992) Potocytosis: sequestration and transport of small molecules by caveolae. *Science*, **255**, 410.
140. Mayor, S., Sabharanjak, S., and Maxfield, F. R. (1998) Cholesterol-dependent retention of GPI-anchored proteins in endosomes. *EMBO J.*, **17**, 4626.
141. Mayor, S., Rothberg, K. G., and Maxfield, F. R. (1994) Sequestration of GPI-anchored proteins in caveolae triggered by cross-linking. *Science*, **264**, 1948.
142. Schnitzer, J. E., Liu, J., and Oh, P. (1995) Endothelial caveolae have the molecular transport machinery for vesicle budding, docking, and fusion including VAMP, NSF, SNAP, annexins, and GTPases. *J. Biol. Chem.*, **270**, 14399.
143. Schnitzer, J. E., McIntosh, D. P., Dvorak, A. M., Liu, J., and Oh, P. (1995) Separation of caveolae from associated microdomains of GPI-anchored proteins. *Science*, **269**, 1435.
144. Gilbert, A., Paccaud, J. P., Foti, M., Porcheron, G., Balz, J., and Carpentier, J. L. (1999) Direct demonstration of the endocytic function of caveolae by a cell-free assay. *J. Cell Sci.*, **112**, 1101.
145. Lobie, P. E., Sadir, R., Graichen, R., Mertani, H. C., and Morel, G. (1999) Caveolar internalization of growth hormone. *Exp. Cell Res.*, **246**, 47.
146. Benlimame, N., Le, P. U., and Nabi, I. R. (1998) Localization of autocrine motility factor receptor to caveolae and clathrin-independent internalization of its ligand to smooth endoplasmic reticulum. *Mol. Biol. Cell*, **9**, 1773.
147. Anderson, H. A., Chen, Y., and Norkin, L. C. (1996) Bound simian virus 40 translocates to caveolin-enriched membrane domains, and its entry is inhibited by drugs that selectively disrupt caveolae. *Mol. Biol. Cell*, **7**, 1825.
148. Stang, E., Kartenbeck, J., and Parton, R. G. (1997) Major histocompatibility complex class I molecules mediate association of SV40 with caveolae. *Mol. Biol. Cell*, **8**, 47.
149. Parton, R. G. and Lindsay, M. (1999) Exploitation of major histocompatibility complex class I molecules and caveolae by simian virus 40. *Immunol. Rev.*, **168**, 23.
150. Norkin, L. C. (1999) Simian virus 40 infection via MHC class I molecules and caveolae. *Immunol. Rev.*, **168**, 13.
151. Harder, T., Scheiffele, P., Verkade, P., and Simons, K. (1998) Lipid domain structure of the plasma membrane revealed by patching of membrane components. *J. Cell Biol.*, **141**, 929.

152. Verkade, P., Harder, T., Lafont, F., and Simons, K. (1999) Induction of caveolae in the apical plasma membrane of Madin-Darby Canine Kidney cells. *J. Cell Biol.*, **148**, 727.
153. Lencer, W. I., Hirst, T. R., and Holmes, R. K. (1999) Membrane traffic and the cellular uptake of cholera toxin. *Biochim. Biophys. Acta*, **1450**, 177.
154. Orlandi, P. A. and Fishman, P. H. (1998) Filipin-dependent inhibition of cholera toxin: evidence for toxin internalization and activation through caveolae-like domains. *J. Cell Biol.*, **141**, 905.
155. Henley, J. R., Cao, H., and McNiven, M. A. (1999) Participation of dynamin in the biogenesis of cytoplasmic vesicles. *FASEB J.*, **13**, S243.
156. Llorente, A., Garred, O., Holm, P. K., Eker, P., Jacobsen, J., van Deurs, B., *et al.* (1996) Effect of calmodulin antagonists on endocytosis and intracellular transport of ricin in polarized MDCK cells. *Exp. Cell Res.*, **227**, 298.
157. Eker, P., Holm, P. K., van Deurs, B., and Sandvig, K. (1994) Selective regulation of apical endocytosis in polarized Madin-Darby canine kidney cells by mastoparan and cAMP. *J. Biol. Chem.*, **269**, 18607.
158. Rodal, S. K., Skretting, G., Øystein, G., Vilhardt, F., van Deurs, B., and Sandvig, K. (1999) Extraction of cholesterol with methyl-β-cyclodextrin perturbs formation of clathrin-coated endocytic vesicles. *Mol. Biol. Cell*, **10**, 961.
159. Subtil, A., Gaidarov, I., Kobylarz, K., Lampson, M. A., Keen, J. H., and McGraw, T. E. (1999) Acute cholesterol depletion inhibits clathrin-coated pit budding. *Proc. Natl. Acad. Sci. USA*, **96**, 6775.
160. Skretting, G., Torgersen, M. L., van Deurs, B., and Sandvig, K. (1999) Endocytic mechanisms responsible for uptake of GPI-linked diphtheria toxin receptor. *J. Cell Sci.*, **112**, 3899.
161. Vilhardt, F., Nielsen, M., Sandvig, K., and van Deurs, B. (1999) Urokinase-type plasminogen activator receptor is internalized by different mechanisms in polarized and non-polarized Madin-Darby canine kidney epithelial cells. *Mol. Biol. Cell*, **10**, 179.
162. Lee, S., Zhao, Y., and Anderson, W. F. (1999) Receptor-mediated Moloney murine leukemia virus entry can occur independently of the clathrin-coated-pit-mediated endocytic pathway. *J. Virol.*, **73**, 5994.
163. Zhang, J., Ferguson, S. S. G., Barak, L. S., Menard, L., and Caron, M. G. (1996) Dynamin and beta-arrestin reveal distinct mechanisms for G protein-coupled receptor internalization. *J. Biol. Chem.*, **271**, 18302.
164. Pals-Rylaarsdam, R., Gurevich, V. V., Lee, K. B., Ptasienski, J. A., Benovic, J. L., and Hosey, M. M. (1997) Internalization of the m2 muscarinic acetylcholine receptor. Arrestin-independent and -dependent pathways. *J. Biol. Chem.*, **272**, 23682.
165. Lee, K. B., Pals-Rylaarsdam, R., Benovic, J. L., and Hosey, M. M. (1998) Arrestin-independent internalization of the m1, m3, and m4 subtypes of muscarinic cholinergic receptors. *J. Biol. Chem.*, **273**, 12967.
166. Vickery, R. G. and von Zastrow, M. (1999) Distinct dynamin-dependent and -independent mechanisms target structurally homologous dopamine receptors to different endocytic membranes. *J. Cell Biol.*, **144**, 31.
167. Subtil, A. and Dautry-Varsat, A. (1998) Several weak signals in the cytosolic and transmembrane domains of the interleukin 2 receptor β chain allow for its efficient endocytosis. *Eur. J. Biochem.*, **253**, 525.
168. Morelon, E. and Dautry-Varsat, A. (1998) Endocytosis of the common cytokine receptor γc chain: identification of sequences involved in internalization and degradation. *J. Biol. Chem.*, **273**, 22044.

169. Miller, K., Shipman, M., Trowbridge, I. S., and Hopkins, C. R. (1991) Transferrin receptors promote the formation of clathrin lattices. *Cell*, **65**, 621.
170. Hansen, S. H., Sandvig, K., and Van Deurs, B. (1992) Internalization efficiency of the transferrin receptor. *Exp. Cell Res.*, **199**, 19.
171. Mukherjee, S. and Maxfield, F. R. (2000) Role of membrane organization and membrane domains in endocytic lipid trafficking. *Traffic*, **1**, 203.
172. Maxfield, F. R. and Mayor, S. (1997) Cell surface dynamics of GPI-anchored proteins. *Adv. Exp. Med. Biol.*, **419**, 355.
173. Mukherjee, S., Soe, T. T., and Maxfield, F. R. (1999) Endocytic sorting of lipid analogues differing solely in the chemistry of their hydrophobic tails. *J. Cell Biol.*, **144**, 1271.
174. Wilkening, G., Linke, T., and Sandhoff, K. (1998) Lysosomal degradation on vesicular membrane surfaces. Enhanced glucosylceramide degradation by lysosomal anionic lipids and activators. *J. Biol. Chem.*, **273**, 30271.
175. Kobayashi, T., Stang, E., Fang, K. S., de Moerloose, P., Parton, R. G., and Gruenberg, J. (1998) A lipid associated with the antiphospholipid syndrome regulates endosome structure and function. *Nature*, **392**, 193.
176. Kobayashi, T., Gu, F., and Gruenberg, J. (1998) Lipids, lipid domains and lipid-protein interactions in endocytic membrane traffic. *Semin. Cell Dev. Biol.*, **9**, 517.
177. Farge, E., Ojcius, D. M., Subtil, A., and Dautry-Varsat, A. (1999) Enhancement of endocytosis due to aminophospholipid transport across the plasma membrane of living cells. *Am. J. Physiol.*, **276**, C725.
178. Hémar, A., Subtil, A., Lieb, M., Morelon, E., Hellio, R., and Dautry-Varsat, A. (1995) Endocytosis of interleukin 2 receptors in human T lymphocytes: distinct intracellular localization and fate of the receptor α, β and γ chains. *J. Cell Biol.*, **129**, 55.
179. Subtil, A., Delepierre, M., and Dautry-Varsat, A. (1997) An α-helical signal in the cytosolic domain of the interleukin 2 receptor β chain mediates sorting towards degradation after endocytosis. *J. Cell Biol.*, **136**, 583.
180. Subtil, A., Rocca, A., and Dautry-Varsat, A. (1998) Molecular characterization of the signal responsible for the targeting of the IL2 receptor β chain toward intracellular degradation. *J. Biol. Chem.*, **273**, 29424.
181. Thomas, P., Lee, A. K., Wong, J. G., and Almers, W. (1994) A triggered mechanism retrieves membrane in seconds after Ca(2+)-stimulated exocytosis in single pituitary cells. *J. Cell Biol.*, **124**, 667.
182. Artalejo, C. R., Henley, J. R., McNiven, M. A., and Palfrey, H. C. (1995) Rapid endocytosis coupled to exocytosis in adrenal chromaffin cells involves Ca2+, GTP, and dynamin but not clathrin. *Proc. Natl. Acad. Sci. USA*, **92**, 8328.
183. Smart, E. J., Graf, G. A., McNiven, M. A., Sessa, W. C., Engelman, J. A., Scherer, P. E., *et al.* (1999) Caveolins, liquid-ordered domains, and signal transduction. *Mol. Cell. Biol.*, **19**, 7289.

3 | Phagocytosis

EMMANUELLE CARON and ALAN HALL

1. Introduction

Phagocytosis is the process by which cells bind and internalize particulate matter larger than around 0.75 μm in diameter, e.g. from small-sized dust particles and cell debris to micro-organisms, parasites, and apoptotic cells. The terms 'phagocytosis' (the eating process) and 'phagocytes' (eating cells) were coined in the 1880s by Elie Metchnikoff, but the existence of such eating cells had been described even in the previous century, in protozoans and in leukocytes from invertebrates (1).

In protozoans, phagocytosis is probably only used for feeding, but in higher eukaryotes it serves two quite different functions, which contribute to maintain homeostasis: the removal of foreign particles, such as micro-organisms or parasites, and the removal of host cells, such as apoptotic cells generated during development, wound healing, or inflammation. Phagocytosis can be seen as a restricted cellular process limited to the binding and engulfment of targets. However, using a broader definition, which takes into account the uptake process and its consequences, phagocytosis is necessary for non-specific host defence (against micro-organisms or exceeding cells), for the induction of specific immunity (through antigen presentation and cytokine release), and as a component of specific immunity (against microbes or tumour cells in an immune organism). Much has been learnt about phagocytes and phagocytosis in the past 120 years, and exciting new studies and approaches have shed light on the intimate details of the process. This chapter will summarize the current state of knowledge concerning the cellular and molecular mechanisms of phagocytosis and due to space restrictions, will focus primarily on the work done in mammalian macrophages.

2. Phagocytes

Anyone who is familiar with tissue culture will have noticed cell debris or particulate material inside cultured cells, regardless of their lineage or origin. In fact, all eukaryotic cells with the noticeable exception of yeasts, can phagocytose. The first description of phagocytic behaviour was observed in single-celled Infusoria (1), and many other protozoan species, including important human pathogens such as *Entamoeba histolytica*, have been used as model systems to study phagocytosis (2). The slime mould

Dictyostelium discoideum has provided a genetically tractable model system (see Chapter 4) (3). In higher Metazoa, however, specific cell types have evolved with the task of removing debris, apoptotic cells, and micro-organisms by phagocytosis. These highly specialized cells are often referred to as 'professional' phagocytes or 'true' phagocytes.

Metchnikoff himself proposed the existence of two types of dedicated phagocyte: macrophages and microphages (now called granulocytes or polymorphonuclear cells). Macrophages are present in all organs and all epithelia that line contact surfaces between the outside world and the tissues (e.g. skin, gut, lung, mucosae). They are particularly concentrated in organs and tissues that filter blood and lymph and in fact neutrophils and the macrophage precursors (monocytes) are carried by the blood around the body and can be quickly recruited to sites of infection/inflammation. The importance of specific cell populations devoted to phagocytosis is clearly demonstrated through genetic mutations in a variety of organisms. Thus, patients suffering from Leukocyte Adhesion Deficiency (LAD) (4), mice lacking the transcription factor *PU.1* (5), and *domino* mutants of *Drosophila* (6) all have problems clearing micro-organisms due to the functional absence of professional phagocytes, i.e. macrophages and neutrophils in mammals, or haemocytes in insects.

Cells other than neutrophils and macrophages can however, also, take up particles, both *in vitro* and *in vivo*. *In vivo*, this is generally observed in situations where inert material or large numbers of apoptotic or senescent cells need to be removed. For example, bladder epithelial cells phagocytose damaged erythrocytes (7) while in the retina, cells from the pigment epithelium scavenge by phagocytosis the effete tips of retinal rods and cones (8, 9). Endothelial cells can internalize apoptotic lymphocytes both *in vitro* and *in vivo* (10), and there is evidence that dermal fibroblasts can phagocytose apoptotic cells (11). Compared to professional phagocytes however, the non-professional cells are far less efficient and are unable to deal with as large a variety of targets. Moreover the responses that accompany phagocytosis (such as activation of the NADPH oxidase or cytokine production, see Section 6) seem restricted to neutrophils and macrophages. Finally, macrophages and related dendritic cells are able to process and present antigens, derived from phagocytized material, thereby bridging the innate and acquired immunity host defence systems.

3. Phagocytic receptors

The three major steps of phagocytosis are:

(a) Binding of the phagocytic target to cell surface receptors.

(b) Actin-dependent clathrin-independent enclosure of the particle within a membrane-bound phagosome.

(c) Depolymerization of F-actin and maturation of the phagosome.

The first requirement for phagocytosis is that of particle recognition and this involves phagocyte receptors binding directly, or indirectly through opsonins (host proteins

Table 1 Macrophage phagocytic receptors

Receptor		Ligand	Confers phagocytic potential upon transfection into non-professional phagocytes
Direct recognition of target			
Mannose receptor		Mannose, fucose (zymosan)	+ (12)
DEC 205		Mannans[a]	
CR3		LPS, β-glucans	+ (13)
Scavenger Rs		LPS	
	class A I and II,	polyanions, AC[b]	
	Marco	Gram + and Gram − bacteria	+ (14)
	class B CD36 (+VnR)	AC	+ (15)
CD14		AC	+ (16)
Indirect recognition (through opsonization)			
Fcγ receptors		IgG	
	FcγRI		+ (17)
	FcγRIIA (CD32)		+ (18)
	FcγRIIIA (CD16)		+ (19)
Complement receptors (CR)			
	CR3 (CD11b/CD18)	C3bi, fibrinogen	+ (20)
	CR1	C3b, C4b, C1q, C3bi[a]	
	C1qR	C1q, MBP, SPA	
	CR4 ($\alpha_x\beta_2$)	C3bi	
CD14		LPS/LBP complex[a]	
SPR210		SPA[a]	
$\alpha_5\beta_1$		Fibronectin	

[a] May be involved in particle binding only, and not mediate phagocytosis *per se*.
[b] Abbreviations: AC, apoptotic cell; VnR, vitronectin receptor ($\alpha_v\beta_3$).

that uniformly coat the particle). Around 12 distinct types of phagocytic receptor have been identified on the surface of mammalian macrophages (see Table 1). These belong to quite distinct gene families, they share no obvious homology in their intracellular domain, and they recognize very different structures on the phagocytic targets.

The identification of phagocytic receptors has been achieved in a variety of ways such as blocking or competition studies, or by expression of cDNAs in CHO or COS cells. A limitation of ectopic expression type experiments, however, is that some receptors may bind targets without triggering phagocytosis. For instance, expression of the GPI-linked CD14 molecule confers recognition but not phagocytic properties (21); two FcγR subtypes, FcγRIA and FcγRIIIA, only confer phagocytic potential to COS cells when co-transfected with accessory γ chains (17, 19).

For opsonin-independent uptake, specific motifs or molecular patterns are recognized on the target surface. These motifs appear to be general signatures associated with the surface of a micro-organism or an apoptotic cell, for example, certain combinations of sugar residues found on the outer membrane of yeasts are recognized by the lectin-like carbohydrate recognition domains of the mannose receptor, while polyanionic motifs, a common feature of apoptotic cells, are recognized by several classes of phagocytic receptors (e.g. scavenger receptors, CD14). Opsonization (to opsonize literally means 'to season') extends the range of particles that can be phagocytosed and increases the rate of uptake. There are several categories of opsonins, including the mannose-binding protein and surfactant protein A, but the most important are fragments of the complement system and immunoglobulins.

Finally, apoptotic cells and micro-organisms are rarely taken up through a unique receptor *in vivo* and results obtained using knock-out mice (22) or *in vitro* data (23, 24) show several receptor/ligand pairs are often involved. Apoptotic cells for example can activate the complement cascade so that the complement receptors CR3 and CR4, most often associated with the uptake of micro-organisms, can also mediate uptake of apoptotic cells (25). Similarly, if a Gram negative bacterium makes its way to tissue of a non-immune host, macrophages are able to recognize it through multiple recognition systems, including CD14 (26), scavenger receptors, CR3 (potentially both by opsonin-dependent and -independent recognition), and C1qR (27).

4. Cell biology of phagocytosis

Binding of a phagocytic particle to a macrophage is necessary but not always sufficient for phagocytosis: for example, when macrophages are fed with phagocytic targets in the cold (28), when a phagocytic target is not fully opsonized (29), or when an additional activation signal is needed to trigger phagocytosis, e.g. for the CR3 receptor (30). In this section, we will examine the cellular aspects of phagocytosis, i.e. how the phagocytic target is internalized and what happens to the phagosome once formed.

4.1 Ultrastructural changes associated with internalization

A series of pioneering experiments form the basis of our understanding of how phagocytosis proceeds. After initial interaction between the IgG- or complement-opsonized targets and macrophage Fcγ or CR3 receptors respectively, particle uptake follows a zipper-like process, schematized in Fig. 1. Portions of plasma membrane are seen to glide circumferentially around the target and progressively cover its entire surface as more and more receptor/ligand pairs form (31). Zippering is confined to the segment of plasma membrane immediately adjacent to the phagocytic target, it requires receptors that were not initially involved in ligand recognition, and ligand must be available all around the phagocytic particle (29). Phagocytosis is therefore a highly localized, segmented process rather than a general response of the macrophage to a phagocytic stimulus.

It has long been known from ultrastructural studies that the morphology of nascent phagosomes differs depending on which phagocytic receptor is mediating internalization. During FcγR-mediated phagocytosis, sequential interaction of IgG-coated particles with receptors stimulates the extension of pseudopods, rising from the macrophage surface, and protrusion of a localized cup-shaped lamellipodium around the particle. By contrast, few protrusions form during phagocytosis of complement-

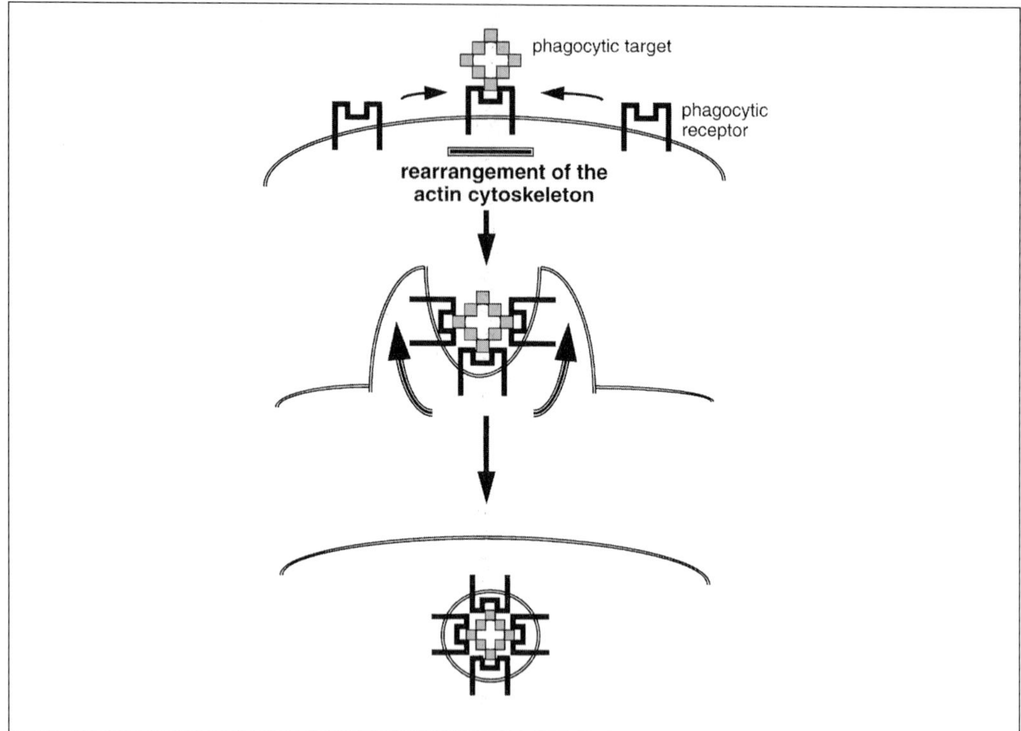

Fig. 1 The zipper model of phagocytosis. The initial interaction of ligands on the target with specific receptors on the phagocyte surface triggers the recruitment of additional receptors and the initiation of a localized cytoskeletal response. F-actin polymerization and recruitment of F-actin and other cytoskeletal proteins proceed towards the rims of the ligand binding zone (zippering), as more and more receptors become engaged in target recognition. When the phagocytic target is circumferentially bound to the cell, membrane fusion occurs, giving rise to the early phagosome. After losing its F-actin and cytoskeletal shell, the phagosome can then mature.

coated particles, associating with the CR3 receptor; rather the particles appear to sink into the cells (32, 33). These early observations suggested that different mechanisms of phagocytosis might operate in the same cell type.

FcγR-mediated phagocytosis also leads to general morphological changes in macrophages. Resident peritoneal macrophages show a very ruffly surface in scanning electron microscopy; addition of IgG-coated targets induces a time-dependent disappearance of membrane folds, accompanying an increase in cell volume (34). Recent results show that phagocytes that have internalized IgG-coated targets have a higher surface area compared to control cells, the difference correlating with the number of particles taken up (35). There is little doubt that some extensive membrane remodelling must occur for phagocytosis to proceed. Moreover, the area of membrane internalized is considerable, estimated to be up to 100% of the total plasma membrane area (34). This calculation is well in line with data indicating that within 30 minutes of interaction with IgG-coated targets, a macrophage can ingest the equivalent of 1.5 times its diameter or 3 times its volume (36).

Currently it is unclear where the required extra-membrane comes from during phagocytosis. There are at least four possibilities:

(a) Internal membranes could be brought to the surface.

(b) Unfolding of plasma membrane folds could provide the extra-membrane.

(c) Membrane could be recycled from other parts of the plasma membrane.

(d) Membrane could be recruited from the phagolysosomal compartment itself.

Experimental evidence has so far not distinguished these possibilities. Professional phagocytes are known to contain mobilizable internal membranes: for example, upon stimulation with phorbol esters, macrophages undergo a 66% increase in surface area, which correlates with the appearance of endosomal markers on the plasma membrane, and flattening of the cell (37). In fact, electron-lucent vesicles can be seen in the region of the phagocytic cups, and these accumulate when phagocytosis is inhibited (38). Finally, there is evidence for a continuous, bidirectional flow of membrane between the plasma membrane and the lysosomal compartment in phagocytizing macrophages (39), and clathrin-coated vesicles have been observed budding off or fusing with phagosomes (40).

Once the phagocytic target is totally surrounded by membrane, a fusion event isolates it in a morphologically distinct membrane-bound phagosome, which then matures by fusion with other intracellular compartments (see Section 4.3)

4.2 Cytoskeletal reorganization during phagocytosis

Despite the great variety of targets and phagocytic receptors, two features of receptor-mediated phagocytosis are always conserved: the zipper-like mechanism of uptake and the dependence on the actin cytoskeleton. Phagocytosis, but not particle binding to phagocytic receptors, is sensitive to drugs that prevent actin polymerization, such as cytochalasins. This is true for all pairs of phagocytic ligand/receptors that have been looked at, in professional as well as non-professional phagocytes (18, 32, 41, 42).

Actin polymerization underneath bound particles is a very early marker of ongoing phagocytosis: for example, it does not take place when cells are fed with IgG-coated particles at 4 °C, but can be observed as early as one minute after warming to 37 °C (28, 43). Numerous cytoskeletal molecules are recruited to the sites of particle binding, and most are conserved in phagocytosis mediated by several distinct receptors (Table 2). However there are some differences: paxillin and vinculin are recruited to IgG- as well as complement-dependent phagosomes for example, but are absent from phagosomes forming around unopsonized zymosan. Moreover, whereas these two cytoskeletal proteins accumulate diffusely underneath IgG targets, they show a punctate staining during CR3-mediated phagocytosis (33). The function of paxillin and vinculin during phagocytosis and the meaning of their differential distribution remain unknown. Most cytoskeletal molecules associated with forming phagosomes are lost shortly after enclosure of the particle, possibly to allow fusion events associated with phagosome maturation.

Table 2 Cytoskeletal proteins recruited to nascent phagosomes

Cytoskeletal component		Phagocytic target			Presence on mature phagosomes	Reference
		IgG-target	C3bi-target	Zymosan		
F-actin		+	+	+	−	28, 33
Myosins	I	+		+	−	44, 45
	II	+		+		45, 46
	V	−			+	45
	VI	−			−	45
	VII	−			−	45
	IXb	+			−	45
Talin		+	+	+	−	28, 33
Vinculin		+	+	−		33
Paxillin		+	+	−		33, 47
α-actinin		+	+			33
MARCKS, MacMARCKS		+	+	+	−	44, 48
Pleckstrin		+			−	49

There are numerous reports suggesting a role for myosin motors during phagocytosis. For example, 2,3-butanedione monoxime (BDM), an uncompetitive inhibitor of myosin II and perhaps other myosins, inhibits phagocytosis, though not the formation of F-actin-rich phagocytic cups underneath bound targets (45). Also, distinct myosin classes have been found to accumulate on phagosomes as they form and mature (44–46, and see Table 2). However, there is no single class of myosin found on all phagosomes, which underlines the complexity of the cytoskeletal remodelling that occurs during uptake. Finally, the contractile force within a phagocyte cell body increases during phagocytosis of zymosan, although this occurs after engulfment (50).

Another interesting issue is the role of the microtubule network during phagocytosis and different phagocytic receptors appear to behave differently in this regard. For example, CR3-mediated phagocytosis is sensitive to colchicine (which disrupts the microtubule network), but FcγR-mediated phagocytosis and the uptake of latex beads are not (31). This has led to the suggestion that phagocytosis through complement receptors resembles the microtubule- and actin-dependent process of cell spreading seen on extracellular matrix (51). By contrast, macrophages can spread on surfaces coated with IgG in the presence of microtubule poisons (30, 36).

4.3 Phagosome maturation

After internalization is complete, the actin-based machinery is shed from the phagosome, through a completely unknown mechanism. Phagosomes then mature, and plasma membrane proteins are recycled back to the cell surface (39), while phagocytic receptors are rapidly and selectively degraded (52), as they sequentially fuse with endosomes and lysosomes. The process culminates in the formation of a mature, acidified phago-lysosome (53–55).

Maturation is, therefore, characterized by changes in both the protein and the phospholipid composition of the phagosome (56) in a process that resembles the maturation of early endosomes into late endosomes (57). Cell-free assays suggest that maturing phagosomes acquire the ability to fuse with different populations of endocytic vesicles (58). A number of proteins participating in membrane trafficking along the endocytic pathway associate transiently with maturing phagosomes, such as heterotrimeric G-proteins, annexins, and small GTP-binding proteins of the Rab family (55, 56, 59).

The fusion of lysosomes with phagosomes does not seem to be greatly influenced by the nature of the receptor/ligand pair originally initiating phagocytosis (60), though the delivery of lysosomes occurs concomitantly with phagocytosis of IgG-coated bacteria but it is delayed during phagocytosis of non-particulate material (61). Furthermore, even though uptake and fusion both happen shortly after particle binding, the uptake process *per se* is independent of phagosome maturation and phagocytosis is unimpaired by inhibitors of phago-lysosome fusion (e.g. ammonium chloride) (62).

Phagosome acidification involves the insertion of the Na^+/H^+ exchanger (NHE), present on the plasma membrane, and of the vacuolar H^+-ATPase (V-ATPase), acquired through fusion with endosomes and lysosomes, into the phagosome (63, 64). Over time, macrophages gradually accumulate the 39 kDa subunit of the V-ATPase on IgG-dependent phagosomes (49), and the kinetics of acidification of phagosomes originating from other phagocytic receptors appears to be similar (60). Finally, the rates of phagosome maturation, acidification, and fusion with lysosomes are very similar in macrophages and non-professional phagocytes. For example, the kinetics of phagosome acidification or of the acquisition of the markers CD63, LAMP1 and 2 is comparable in macrophages and in FcγR-transfected COS cells (41, 65).

4.4 Phagocytosis and endocytosis

In 1963, Christian de Duve proposed the term endocytosis to cover both notions of pinocytosis and phagocytosis. Indeed, phagocytosis and receptor-mediated endocytosis share many characteristics (57). However, several features distinguish the two processes.

First, unlike endocytic receptors, there are no obvious motifs shared by the intracellular domains of phagocytic receptors. Actually, within the FcγR family, those receptors that are particularly efficient for endocytosis are poor at phagocytosis (66), and it has been shown that different subdomains of FcγRI and FcγRIIA signal phagocytosis and endocytosis (67, 68). Secondly, even though coated pits, clathrin, and adaptins have been observed in newly formed phagosomes (40, 53, 69), there is as yet no proven role for clathrin in the phagocytic process. Finally, the absolute dependence of phagocytosis upon actin polymerization (see Section 4.2) is not a universal requirement during receptor-mediated endocytosis (70, 71).

These differences are well illustrated in a recent study in macrophages where uptake of IgG-coated beads of different sizes was compared to internalization of mannosylated BSA (61). Even though all targets were internalized with similar kinetics, small beads (0.2–0.5 μm in diameter) behaved like BSA and internalization was not sensitive to cytochalasin, there was no actin cup formation. Beads of 0.75 μm or more on the other hand triggered the formation of cytochalasin-sensitive, actin-rich phagocytic cups, and delivery of lysosomes was concomitant with internalization.

5. Signal transduction of phagocytosis

Over the last ten years a lot of effort has been put into understanding the molecular events linking the ligation of phagocytic receptors to particle uptake. In line with the zipper mechanism, the current model is that a multimolecular signalling complex assembles beneath the membrane at the site of particle attachment and that this induces the remodelling of the actin cytoskeleton required to complete internalization. A combination of pharmacological, biochemical, cell biology, and genetic approaches has contributed to an increased understanding of the signal transduction pathways activated downstream of the two best-characterized phagocytic receptors: the Fcγ and CR3 receptors. The main features of phagocytosis mediated by these receptors are summarized in Fig. 2.

5.1 Tyrosine kinases

Engagement by a phagocytic target of either Fcγ or CR3 receptors leads to the recruitment of tyrosine-phosphorylated proteins at nascent phagosomes, even though

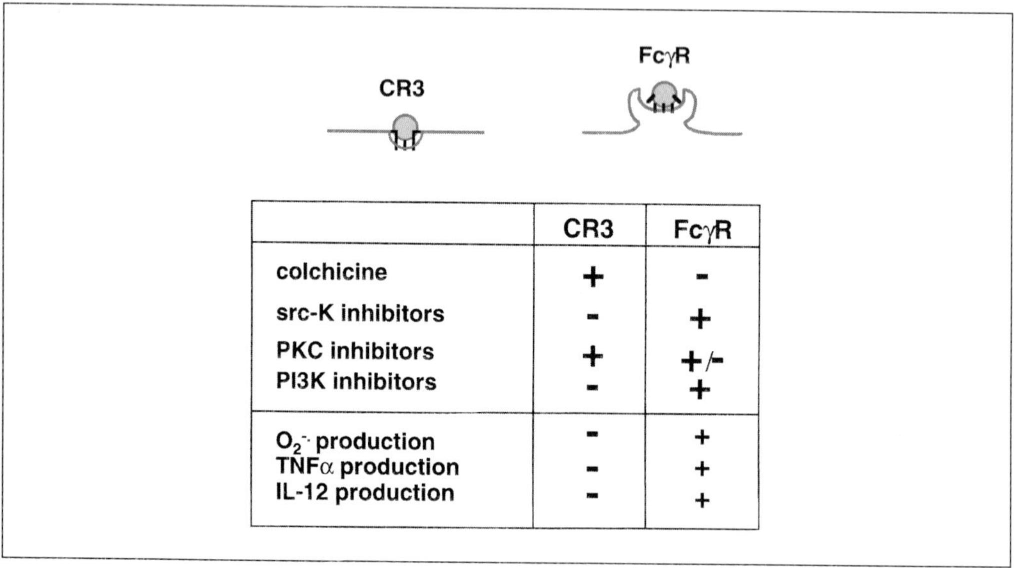

	CR3	FcγR
colchicine	+	-
src-K inhibitors	-	+
PKC inhibitors	+	+/-
PI3K inhibitors	-	+
$O_2^{\cdot-}$ production	-	+
TNFα production	-	+
IL-12 production	-	+

Fig. 2 Ultrastructural, pharmacological, and functional differences between CR3- and FcγR-mediated phagocytosis. The ultrastructures of ongoing CR3- and FcγR-mediated phagocytic events, as they appear on electron micrographs, are schematized at the top. In the upper part of the table are shown the effects of various inhibitors on phagocytosis through the two types of receptors (+, inhibition; –, no inhibition). The bottom part of the table indicates the production of oxygen radicals and two cytokines important for inflammation and induction of cell-mediated immunity in response to phagocytosis.

neither of the receptors harbours a kinase domain. However, only FcγR- and not CR3-mediated phagocytosis (or particle binding) is sensitive to tyrosine kinase inhibitors (33, 43, 72).

The earliest signal identifiable after FcγR clustering is tyrosine kinase activity and this is necessary for phagocytosis to proceed and for F-actin to assemble underneath bound targets (43, 73). Importantly, inhibiting actin polymerization does not affect the pattern of proteins that are phosphorylated on tyrosine residues following FcγR ligation (47). Amongst the tyrosine-phosphorylated proteins are the Fcγ receptors themselves, which harbour ITAM (immunoreceptor tyrosine activation motif) motifs in their intracellular domain or on their accessory chains. ITAM phosphorylation is necessary for phagocytosis, as shown by mutational analysis of FcγR and COS cell transfection studies (74). It was thought for a long time that src-like kinases were responsible for ITAM phosphorylation, because c-src can be co-precipitated with FcγR on unchallenged phagocytes (75) and c-src activity increases following FcγR cross-linking (76). However, phagocytosis in macrophages from mice deficient in three of the src-like kinases is only marginally affected (77).

Another candidate kinase is p72Syk. This tyrosine kinase associates with FcγRII (78), the γ chain associated with FcγRIA (79) and with FcγRIIIA (80), even in the absence of a phagocytic challenge. When co-transfected with FcγRIA or FcγRIIIA (and γ chain), Syk increases the phagocytic potential of COS cells towards IgG-coated particles; however, it has no effect on FcγRIIA-mediated phagocytosis (72). Syk-deficient macrophages are unable to phagocytose (78, 81), but the specific role of p72Syk downstream of the Fcγ receptors is still controversial: one report claims that it is needed for γ chain phosphorylation (81), while another claims that Syk-deficient macrophages can still polymerize F-actin into phagocytic cups (77).

5.2 Phosphatidylinositol 3-kinase (PI3K)

Two PI3K inhibitors, wortmannin at low concentrations and LY294002 both inhibit FcγR-mediated phagocytosis (see Chapter 4). This is found in macrophages as well as in transfected COS cells (82). Despite this, inhibition of PI3K does not affect target binding, the pattern of tyrosine-phosphorylated proteins (83, 84), or F-actin polymerization (84, 85). Phagocytosis of other targets, such as latex beads, bacteria, or yeasts is not sensitive to PI3K inhibitors (77). Cross-linking of FcγR increases PI3K activity (83), and causes PI3K to associate with receptor complexes, except in Syk-deficient macrophages (77). PI3K activity seems to be needed for the final steps of FcγR-mediated phagocytosis, as suggested by scanning electron microscopy which shows that wortmannin-treated macrophages can still almost completely surround IgG-targets in membrane protrusions (84). However, since PI3K involvement is restricted to FcγR-mediated internalization, this underlines the existence of distinct signalling requirements for uptake mediated by different receptors.

5.3 Small GTP-binding proteins

A lot of attention has been given in recent years to the superfamily of small GTP-binding proteins (GTPases), which act as molecular switches to regulate many signal

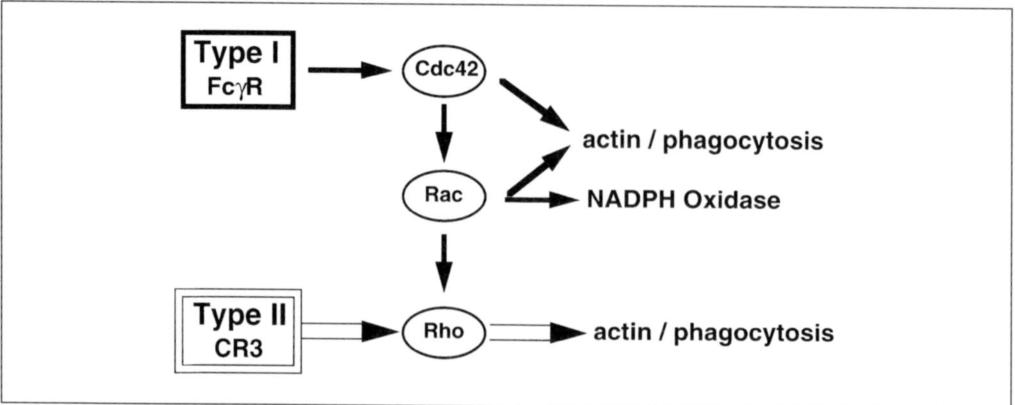

Fig. 3 Rho GTPases define two distinct mechanisms for FcγR- and CR3-mediated phagocytosis. Ligation of Fcγ receptors leads to the activation of Cdc42, Rac (necessary for activation of the NADPH oxidase) and Rho, but only Cdc42 and Rac are required for phagocytosis to occur. By contrast, Rho alone is activated and is required for CR3-mediated phagocytosis.

transduction pathways involved in various aspects of cell biology. Of particular interest to the problem of phagocytosis are members of the Rho family of small GTPases, which control the reorganization of the actin cytoskeleton in all eukaryotic cells (86). Toxin B from *Clostridium difficile*, which inactivates all Rho-family members including Rho, Rac, and Cdc42, blocks phagocytosis through both Fcγ and CR3 receptors without affecting particle binding. Moreover FcγR-mediated phagocytosis is abolished in mouse macrophages or in FcγRIIA-transfected COS cells by dominant negative forms of Cdc42 and Rac, but not by interfering specifically with Rho function (20, 87). By contrast, specific inactivation of Rho in macrophages or CR3-transfected COS cells blocks complement-mediated phagocytosis, but dominant negative Rac and Cdc42 have no effect (20). Finally, when overexpressed in COS cells, the wild-type versions of Cdc42, Rac, and Rho are recruited along with F-actin underneath IgG-opsonized particles, whereas only Rho is recruited to F-actin rich phagosomes formed during CR3-mediated phagocytosis. These results led us to propose the existence of two distinct mechanisms for phagocytosis, controlled by different Rho GTPases (Fig. 3).

The evidence for the involvement of Cdc42 and Rac downstream of Fc receptors is now compelling, even though the data and conclusions do not always entirely agree. FcεRI is functionally and structurally very closely related to the FcγRI/III receptors and mediates the toxin B-sensitive uptake of IgE-opsonized particles by mast cells (88). Using mast cells transfected with dominant negative versions of Cdc42 or Rac, it was reported that although both dominant negative GTPases inhibit phagocytosis by more than 50% (without affecting particle binding), neither Cdc42 nor Rac were needed for the recruitment of F-actin around particles. This is quite surprising considering that toxin B blocks all F-actin deposition underneath IgE-opsonized particles and could indicate either that another Rho-family member (i.e. not Rac or Cdc42) controls actin polymerization downstream of FcεRI, or that the level of inhibition achieved is greater with the toxin than with the dominant negative constructs in these experiments.

Two other small GTPases have been linked to FcγR-mediated phagocytosis. Microinjection of macrophages with C3 exoenzyme, which ADP-ribosylates and inhibits Rho, blocked FcγR-mediated phagocytosis, but in this report it also triggered a 80% drop in the number of IgG-opsonized particles able to bind Fcγ receptors and impaired capping of the Fc receptors (89). This raises the possibility that the inhibitory effects seen on FcγR-mediated phagocytosis might be due to a more indirect effect on cell behaviour. Finally, expression of dominant negative or constitutively active constructs of the small GTPase ARF6 abolishes FcγR-mediated phagocytosis, and inhibits the accumulation of F-actin at nascent phagosomes (90).

Interestingly, in *C. elegans*, genetic analysis suggests a role for Rho GTPases in apoptotic cell uptake. *ced5*, one of the six *ced* (cell death abnormal) genes required for the removal of cell corpses (91) is thought to be necessary for membrane gliding around the phagocytic target (92). The *Drosophila* and mammalian homologues of Ced-5, Mbc (myoblast city) and DOCK180 respectively, have been shown to play a role in integrin signalling and in particular in the activation of Rac (93, 94).

In the past two years, small GTP-binding proteins, especially in the Rho family, have emerged as key players in controlling the early stages of phagocytosis and have been found to define two distinct mechanisms for phagocytosis; Type I, Cdc42- and Rac- mediated, Type II Rho-mediated. A major task now is to see how general this requirement for Rho GTPases is, by looking at other phagocytic receptors. In particular, investigating the role of Rho-like small GTPases during phagocytosis of apoptotic cells should be very informative; the ultrastructural and functional similarities in the uptake of complement-opsonized targets and dying cells led us to speculate that apoptotic cell uptake could use a Type II, Rho-mediated mechanism in mammalian cells (20, and see Section 6). For FcγR and CR3, major outstanding questions remain: how are Cdc42/Rac and Rho, respectively, activated, where exactly do they lie downstream of phagocytic receptors, and how do they relay a signal to actin polymerization.

5.4 Other signalling pathways

Many other signalling elements are likely to participate in the control of phagocytosis. Pharmacological studies have pointed towards a role for protein kinase C (PKC) activation, transient Ca^{2+} rises, and phospholipases in FcγR- or CR3-mediated phagocytosis (38, 42). However, the data are often contradictory and the definitive assessment of the role of these signalling intermediates will have to await more specific inhibitors or better defined systems. This is particularly true in the case of protein kinase C involvement during phagocytosis.

Chelerythrine, a broad PKC inhibitor, inhibits by 90% both IgG- and complement-dependent phagocytosis in resident peritoneal macrophages and impairs the recruitment of F-actin and other cytoskeletal elements to nascent phagosomes (33). When combined to the observation that PKCα is recruited underneath opsonized particles

(33), this strongly suggests an important role for PKCα and maybe other PKC isoforms in early phagocytic signalling downstream of both FcγR and in activated macrophages, CR3. However, although cytoskeleton-associated PKC activity increases during FcγR-mediated phagocytosis, there is no increase in PKC activity during CR3-mediated uptake (95). Furthermore, a role for PKC downstream of the CR3 receptor is difficult to assess in most systems because of the need for receptor activation—thought to be PKC-dependent—in macrophages (30, 96).

As far as FcγR-mediated phagocytosis is concerned, the requirement for PKC is hotly debated. On the one hand, PKC is activated upon ligation of the Fcγ receptor and necessary for phagocytosis in monocytes (97). On the other hand, none of several known PKC inhibitors are able to affect actin polymerization or impair phagocytosis in elicited peritoneal macrophages (43). In order to reconcile these conflicting results, one would have to postulate either the existence of two distinct signalling pathways downstream of Fc receptors depending on the differentiation stage of the cells used or, if we assume that a unique signalling pathway is activated upon FcγR ligation, that PKC activity is not a universal requirement for IgG-dependent phagocytosis in professional phagocytes.

6. Functional, physiological, and pathological consequences of phagocytosis

The signals and molecules involved in actin polymerization are being characterized, and this will help define the molecular mechanisms of receptor-dependent uptake. But, as already alluded to in the introduction, phagocytosis is much more than just an uptake process.

As a consequence of phagocytosis, professional phagocytes can secrete a variety of important molecules, including cytotoxic radicals of nitrogen and oxygen, enzymes that degrade the extracellular matrix, lipid mediators of inflammation, and cytokines that can modify the behaviour of phagocytes and several other cell types. These mediators are important effectors in mediating anti-bacterial defences *in vivo*.

The exact nature of the secretory response, however, depends on the phagocytic stimulus. Activation of the NADPH oxidase, responsible for the production of oxygen radicals and known to be dependent upon Rac activation *in vitro* (98), accompanies the cross-linking of Fcγ but not CR3 receptors in macrophages (95, 99, 100, and see Figs 2 and 3). Moreover, binding and phagocytosis via FcγR, but not CR3, trigger leukotriene release (101). Finally, FcγR (not CR3) cross-linking leads to TNF-α (102) and IL-12 production. In fact, it is even thought that ligation of CR3 and other complement receptors may be immunosuppressive by down-regulating cytokine production by human monocytes (103, 104). This is reminiscent of the effect of apoptotic cell uptake on cytokine production by macrophages (16, 105).

Phagocytosis is required for presentation of antigens by MHC class II molecules on the macrophage surface, thereby enhancing T cell responses to bacterial infections (106, 107). The apoptotic material taken up by macrophages would be extensively degraded (108), but there is as yet no evidence for—or against—presentation of peptides derived from apoptotic cells on the macrophage surface.

Paradoxically, phagocytosis is also the common mechanism by which intracellular pathogens invade non-professional phagocytes (109) or colonize macrophages (see Chapter 11) (110). In any case, this deleterious aspect of phagocytosis means that the micro-organisms have found shelter both from the lethal combination of antibodies and complement and from the bactericidal phagocytes.

7. Concluding remarks

More than a century after Metchnikoff's initial observations, phagocytosis is still a poorly understood process. XXth-century scientists have discovered that phagocytosis is a localized cytoskeletal and signalling response to particle binding, that can be associated with other effector responses such as inflammatory mediators and cytokine production to amplify host defence mechanisms. Converging ultrastructural, biological, pharmacological, and signalling approaches indicate that there is one basic zipper-like model for phagocytosis, but two functional outcomes in terms of inflammation and cytokine production and at least two molecular mechanisms accounting for the progression from receptor binding to actin-driven engulfment. The fact that most cell types can perform phagocytosis has facilitated the dissection of the molecular details underlying the uptake process, and combined with the growing number of molecular and cellular tools that are becoming available, this promises a tremendously exciting research area for the years to come.

Acknowledgements

E. C. is supported by the Wellcome Trust.

References

1. Stossel, T. P. (1999) The early history of phagocytosis. In *Advances in cell and molecular biology of membranes and organelles* (ed. A. M. Tartakoff). JAI Press Inc., Stamford, Connecticut. Vol. 5 (Phagocytosis: the host, ed. S. Gordon), p.3.
2. Guillèn, N. (1996) Role of signalling and cytoskeletal rearrangements in the pathogenesis of *Entamoeba histolytica*. *Trends Microbiol.*, **4**, 191.
3. Cohen, C. J., Bacon, R., Clarke, M., Joiner, K., and Mellman, I. (1994) *Dictyostelium discoideum* mutants with conditional defects in phagocytosis. *J. Cell Biol.*, **126**, 955.
4. Repo, H. and Harlan, J. M. (1999) Mechanisms and consequences of phagocyte adhesion to endothelium. *Ann. Med.*, **31**, 156.
5. McKercher, S. R., Torbett, B. E., Anderson, K. L., Henkel, G. W., Vestal, D. J., Baribault, H., *et al.* (1996) Targeted disruption of the PU.1 gene results in multiple hematopoietic abnormalities. *EMBO J.*, **15**, 5647.
6. Braun, A., Hoffmann, J. A., and Maister, M. (1998) Analysis of the Drosophila host defense in domino mutant larvae, which are devoid of hemocytes. *Proc. Natl. Acad. Sci. USA*, **95**, 14337.

7. Wakefield, J. and Hicks, R. M. (1974) Erythrophagocytosis by the epithelial cells of the bladder. *J. Cell Sci.*, **15**, 555.
8. Young, R. W. and Bok, D. (1969) Participation of the retinal pigment epithelium in the rod outer segment renewal process. *J. Cell Biol.*, **42**, 392.
9. Hogan, M. J., Wood, I., and Steinberg, R. H. (1974) Phagocytosis by pigment epithelium of human retinal cones. *Nature*, **252**, 305.
10. Hess, K. L., Tudor, K. S., Johnson, J. D., Osati-Ashtiani, F., Askew, D. S., and Cook-Mills, J. M. (1997) Human and murine high endothelial venule cells phagocytose apoptotic leukocytes. *Exp. Cell Res.*, **236**, 404.
11. Hall, S. E., Savill, J. S., Henson, P. M., and Haslett, C. (1994) Apoptotic neutrophils are phagocytosed by fibroblasts with participation of the fibroblast vitronectin receptor and involvement of a mannose/fucose-specific lectin. *J. Immunol.*, **153**, 3218.
12. Ezekowitz, R. A. B., Sastry, K., Bailly, P., and Warner, A. (1990) Molecular characterization of the human macrophage mannose receptor: demonstration of multiple carbohydrate recognition-like domains and phagocytosis of yeasts in Cos-1 cells. *J. Exp. Med.*, **172**, 1785.
13. Krauss, J. C., Poo, H., Xue, W., Mayo-Bond, L., Todd III, R. F., and Petty, H. R. (1994) Reconstitution of antibody-dependent phagocytosis in fibroblasts expressing Fcγ receptor IIIB and the complement receptor type 3. *J. Immunol.*, **153**, 1769.
14. van der Laan, L. J. W., Dopp, E. A., Haworth, R., Pikkarainen, T., Kangas, M., Elomaa, O., *et al.* (1999) Regulation and functional involvement of macrophage scavenger receptor MARCO in clearance of bacteria *in vivo*. *J. Immunol.*, **162**, 939.
15. Ren, Y., Silverstein, R. L., Allen, J., and Savill, J. (1995) CD36 gene transfer confers capacity for phagocytosis of cells undergoing apoptosis. *J. Exp. Med.*, **181**, 1857.
16. Devitt, A., Moffatt, O. D., Raykundalia, C., Capra, J. D., Simmons, D. L., and Gregory, C. D. (1998) Human CD14 mediates recognition and phagocytosis of apoptotic cells. *Nature*, **392**, 505.
17. Indik, Z. K., Hunter, S., Huang, M. M., Pan, X. Q., Chien, P., Kelly, C., *et al.* (1994) The high affinity Fcγ receptor (CD64) induces phagocytosis in the absence of its cytoplasmic domain: the γ subunit of FcγRIIIA imparts phagocytic function to FcγRI. *Exp. Haematol.*, **22**, 599.
18. Indik, Z. K., Kelly, C., Chien, P., Levinson, A. I., and Schreiber, A. D. (1991) Human FcγRII, in the absence of any other Fcγ receptors, mediates a phagocytic signal. *J. Clin. Invest.*, **88**, 1766.
19. Park, J.-G., Isaacs, R. E., Chien, P., and Schreiber, A. D. (1993) In the absence of any other Fc receptors, FcγRIIIA transmits a phagocytic signal that requires the cytoplasmic domain of its γ subunit. *J. Clin. Invest.*, **92**, 1967.
20. Caron, E. and Hall, A. (1998) Identification of two distinct mechanisms of phagocytosis controlled by different Rho GTPases. *Science*, **282**, 1717.
21. Wright, S. D., Tobias, P. S., Ulevitch, R. J., and Ramos, R. A. (1989) Lipopolysaccharide (LPS) binding protein opsonizes LPS-bearing particles for recognition by a novel receptor on macrophages. *J. Exp. Med.*, **170**, 1231.
22. Platt , N., Suzuki, H., Kurihara, Y., Kodama, T., and Gordon, S. (1996) Role for the class A macrophage scavenger receptor in the phagocytosis of apoptotic thymocytes *in vitro*. *Proc. Natl. Acad. Sci. USA*, **93**, 12456.
23. Fadok, V. A., Savill, J. S., Haslett, C., Bratton, D. L., Doherty, D. E., Campbell, P. A., *et al.* (1992) Different populations of macrophages use either the vitronectin receptor or the phosphatidylserine receptor to recognize and remove apoptotic cells. *J. Immunol.*, **149**, 4029.
24. Fadok, V. A., Warner, M. L., Bratton, D. L., and Henson, P. M. (1998) CD36 is required for

phagocytosis of apoptotic cells by human macrophages that use either a phosphatidylserine receptor or the vitronectin receptor ($\alpha_v\beta_3$). *J. Immunol.*, **161**, 6250.

25. Mevorach, D., Mascarenhas, J. O., Gershov, D., and Elkon, K. B. (1998) Complement-dependent clearance of apoptotic cells by human macrophages. *J. Exp. Med.*, **188**, 2313.
26. Muro, M., Koseki, T., Akifusa, S., Kato, S., Kowashi, Y., Ohsaki, Y., *et al.* (1997) Role of CD14 molecules in internalization of *Actinobacillus actinomycetemcomitans* by macrophages and subsequent induction of apoptosis. *Infect. Immun.*, **65**, 1147.
27. Kuhlman, M., Joiner, K., and Ezekowitz, R. A. (1989) The human mannose-binding protein functions as an opsonin. *J. Exp. Med.*, **169**, 1733.
28. Greenberg, S., Burridge, K., and Silverstein, S. C. (1990) Colocalization of F-actin and talin during Fc receptor-mediated phagocytosis in mouse macrophages. *J. Exp. Med.*, **172**, 1853.
29. Griffin, F. M., Griffin, J. A., Leider, J. E., and Silverstein, S. D. (1975) Studies on the mechanism of phagocytosis. I. Requirement for circumferential attachment of particle-bound ligands to specific receptors on the macrophage plasma membrane. *J. Exp. Med.*, **142**, 1263.
30. Wright, S. D. and Silverstein, S. C. (1982) Tumor-promoting phorbol esters stimulate C3b and C3b′ receptor-mediated phagocytosis in cultured human monocytes. *J. Exp. Med.*, **156**, 1149.
31. Griffin, F. M. and Silverstein, S. C. (1974) Segmental response of the macrophage plasma membrane to a phagocytic stimulus. *J. Exp. Med.*, **139**, 323.
32. Kaplan, G. (1977) Differences in the mode of phagocytosis with Fc and CR3 receptors in macrophages. *Scand. J. Immunol.*, **6**, 797.
33. Allen, L.-A. and Aderem, A. (1996) Molecular definition of distinct cytoskeletal structures involved in complement- and Fc receptor-mediated phagocytosis in macrophages. *J. Exp. Med.*, **184**, 627.
34. Petty, H. R., Hafeman, D. G., and McConell, H. M. (1981) Disappearance of macrophage surface folds after antibody-dependent phagocytosis. *J. Cell Biol.*, **89**, 223.
35. Hackam, D. J., Rotstein, O. D., Sjolin, C., Schreiber, A. D., Trimble, W. S., and Grinstein, S. (1998) v-SNARE-dependent secretion is required for phagocytosis. *Proc. Natl. Acad. Sci. USA*, **95**, 11691.
36. Cannon, G. J. and Swanson, J. A. (1992) The macrophage capacity for phagocytosis. *J. Cell Sci.*, **101**, 907.
37. Buys, S. S., Keogh, E. A., and Kaplan, J. (1984) Fusion of intracellular membrane pools with cell surfaces of macrophages stimulated by phorbol esters and calcium ionophores. *Cell*, **38**, 569.
38. Lennartz, M. R., Yuen, A. F. C., Mc Kenzie Masi, S., Russell, D. G., Buttle, K. F., and Smith, J. J. (1997) Phospholipase A_2 inhibition results in sequestration of plasma membrane into electron-lucent vesicles during IgG-mediated phagocytosis. *J. Cell Sci.*, **110**, 2041.
39. Muller, W. A., Steinman, R. M., and Cohn, Z. A. (1980) The membrane proteins of the vacuolar system. II. Bidirectional flow between secondary lysosomes and plasma membrane. *J. Cell Biol.*, **86**, 304.
40. Aggeler, J. and Werb, Z. (1982) Initial events during phagocytosis by macrophages viewed from outside and inside the cell: membrane-particle interactions and clathrin. *J. Cell Biol.*, **94**, 613.
41. Lowry, M. B., Duchemin, A.-M., Robinson, J. M., and Anderson, C. L. (1998) Functional separation of pseudopod extension and particle internalization during Fcγ receptor-mediated phagocytosis. *J. Exp. Med.*, **187**, 161.
42. Aderem, A. and Underhill, D. M. (1999) Mechanisms of phagocytosis in macrophages. *Annu. Rev. Immunol.*, **17**, 593.

43. Greenberg, S., Chang, P., and Silverstein, S. C. (1993) Tyrosine phosphorylation is required for Fc receptor-mediated phagocytosis in mouse macrophages. *J. Exp. Med.*, **177**, 529.
44. Allen, L.-A. and Aderem, A. (1995) A role for MARCKS, the α isozyme of protein kinase C and myosin I in zymosan phagocytosis by macrophages. *J. Exp. Med.*, **182**, 829.
45. Swanson, J. A., Johnson, M., Beningo, K., Post, P., Mooseker, M., and Araki, N. (1999) A contractile activity that closes phagosomes in macrophages. *J. Cell Sci.*, **112**, 307.
46. Stendahl, O. I., Hartwig, J. H., Brotschi, E. A., and Stossel, T. P. (1980) Distribution of actin-binding protein and myosin in macrophages during spreading and phagocytosis. *J. Cell Biol.*, **84**, 215.
47. Greenberg, S., Chang, P., and Silverstein, S. C. (1994) Tyrosine phosphorylation of the gamma subunit of Fc gamma receptors, p72Syk, and paxillin during Fc receptor-mediated phagocytosis. *J. Biol. Chem.*, **269**, 3897.
48. Zhu, Z., Bao, Z., and Li, J. (1995) MacMARCKS mutation blocks macrophage phagocytosis of zymosan. *J. Biol. Chem.*, **270**, 17652.
49. Brumell, J. H., Howard, J. C., Craig, K., Grinstein, S., Schreiber, A. D., and Tyers, M. (1999) Expression of the protein kinase C substrate pleckstrin in macrophages: association with phagosomal membranes. *J. Immunol.*, **163**, 3388.
50. Evans, E., Leung, A., and Zhelev, D. (1993) Synchrony of cell spreading and contraction force as phagocytes engulf large pathogens. *J. Cell Biol.*, **122**, 1295.
51. Grinnell, F. (1984) Fibroblast spreading and phagocytosis: similar cell responses to different-sized substrata. *J. Cell. Physiol.*, **119**, 58.
52. Mellman, I. S., Plutner, H., Steinman, R. M., Unkeless, J. C., and Cohn, Z. A. (1983) Internalization and degradation of macrophage Fc receptors during receptor-mediated phagocytosis. *J. Cell Biol.*, **96**, 887.
53. Pitt, A., Mayorga, L. S., Schwartz, A. L., and Stahl, P. D. (1992) Transport of phagosomal components to an endosomal compartment. *J. Biol. Chem.*, **267**, 126.
54. Mayorga, L. S., Bertini, F., and Stahl, P. D. (1991) Fusion of newly formed phagosomes with endosomes in intact cells and in a cell-free system. *J. Biol. Chem.*, **266**, 6511.
55. Desjardins, M., Huber, L. A., Parton, R. G., and Griffiths, G. (1994) Biogenesis of phagolysosomes proceeds through a sequential series of interactions with the endocytic apparatus. *J. Cell Biol.*, **124**, 677.
56. Desjardins, M., Celis, J. E., van Meer, G., Dieplinger, H., Jahraus, A., Griffiths, G., *et al.* (1994) Molecular characterization of phagosomes. *J. Biol. Chem.*, **269**, 32194.
57. Beron, W., Alvarez-Dominguez, C., Mayorga, L., and Stahl, P. D. (1995) Membrane trafficking along the phagocytic pathway. *Trends Cell Biol.*, **5**, 100.
58. Desjardins, M., Nzala, N. N., Corsini, R., and Rondeau, C. (1997) Maturation of phagosomes is accompanied by changes in their fusion properties and size-selective acquisition of solute materials from endosomes. *J. Cell Sci.*, **110**, 2303.
59. Beron, W., Colombo, M. I., Mayorga, L. S., and Stahl, P. D. (1995) *In vitro* reconstitution of phagosome-endosome fusion: evidence for regulation by heterotrimeric GTPases. *Arch. Biochem. Biophys.*, **317**, 337.
60. Bouvier, G., Benoliel, A.-M., Foa, C., and Bongrand, P. (1994) Relationship between phagosome acidification, phagosome-lysosome fusion, and mechanism of particle ingestion. *J. Leukoc. Biol.*, **55**, 729.
61. Koval, M., Preiter, K., Adles, C., Stahl, P. D., and Steinberg, T. H. (1998) Size of IgG-opsonized particles determines macrophage response during internalization. *Exp. Cell Res.*, **242**, 265.

62. Gordon, A. H., Hart, P. D., and Young, M. R. (1980) Ammonia inhibits phagosome-lysosome fusion in macrophages. *Nature*, **286**, 79.
63. Hackam, D. J., Rotstein, O. D., Zhang, W.-J., Demaurex, N., Woodside, M., Tsai, O., *et al.* (1997) Regulation of phagosomal acidification. Differential targeting of Na^+/H^+ exchangers, Na^+/K^+/ATPases, and vacuolar-type H^+-ATPases. *J. Biol. Chem.*, **272**, 29810.
64. Lukacs, G. L., Rotstein, O. D., and Grinstein, S. (1990) Phagosomal acidification is mediated by a vacuolar-type H(+)-ATPase in murine macrophages. *J. Biol. Chem.*, **265**, 21099.
65. Downey, G. P., Botelho, R. J., Butler, J. R., Moltyaner, Y., Chien, P., Schreiber, A. D., *et al.* (1999) Phagosomal maturation, acidification, and inhibition of bacterial growth in nonphagocytic cells transfected with FcγRIIA receptors. *J. Biol. Chem.*, **274**, 28436.
66. Miettinen, H. M., Matter, K., Hunziker, W., Rose, J. K., and Mellman, I. (1992) Fc receptor endocytosis is controlled by a cytoplasmic domain determinant that actively prevents coated pit localization. *J. Cell Biol.*, **116**, 875.
67. Davis, W., Harrison, P. T., Hutchinson, M. J., and Allen, J. M. (1995) Two distinct regions of Fc gamma RI initiate separate signalling pathways involved in endocytosis and phagocytosis. *EMBO J.*, **14**, 432.
68. Odin, J. A., Edberg, J. C., Painter, C. J., Kimberly, R. P., and Unkeless, J. C. (1991) Regulation of phagocytosis and $[Ca^{2+}]_i$ flux by distinct regions of an Fc receptor. *Science*, **254**, 1785.
69. Pitt, A., Mayorga, L. S., Stahl, P. D., and Schwartz, A. L. (1992) Alterations in the protein composition of maturing phagosomes. *J. Clin. Invest.*, **90**, 1978.
70. Sandvig, K. and van Deurs, B. (1990) Selective modulation of the endocytic uptake of ricin and fluid phase markers without alteration in transferrin endocytosis. *J. Biol. Chem.*, **265**, 6382.
71. Lamaze, C., Fujimoto, L. M., Yin, H. L., and Schmid, S. L. (1997) The actin cytoskeleton is required for receptor-mediated endocytosis in mammalian cells. *J. Biol. Chem.*, **272**, 20332.
72. Indik, Z. K., Park, J.-G., Pan, X. Q., and Schreiber, A. D. (1995) Induction of phagocytosis by a protein tyrosine kinase. *Blood*, **86**, 4389.
73. Hutchinson, M. J., Harrison, P. T., Floto, R. A., and Allen, J. M. (1995) Fcγ receptor-mediated phagocytosis requires tyrosine kinase activity and is ligand independent. *Eur. J. Immunol.*, **25**, 481.
74. Mitchell, M. A., Huang, M. M., Chien, P., Indik, Z. K., Pan, X. Q., and Schreiber, A. D. (1994) Substitutions and deletions in the cytoplasmic domain of the phagocytic receptor Fc gamma RIIA: effect on receptor tyrosine phosphorylation and phagocytosis. *Blood*, **84**, 1753.
75. Duchemin, A.-M. and Anderson, C. L. (1997) Association of non-receptor protein kinases with the Fc gamma RI/gamma-chain complex in monocytic cells. *J. Immunol.*, **158**, 865.
76. Durden, D. L., Kim, H. M., Calore, B., and Liu, Y. (1995) The Fc gamma RI receptor signals through the activation of hck and MAP kinase. *J. Immunol.*, **154**, 4039.
77. Crowley, M. T., Costello, P. S., Fitzer-Attas, C. J., Turner, M., Meng, F., Lowell, C., *et al.* (1997) A critical role for syk in signal transduction and phagocytosis mediated by Fcγ receptors on macrophages. *J. Exp. Med.*, **186**, 1027.
78. Kiener, P. A., Rankin, B. M., Burkhardt, A. L., Schieven, G. L., Gililand, L. K., Rowley, R. B., *et al.* (1993) Cross-linking of Fc gamma receptor I (Fc gamma RI) and receptor II (Fc gamma RII) on monocytic cells activates a signal transduction pathway common to both Fc receptors that involves the stimulation of p72 Syk protein tyrosine kinase. *J. Biol. Chem.*, **268**, 24442.
79. Durden, D. L. and Liu, Y. B. (1994) Protein-tyrosine kinase p72Syk in Fc gamma RI receptor signalling. *Blood*, **84**, 2102.

80. Darby, C., Geahlen, R. L., and Schreiber, A. D. (1994) Stimulation of macrophage FcγRIIIA activates the receptor-associated protein tyrosine kinase syk and induces tyrosine phosphorylation of multiple proteins including p95Vav and p62/GAP-associated protein. *J. Immunol.*, **152**, 5429.
81. Kiefer, F., Brumell, J., Al-Alawi, N., Latour, S., Cheng, A., Veillette, A., *et al.* (1998) The syk tyrosine kinase is essential for Fcγ receptor signaling in macrophages and neutrophils. *Mol. Cell. Biol.*, **18**, 4209.
82. Lowry, M. B., Duchemin, A.-M., Coggeshall, K. M., Robinson, J. M., and Anderson, C. L. (1998) Chimeric receptors composed of phosphoinositide 3-kinase domains and Fcγ receptor ligand-binding domains mediate phagocytosis in COS fibroblasts. *J. Biol. Chem.*, **273**, 24513.
83. Ninomiya, N., Hazeki, K., Fukui, Y., Seya, T., Okada, T., Hazeki, O., *et al.* (1994) Involvement of phosphatidylinositol 3-kinase in Fcγ receptor signaling. *J. Biol. Chem.*, **269**, 22732.
84. Araki, N., Johnson, M. T., and Swanson, J. A. (1996) A role for phosphoinositide 3-kinase in the completion of macropinocytosis and phagocytosis in macrophages. *J. Cell Biol.*, **135**, 1249.
85. Cox, D., Tseng, C.-C., Bjekic, G., and Greenberg, S. (1999) A requirement for phosphatidylinositol 3-kinase in pseudopod extension. *J. Biol. Chem.*, **274**, 1240.
86. Hall, A. (1999) Rho GTPases and the actin cytoskeleton. *Science*, **279**, 509.
87. Cox, D., Chang, P., Zhang, Q., Reddy, P. G., Bokoch, G. M., and Greenberg, S. (1997) Requirements for both Rac1 and Cdc42 in membrane ruffling and phagocytosis in leukocytes. *J. Exp. Med.*, **186**, 1487.
88. Massol, P., Montcourrier, P., Guillemot, J.-C., and Chavrier, P. (1998) Fc receptor-mediated phagocytosis requires Cdc42 and Rac1. *EMBO J.*, **17**, 6219.
89. Hackam, D. J., Rotstein, O. D., Schreiber, A., Zhang, W.-J., and Grinstein, S. (1997) Rho is required for the initiation of calcium signaling and phagocytosis by Fcγ receptors in macrophages. *J. Exp. Med.*, **186**, 955.
90. Zhang, Q., Cox, D., Tseng, C.-C., Donaldson, J. G., and Greenberg, S. (1998) A requirement for ARF6 in Fcγ receptor-mediated phagocytosis in macrophages. *J. Biol. Chem.*, **273**, 19977.
91. Ellis, R., Jacobson, D., and Horvitz, H. R. (1991) Genes required for the engulfment of cell corpses during programmed cell death in *Caenorhabditis elegans*. *Genetics*, **129**, 79.
92. Wu, Y. C. and Horvitz, H. R. (1998) *C. elegans* phagocytosis and cell-migration protein CED-5 is similar to human DOCK180. *Nature*, **392**, 501.
93. Nolan, K., Barrett, K., Lu, Y., Hu, K.-Q., Vincent, S., and Settleman, J. (1998) Myoblast city, the *Drosophila* homolog of DOCK180/CED-5 is required in a Rac signaling pathway utilized for multiple developmental processes. *Genes Dev.*, **12**, 3337.
94. Kiyokawa, E., Hashimoto, Y., Kobayashi, S., Sugimura, S., Kurata, T., and Matsuda, M. (1998) Activation of Rac1 by a Crk SH3-binding protein, DOCK180. *Genes Dev.*, **12**, 3331.
95. Brozna, J. P., Hauf, N. F., Phillips, W. A., and Johnston, R. B. (1988) Activation of the respiratory burst in macrophages. Phosphorylation specifically associated with Fc receptor-mediated stimulation. *J. Immunol.*, **141**, 1642.
96. Hmama, Z., Knutson, K. L., Herrera-Velit, P., Nandan, D., and Reiner, N. E. (1999) Monocyte adherence induced by lipopolysaccharide involves CD14, LFA-1, and cytohesin-1. *J. Biol. Chem.*, **274**, 1050.
97. Zhelezniak, A. and Brown, E. J. (1992) Immunoglobulin-mediated phagocytosis by human monocytes requires protein kinase C activation. *J. Biol. Chem.*, **267**, 12042.
98. Diekmann, D., Abo, A., Johnston, C., Segal, A. W., and Hall, A. (1994) Interaction of rac with p67phox and regulation of phagocytic NADPH oxidase activity. *Science*, **265**, 531.

99. Wright, S. D. and Silverstein, S. C. (1983) Receptors for C3b and C3bi promote phagocytosis but not the release of toxic oxygen from human phagocytes. *J. Exp. Med.*, **158**, 2016.
100. Yamamoto, K. and Johnston, R. B. (1984) Dissociation of phagocytosis from stimulation of the oxidative metabolic burst in macrophages. *J. Exp. Med.*, **159**, 405.
101. Aderem, A. A., Wright, S. D., Silverstein, S. C., and Cohn, Z. A. (1985) Ligated complement receptors do not activate the arachidonic acid cascade in resident peritoneal macrophages. *J. Exp. Med.*, **161**, 617.
102. Stein, M. and Gordon, S. (1991) Regulation of tumor necrosis factor (TNF) release by murine peritoneal macrophages: role of cell stimulation and specific phagocytic plasma membrane receptors. *Eur. J. Immunol.*, **21**, 431.
103. Marth, T. and Kelsall, B. L. (1997) Regulation of interleukin-12 by complement receptor 3 signalling. *J. Exp. Med.*, **185**, 1987.
104. Karp, C. L., Wysocka, M., Wahl, L. M., Ahearn, J. M., Cuomo, P. J., Sherry, B., *et al.* (1996) Mechanism of suppression of cell-mediated immunity by measles virus. *Science*, **273**, 228.
105. Fadok, V. A., Bratton, D. L., Konowal, A., Freed, P. W., Westcott, J. Y., and Henson, P. M. (1998) Macrophages that have ingested apoptotic cells *in vitro* inhibit proinflammatory cytokine production through autocrine/paracrine mechanisms involving TGF-β, PGE2, and PAF. *J. Clin. Invest.*, **101**, 890.
106. Stockinger, B. (1992) Capacity of antigen uptake by B cells, fibroblasts or macrophages determines efficiency of presentation of a soluble self antigen (C5) to T lymphocytes. *Eur. J. Immunol.*, **22**, 1271.
107. Mellman, I., Turley, S. J., and Steinman, R. M. (1998) Antigen processing for amateurs and professionals. *Trends Cell Biol.*, **8**, 231.
108. Brazil, M. I., Weiss, S., and Stockinger, B. (1997) Excessive degradation of intracellular protein in macrophages prevents presentation in the context of major histocompatibility complex class II molecules. *Eur. J. Immunol.*, **27**, 1506.
109. Ireton, K. and Cossart, P. (1998) Interaction of invasive bacteria with host signaling pathways. *Curr. Opin. Cell Biol.*, **10**, 276.
110. Allen, L.-A. and Aderem, A. (1996) Mechanisms of phagocytosis. *Curr. Opin. Immunol.*, **8**, 36.

4 Macropinocytosis

MARKUS MANIAK

1. Introduction

Many decades ago, when discovery in the field of cell biology was limited by the resolution of the light microscope two endocytic pathways could be observed directly: They were classified as phagocytosis (the uptake of particles) and pinocytosis (the uptake of fluid). Despite this early distinction, the two processes are now known to be mechanistically highly related. They are grossly different from the more recently described types of endocytic processes involving coated pits or caveoli.

Since cells can only exist in aqueous environments, any type of endocytosis results in the concomitant internalization of extracellular fluid. Over time, the term pinocytosis has been increasingly used to specify the pathway of fluid phase uptake involving small (0.1 μm diameter) vesicles formed by clathrin-coated pits at the plasma membrane. Stages of this process can be documented only by electron microscopy. Much is known about clathrin-dependent micropinocytosis, a pathway that is thought to operate constitutively in all cell types. In contrast, macropinocytosis, the formation of large (many μm diameter) fluid-filled vesicles originally described in 1931 (1) is a natural feature of professional phagocytic cells but a rare event in other types of tissue culture cells. In many cells, however, the rate of macropinocytosis can be increased by a variety of stimuli. Under these conditions the signal is transmitted to effector molecules that regulate the cytoskeleton and influence vesicle trafficking. Consequently membrane ruffles form in the periphery of the cell, detach from the substratum, and fold back onto the cell body where they gradually disappear. An aliquot of medium trapped during this process ends up as a macropinosome in the cytoplasm. The fate of the macropinosome varies among cell types: its contents can either be recycled to the cell surface or processed in a manner analogous to material endocytosed through different routes. Only in the latter case the biological function of macropinocytosis is immediately obvious. Soluble material is internalized, enzymatically degraded, and in special cases fragments thereof are presented to the immune system. Macropinocytosis is evident in lower eukaryotes like *Dictyostelium* amoebae, which can serve as a model for the mammalian system because many features have been preserved throughout evolution. The subsequent chapters deal with the following main questions:

- Which pathway is used for fluid phase uptake?
- How is macropinocytosis induced?

- What are the mechanisms mediating macropinocytosis?
- Where does the endocytosed fluid and membrane go?
- Why does it occur in some cells only?

2. Macropinocytosis or other pathways?

To experimentally distinguish macropinocytosis from other pathways of fluid phase uptake, the following criteria apply:

On the surface of the cell or its periphery broad lamellar protrusions should be visible, e.g. in the scanning electron microscope (Fig. 1a–d). These structures should

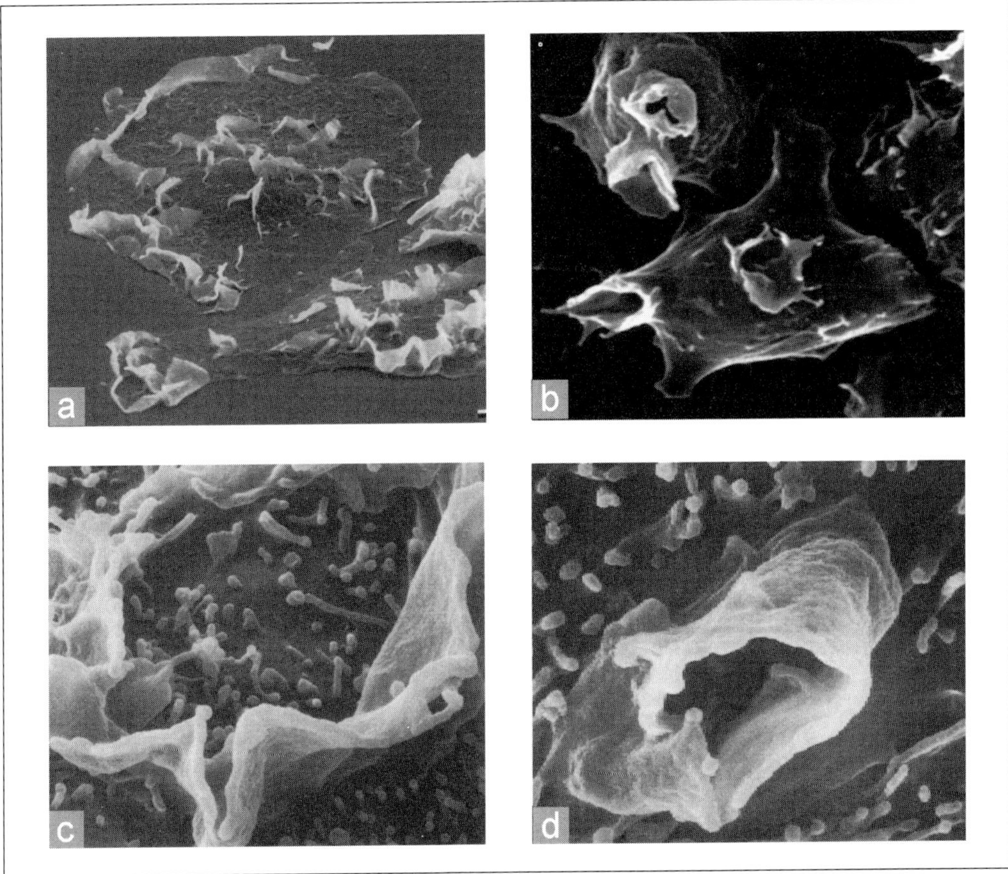

Fig. 1 Morphology and mechanism of macropinocytosis by scanning electron microscopy. (a) Macrophages showing membrane ruffles at the periphery. A cup-shaped ruffle suggestive of macropinocytosis is seen in the lower left corner. Frame is 60 μm wide. Reprinted from ref. 22. (b) *Dictyostelium* cells with a funnel-shaped protrusion at the left edge of the lower cell. Dorsal circular ruffle just prior to closure on the upper cell. Frame is 20 μm wide. Reprinted from ref. 5. (c, d) Details of circular ruffles on MDCK cells during (c, frame 4.3 μm wide) initiation and (d, frame 2.8 μm wide) closure. Reprinted from ref. 15.

be filled with filamentous actin, e.g. as detected by the localization of fluorescent phalloidin. The presence of ruffles only, however, is no indication of their ability to form macropinosomes, e.g. (2).

The uptake of fluid phase markers should be sensitive to drugs acting on intracellular pH or the cytoskeleton. Amiloride, an inhibitor of the Na^+/K^+ transporter, causes the cytoplasm to become more alkaline, as a consequence ruffles disappear and fluid phase uptake is reduced (3). Cytochalasin blocks actin polymerization and decreases fluid phase uptake in many systems (4, 5). This sensitivity, however, is also not a sufficient criterion to define macropinocytosis, because in some way the actin cytoskeleton is involved in other endocytic pathways, e.g. influencing coated pit dynamics (6) and many steps along the endocytic route (7, 8). Moreover, depending on the concentration and other conditions, cytochalasin may also stimulate the uptake of fluid phase marker (9).

Therefore, the best way to monitor macropinocytosis is microscopy of living cells (Plate 3a–h). In an ideal situation, the extracellular fluid should contain a suitable tracer that is internalized (Plate 3e–h). Fluorescent markers are the molecules of choice, because micropinosomes are barely visible in the presence of macropinosomes. The reasons are based on geometry: a tenfold difference in diameter produces a 1000-fold difference in the volume of fluorescent marker, and therefore fluorescence intensity. Extending this argument to include considerations about the surface to volume ratio of a sphere, the mechanism of fluid phase endocytosis can solely be inferred from the ratio of fluid volume taken up to the area of membrane concomitantly internalized (10).

3. Physiological stimulus or cells gone wild?

Membrane ruffling around the edges of the cell and subsequent macropinocytosis is readily seen in professional phagocytes, such as unstimulated macrophages (Plate 3a–d). The addition of M-CSF (macrophage colony stimulating factor) results in the production of more and larger ruffles and the rate of macropinosome formation increases (4) while the rate of clathrin-dependent micropinocytosis remains unchanged (11). EGF (epidermal growth factor) induces the formation of macropinocytosis in various cell types like glia (12) but most prominently in the epidermoid cell line A431. Here, the response is possibly a function of the number of receptors expressed (13). The planar ruffles that are seen at the periphery of cells typically result in the formation of macropinosomes about 1 μm in diameter. In contrast, fibroblasts treated with PDGF (platelet-derived growth factor) show large circular ruffles on dorsal surface (14). MDCK cells (Madin-Darby canine kidney) stimulated with HGF/SF (hepatocyte growth factor / scatter factor) were the first cells for which it was shown that circular ruffles close to form macropinosomes (Fig. 1c, d) (15). Macropinosomes formed by circular ruffles are much larger (often 10 μm in diameter) than ones from peripheral ruffles. In the lower eukaryote *Dictyostelium* circular ruffles appear both on the dorsal surface of the cell and around the edges (Fig. 1b). They develop into funnel-like structures and the size of macropinosomes ultimately formed is not grossly different (5).

The growth factors known to stimulate the formation of membrane ruffles in the mammalian system act through the class of receptor tyrosine kinases (RTK). This signalling pathway is particularly sensitive to mutations, since many of the genes involved are proto-oncogenes. For example, ruffling can be elicited by expression of an activated form of the non-receptor tyrosine kinase *Src* (16). One of the immediate targets of membrane-associated tyrosine kinases is the small GTPase *Ras*. Indeed, microinjection of normal and oncogenic variants of ras elicits membrane ruffling and macropinocytosis (17). In addition a signalling cascade is set off that involves various protein kinases and lipid kinases. One of the central molecules involved in signalling appears to be protein kinase C (PKC) as judged from the increased rate of macropinocytosis in cells treated with phorbol esters and diacylglycerol (18, 19). Modification of lipids, which could create membrane microdomains favouring macropinocytosis is apparently mediated by phosphatidylinositol 3-kinase (PI3K) (20, 21). A very interesting observation is that the drug wortmannin, a specific inhibitor of PI3K allows membrane ruffling in the absence of macropinocytosis, suggesting that PI3K is involved in one of the last steps of this event, namely the closure of the macropinosome (22). PKC and PI3K are signalling molecules implicated to act upstream of small GTPases (23, 24). Among the small GTPases *Rac* may constitute the main effector, as it induces the polymerization of actin filaments and the dramatic formation of ruffles in the cell periphery at the expense of actin stress-fibre bundles in the cytoplasm (23). *Rac* again may act through the p21-activated kinase PAK (25). Another small GTPase involved in the regulation of membrane ruffling is *Arf*6 (26). Both GTPases, *Rac* and *Arf*6 interact with common target molecules (27, 28) and thus jointly regulate cytoskeletal rearrangements that lead to the formation of ruffles and macropinosomes.

The molecules involved in the signalling pathway leading to macropinocytosis are conserved in evolution. Genes homologous to protein kinases, lipid kinases, and small GTPases have been identified in *Dictyostelium*. Mutants carrying targeted gene-disruptions in two of the three p110-related PI3K genes are severely impaired in the uptake of fluid phase marker (29). The genetic situation in the family of small GTPases is complex. Currently 14 genes homologous to *Rac* have been identified. Among the few initially characterized, *Rac*F localizes to nascent macropinosomes but the genetic inactivation of *Rac*F has no effect on fluid phase uptake (30). The over-expression of *Rac*C causes the excessive formation of membrane ruffles, some circular in morphology, but surprisingly macropinocytosis is reduced threefold (31). Similar observations have been made with cells overexpressing *Rap*1, a *Ras*-like GTPase (32, 33) emphasizing that this is not a unique result.

One conclusion that immediately transpires from both the mammalian and the *Dictyostelium* system is that the formation of membrane ruffles does not necessarily result in the formation of a macropinosome. One possible explanation comes from the analysis of phagocytosis: When extracellular particles are used to locally trigger the intracellular accumulation of individual small GTPases using the rapamycin-dependent recruitment system, protrusions resembling filopods or pedestals are formed instead of functional phagocytic cups (34, 35). It is therefore conceivable that the activity driving membrane protrusion needs precise spatial and temporal control in order to form a productive endocytic structure.

4. Reorganization of the cytoskeleton or redirected vesicle traffic?

4.1 The actin cytoskeleton

Protrusions at the plasma membrane are thought to be mainly a consequence of rearrangements within the actin cytoskeleton. Actin-binding proteins serve as targets of signalling pathways and fulfil four main functions.

(a) The organization of actin filaments into higher order structures of a certain geometry.
(b) The coupling of actin to the plasma membrane.
(c) The generation of protrusive force by actin polymerization.
(d) The generation of contractile force by molecular motors of the myosin family.

(a) Among the proteins cross-linking filamentous actin into networks or bundles in the mammalian system, one of the most promising targets is MARCKS (myristoylated, alanine-rich, C kinase substrate), since it is a major target of PKC. In cells lacking the MARCKS protein, however, no effects on macropinocytosis can be observed (36). In *Dictyostelium*, at least three proteins of the acting cross-linking class, namely ABP 120, the 34 kDa bundling protein, and fimbrin are enriched in macropinocytic ruffles, but deletion of the corresponding genes does not affect fluid phase uptake (37–39) (Maniak unpublished data).

(b) In mammalian cells ezrin is the prime candidate for a protein coupling the cytoskeleton to the plasma membrane during macropinocytosis (40). EGF stimulation of cells leads to ezrin phosphorylation and its localization to ruffles (41) in a process that is likely to depend on *Rac* signalling (42). In *Dictyostelium* no members of the ezrin/moesin/radixin (ERM)-family have been found. The protein most closely related to the ERM-family is talin. Genetic deletion of talin results in a decreased rate of phagocytosis while macropinocytosis is unaffected (43), indicating that talin is not a general link between the membrane and the actin cytoskeleton. The probably most abundant actin-binding protein on the *Dictyostelium* plasma membrane is ponticulin. But again, ponticulin mutant cells do not show any defect in the process of fluid phase uptake (44).

(c) Actin assembly is a highly organized process involving many cytoskeletal proteins that are potential targets of signalling cascades. The proposed mechanism for actin polymerization involves the following sequence of events: Nucleation of an actin filament by the Arp2/3 complex, elongation of the filaments to form a branched network underneath the plasma membrane, disassembly of older filaments in the cytoplasm by the activity of cofilin, binding of the actin monomers by profilin, which again delivers actin to the zones of protrusion at the membrane (for a model see ref.

45). Indeed, components of the Arp2/3 complex localize to growth factor-induced membrane ruffles (46). This localization and the activity of the complex is regulated by the Wiscott-Aldrich-Protein (WASP) and the related protein Scar (47). Ruffles formed after stimulation also contain coronin, a potential regulator of actin polymerization (48), and the disassembly-promoting protein cofilin (49, 50). One protein that could stabilize actin filaments against breakdown is hsp27. Overexpression of this protein increases F-actin concentration beneath the plasma membrane and macropinocytic activity (51). The Arp2/3 complex has been originally identified in the lower eukaryote *Acanthamoeba* (52), and most of its constituents are also represented in the *Dictyostelium* Genome database. The regulatory protein Scar was first described in *Dictyostelium* (53), but its role in macropinocytosis remains to be established. Coronin was the first *Dictyostelium* protein associated with crown-shaped circular ruffles (54), later shown to be sites of macropinocytosis (Plate 3e–h) (5). A mutant lacking coronin has defects in the organization of the cortical cytoskeleton (55) and a strongly reduced rate of macropinocytosis (5). *Dictyostelium* cofilin localizes to macropinocytic ruffles (56), but it proved to be encoded by an essential gene (57), so that the effect of its deletion on macropinocytosis could not be studied. The activity of cofilin is positively regulated by the actin-interacting protein 1 (Aip1) (58). In a *Dictyostelium* mutant lacking Aip1, fluid phase uptake is reduced as compared to wild-type cells, indirectly suggesting that actin depolymerization plays some role in macropinocytosis (59). It is, however, important to note that overexpression of Aip1 stimulates particle internalization but not fluid phase uptake, pointing to a subtle difference in the organization of the phagocytic cup versus the macropinocytic ruffle (59).

(d) Another important mechanism for generating force at the plasma membrane is by the action of myosin motors. Especially many of the unconventional myosins (that is other than class II, the muscle-like myosin) have the ability to associate with cellular membranes (60). In the mammalian system, myosins I and V could be detected in circular ruffles induced on RBL cells (2), and myosins V and VI localize to EGF induced ruffles (61) under conditions known to stimulate macropinocytosis. As in the case of the *Rac* genes, the situation in the *Dictyostelium* system is complicated by the large family of myosins. Among the unconventional myosins, myosin I localizes to so-called crowns (62) which correspond to the circular ruffles involved in macropinocytosis (5). In mutants lacking combinations of myosin I isoforms (e.g. myo A^-/B^- and B^-/C^-) the number of crowns on the cell surface increases, but the rate of macropinocytosis decreases (63, 64). This finding is compatible with the view that myosins are involved in the closure of macropinosomes, as has been postulated for phagosome formation (65).

Taken together, there is surprisingly little evidence for higher order organization of the actin cytoskeleton in the formation of the macropinosome. On the other hand a strong case is made for the involvement of actin polymerization in the protrusion of membrane ruffles. The multitude of unconventional myosins found in ruffles may provide the force for the purse-sting closure mechanism of macropinosomes by combining the properties of membrane binding and motor activity.

4.2 Microtubules and vesicle transport

In addition to the contribution of the actin system in the formation of macropinosomes, the role of the microtubule cytoskeleton has to be briefly considered. The importance of the microtubules in this endocytic process is best illustrated by the fact that depolymerizing drugs inhibit macropinocytosis in mammalian cells (4, 11) and *Dictyostelium* (66). The direct interaction between tubulin and *Rac* (67) explains how the microtubular and microfilament systems could communicate and how growing microtubules could determine the sites of membrane ruffle protrusion (68). The appearance of membrane ruffles does not only depend on the interaction of actin and microtubules but also correlates with exocytosis in various systems, e.g. (2, 69). This observation underscores the cell's need to keep the surface area of the plasma membrane constant. To achieve this goal endocytosis and recycling must be at equilibrium (membrane homeostasis). Attempts to track the origin of the recycled membrane have identified lysosomal proteins and recycling receptors at membrane ruffles (70, 71). Thus, the microtubular system could direct macropinocytosis in two steps: First, microtubules could function as tracks directing vesicles to sites on the surface where the inserted membrane forms ruffles. In the second step *Rac* is activated to restructure the cytoskeleton for internalization at this site. Along these lines microtubule dynamic instability could provide the basis for the apparent random localization of macropinocytic events around the circumference of the cell.

5. Recycling or progression?

The macropinosome released into the cytoplasm is surrounded by a coat of filamentous actin. If this coat is homogeneous, no movement occurs. Occasionally, as seen in mast cells, the coat is polarized and as a consequence the macropinosome rockets at 0.2 μm/sec through the cytoplasm (72). On the order of minutes after internalization, the coat dissociates from the macropinosome and makes the membrane available for fusion and fission processes. Depending on the cell type, the contents of the macropinosome can be either recycled to the cell surface or delivered to the endosomal system.

In the macrophage and probably other phagocytic cells, macropinosomes that are formed at the cell periphery move towards the cell centre and acidify over about 15 minutes (73). During this period plasma membrane proteins like the transferrin receptor are recycled to the cell surface and as a consequence the size of the macropinosome decreases. One fascinating observation is, that as the macropinosome matures, a mix of marker molecules is separated according to molecular weight (74). This could be explained by a series of sequential transient interactions with preexisting endocytic compartments like in the case of phagosomes (75). An alternative explanation is based on a fusion event, in which a stable pore of small diameter favours the exit of small molecules and the exchange of membrane components before mixing of the entire contents occurs (Plate 4a, b) (76). Judging from obvious changes in morphology, macropinosomes formed in MDCK cells also deliver their contents to the endosomal system (15).

The fate of macropinosomes formed in A431 cells is different from the pathway described above. Endocytosed marker is neither acidified nor delivered to the endosomal system, but rather recycled to the cell surface (77). A similar observation has been made in v-src transformed fibroblasts where recycling is accelerated by increasing intracellular cAMP levels (78). The precise mechanism how the marker is recycled to the cell surface remains obscure, but it may involve smaller vesicular intermediates since extensive tubulation of macropinosomes has been documented (77). Interestingly, overexpression of the small GTPase Arf6 apparently short-circuits the formation of macropinosomes at the plasma membrane and their subsequent exocytosis (26).

Concerning the fate of the macropinosome, *Dictyostelium* again behaves in a fashion most similar to the professional phagocytes. Time lapse confocal microscopy of the green fluorescent protein (GFP) fused to actin-binding proteins shows that the cytoskeletal coat remains on the macropinosome for less than one minute after internalization (Plate 3e–h) (5, 59). Concomitant with release of the cytoskeletal coat the contents of the macropinosome is acidified within seconds (79) and different classes of lysosomal enzymes are delivered sequentially over the next couple of minutes (80). There is one special feature, however, that distinguishes *Dictyostelium* from the cells of higher eukaryotes. After a period of 30–45 minutes under acidic conditions (Plate 4c), the pH of the endosome returns to neutral values (Plate 4d) and remnants of the endocytosed material are efficiently exocytosed (81–83).

6. Biological function or accidental event?

Given the different fates of material internalized through macropinocytosis, one would expect a serious biological function of the process only in cells where the material progresses to later endocytic compartments. In *Dictyostelium*, therefore macropinocytosis is interpreted as phagocytosis without a particulate stimulus, which solely serves a nutritional purpose under laboratory conditions. In the mammalian system progression of marker into a compartment active in degradation is mainly observed in cells belonging to the immune system. Among these, immature dendritic cells are sampling the environment for soluble antigens by internalizing large amounts of fluid via macropinocytosis. During their maturation, antigens become proteolytically processed, bound to MHC class II molecules, and transported to the cell surface to stimulate cytotoxic T cells (84). Alternatively, processed antigen can escape into the cytosol by means of a TAP-dependent transport. In this case the antigen is presented on the surface in association with MHC class I (85, 86). Interestingly this pathway also operates in macrophages (87).

One cell type where macropinocytosis may turn out to be central to its biological function is the osteoclast. Osteoclasts mediate bone resorption by secreting protons and enzymes to solubilize minerals and protein components of bone. This process takes place in the resorption lacuna, a dip beneath the cell, the edges of which are tightly sealed by close apposition of the osteoclast's membrane. In order to progress deeper into the bone, the resorption products need to be removed from the lacuna by transcytosis (Plate 4e) (88, 89). Since coated pits are frequently seen at dorsal and lateral membrane regions but rarely detected at the interface of the resorption lacuna (90), endocytosis of degraded bone material is likely to be mediated by a clathrin-independent mechanism. The membrane facing the lacuna has a characteristic ruffled appearance, the ruffles are rich in filamentous actin (Plate 4f), and the membrane is derived from the late endosomal compartment (91, 92). Moreover, as judged from the published images, primary vesicles formed at the ruffled border do not contain single defined particles (like a phagosome) but a rather fluid amorphous mass of degraded bone matrix. Endocytic vesicles are variable in size and larger than 0.1 μm (Plate 4g, h). Taken together these observations suggest that the endocytic mechanism involved in bone resorption may be macropinocytosis.

A biological function for macropinocytosis can be easily envisaged if the internalized material is degraded as in the cases discussed above. One example where internalization at the plasma membrane does not result in the delivery to the lysosomal compartment is the macropinocytosis of desmosomal plaques in epithelial cells (93), a process that may play a role during embryogenesis or wound healing. But what may be the function of macropinocytosis in cultured cells stimulated by growth factors? Clearly, macropinocytosis is not the pathway through which the growth factor receptors are internalized (13). For an alternative explanation it is important to consider that some growth factors, e.g. PDGF concomitantly stimulate macropinocytosis and chemotaxis (94). A naïve cell challenged by chemoattractant could respond by the following sequence of events: First, internal membranes could be inserted into the plasma membrane for expansion. As a consequence ruffles form. In protrusions that fail to contact the substratum, membrane is retrieved by macropinocytosis. Instead of delivering the endocytosed fluid to the lysosomal system, the membrane of the macropinosome is available to expand the plasma membrane elsewhere, thus enabling the cell to test a new direction. In the end, randomly localized ruffles would be suppressed at the expense of a stable direction of movement. The above hypothesis that the formation of macropinosomes at sites of ruffling membrane is a by-product of chemotactic movement has not been tested experimentally, but pieces of supportive evidence can be found scattered in the literature (70, 71, 95–99).

7. Conclusions

Superficially, macropinocytosis appears like phagocytosis without a particle. However, its main function is neither to internalize a ligand, nor a receptor, but rather a large aliquot of extracellular fluid. To this end the plasma membrane expands by the insertion of internal membranes, which allows the formation of large surface ruffles supported by the actin cytoskeleton. Subtle differences in the architecture of the cytoskeleton in membrane ruffles as compared to phagocytic cups could be the consequence of triggering different signal transduction pathways. Small GTPases have a central role in orchestrating actin dynamics and vesicle traffic and mediate the cross-talk between the systems. There is increasing evidence that the *Rho* family members once thought to be specific regulators of the cytoskeleton, influence vesicle movements, e.g. (100). Similarly, members of the arf family, thought to be mainly involved in vesicle trafficking, affect the activities of the cytoskeleton, e.g. (26). Among the cell types capable of macropinocytosis are lower eukaryotes like amoebae, phagocytes of the immune system that sample and expose antigens, and osteoclasts resorbing bone material. Many other cell types perform macropinocytosis in culture after stimulation with growth factors and it is possible that this response is a, perhaps necessary, side-product of chemotactic cell movement rather than an attempt to internalize extracellular fluid. Finally, as detailed elsewhere in this book, the pathway of macropinocytosis has been exploited by viruses and pathogenic bacteria as a means of entering the cell.

Acknowledgements

I wish to thank Drs Albert Haas, Thierry Soldati, and Dirk Wienke for critical comments on the manuscript, Harald Rühling for help with the illustrations, and Steve Nesbitt for helpful discussions about osteoclasts. Previous work in my laboratories has been supported by the Max-Planck-Gesellschaft, and the MRC Laboratory for Molecular Cell Biology. Photographs are reproduced with permission of: The Rockefeller University Press, The Company of Biologists Ltd., Urban & Fischer Verlag, and the American Association for the Advancement of Science.

References

1. Lewis, W. (1931) Pinocytosis. *Johns Hopkins Hosp. Bull.*, **49**, 17.
2. Edgar, A. J. and Bennett, J. P. (1997) Circular ruffle formation in rat basophilic leukemia cells in response to antigen stimulation. *Eur. J. Cell Biol.*, **73**, 132.
3. West, M. A., Bretscher, M. S., and Watts, C. (1989) Distinct endocytotic pathways in epidermal growth factor-stimulated human carcinoma A431 cells. *J. Cell Biol.*, **109**, 2731.
4. Racoosin, E. L. and Swanson, J. A. (1989) Macrophage colony-stimulating factor (rM-CSF) stimulates pinocytosis in bone marrow-derived macrophages. *J. Exp. Med.*, **170**, 1635.
5. Hacker, U., Albrecht, R., and Maniak, M. (1997) Fluid-phase uptake by macropinocytosis in Dictyostelium. *J. Cell Sci.*, **110**, 105.
6. Gaidarow, I., Santini, F., Warren, R., and Keen, J. (1999) Spatial control of coated-pit dynamics in living cells. *Nat. Cell Biol.*, **1**, 1.
7. van Deurs, B., Holm, P. K., Kayser, L., and Sandvig, K. (1995) Delivery to lysosomes in the human carcinoma cell-line Hep-2 involves an actin filament-facilitated fusion between mature endosomes and preexisting lysosomes. *Eur. J. Cell Biol.*, **66**, 309.
8. Durrbach, A., Louvard, D., and Coudrier, E. (1996) Actin-filaments facilitate two steps of endocytosis. *J. Cell Sci.*, **109**, 457.
9. Keller, H. and Niggli, V. (1995) Effects of cytochalasin D on shape and fluid pinocytosis in human neutrophils as related to cytoskeletal changes (actin, alpha-actinin and microtubules). *Eur. J. Cell Biol.*, **66**, 157.
10. Thilo, L. and Vogel, G. (1980) Kinetics of membrane internalization and recycling during pinocytosis in Dictyostelium. *Proc. Natl. Acad. Sci. USA*, **77**, 1015.
11. Racoosin, E. L. and Swanson, J. A. (1992) M-CSF-induced macropinocytosis increases

solute endocytosis but not receptor-mediated endocytosis in mouse macrophages. *J. Cell Sci.*, **102**, 867.

12. Brunk, U., Schellens, J., and Westermark, B. (1976) Influence of epidermal growth factor (EGF) on ruffling activity, pinocytosis and proliferation of cultivated human glia cells. *Exp. Cell Res.*, **103**, 295.
13. Willingham, M. C., Haigler, H. T., Fitzgerald, D. J., Gallo, M. G., Rutherford, A. V., and Pastan, I. H. (1983) The morphologic pathway of binding and internalization of epidermal growth factor in cultured cells. Studies on A431, KB, and 3T3 cells, using multiple methods of labelling. *Exp. Cell Res.*, **146**, 163.
14. Mellström, K., Heldin, C., and Westermark, B. (1988) Induction of circular membrane ruffling on human fibroblasts by platelet derived growth factor. *Exp. Cell Res.*, **177**, 347.
15. Dowrick, P., Kenworthy, P., McCann, B., and Warn, R. (1993) Circular ruffle formation and closure lead to macropinocytosis in hepatocyte growth factor/scatter factor-treated cells. *Eur. J. Cell Biol.*, **61**, 44.
16. Veithen, A., Cupers, P., Baudhuin, P., and Courtoy, P. J. (1996) v-Src induces constitutive macropinocytosis in rat fibroblasts. *J. Cell Sci.*, **109**, 2005.
17. Bar-Sagi, D. and Feramisco, J. R. (1986) Induction of membrane ruffling and fluid-phase pinocytosis in quiescent fibroblasts by ras proteins. *Science*, **233**, 1061.
18. Swanson, J. A. (1989) Phorbol esters stimulate macropinocytosis and solute flow through macrophages. *J. Cell Sci.*, **94**, 135.
19. Keller, H. U. (1990) Diacylglycerols and PMA are particularly effective stimulators of fluid pinocytosis in human neutrophils. *J. Cell. Physiol.*, **145**, 465.
20. Wennström, S., Siegbahn, A., Yokote, K., Arvidsson, A., Heldin, C., Mori, S., *et al.* (1994) Membrane ruffling and chemotaxis transduced by the PDGF β-receptor require the binding site for phosphatidylinositol 3 kinase. *Oncogene*, **9**, 651.
21. Barker, S. A., Caldwell, K. K., Hall, A., Martinez, A. M., Pfeiffer, J. R., Oliver, J. M., *et al.* (1995) Wortmannin blocks lipid and protein kinase activities associated with PI 3-kinase and inhibits a subset of responses induced by Fc epsilon R1 cross-linking. *Mol. Biol. Cell*, **6**, 1145.
22. Araki, N., Johnson, M. T., and Swanson, J. A. (1996) A role for phosphoinositide 3-kinase in the completion of macropinocytosis and phagocytosis by macrophages. *J. Cell Biol.*, **135**, 1249.
23. Ridley, A. J., Paterson, H. F., Johnston, C. L., Diekmann, D., and Hall, A. (1992) The small GTP-binding protein rac regulates growth factor-induced membrane ruffling. *Cell*, **70**, 401.
24. Kotani, K., Hara, K., Yonezawa, K., and Kasuga, M. (1995) Phosphoinositide 3-kinase as an upstream regulator of the small GTP-binding protein Rac in the insulin signaling of membrane ruffling. *Biochem. Biophys. Res. Commun.*, **208**, 985.
25. Dharmawardhane, S., Sanders, L. C., Martin, S. S., Daniels, R. H., and Bokoch, G. M. (1997) Localization of p21-activated kinase 1 (PAK1) to pinocytic vesicles and cortical actin structures in stimulated cells. *J. Cell Biol.*, **138**, 1265.
26. Radhakrishna, H., Klausner, R. D., and Donaldson, J. G. (1996) Aluminum fluoride stimulates surface protrusions in cells overexpressing the ARF6 GTPase. *J. Cell Biol.*, **134**, 935.
27. D'Souza-Schorey, C., Boshans, R. L., McDonough, M., Stahl, P. D., and Van Aelst, L. (1997) A role for POR1, a Rac1-interacting protein, in ARF6-mediated cytoskeletal rearrangements. *EMBO J.*, **16**, 5445.
28. Franco, M., Peters, P. J., Boretto, J., van Donselaar, E., Neri, A., D'Souza-Schorey, C., *et al.*

(1999) EFA6, a sec7 domain-containing exchange factor for ARF6, coordinates membrane recycling and actin cytoskeleton organization. *EMBO J.*, **18**, 1480.

29. Buczynski, G., Grove, B., Nomura, A., Kleve, M., Bush, J., Firtel, R. A., *et al.* (1997) Inactivation of two Dictyostelium discoideum genes, DdPIK1 and DdPIK2, encoding proteins related to mammalian Phosphatidylinositide 3-kinases, results in defects in endocytosis, lysosome to post-lysosome transport and actin cytoskeleton organisation. *J. Cell Biol.*, **136**, 1271.
30. Rivero, F., Albrecht, R., Dislich, H., Bracco, E., Graciotti, L., Bozzaro, S., *et al.* (1999) RacF1, a novel member of the Rho protein family in Dictyostelium discoideum, associates transiently with cell contact areas, macropinosomes, and phagosomes. *Mol. Biol. Cell*, **10**, 1205.
31. Seastone, D., Lee, E., Bush, J., Knecht, D., and Cardelli, J. (1998) Overexpression of a novel rho family GTPase, racC, induces unusual actin-based structures and positively affects phagocytosis in Dictyostelium discoideum. *Mol. Biol. Cell*, **9**, 2891.
32. Rebstein, P. J., Weeks, G., and Spiegelman, G. B. (1993) Altered morphology of vegetative amebas induced by increased expression of the Dictyostelium discoideum ras-related gene rap1. *Dev. Genet.*, **14**, 347.
33. Seastone, D. J., Zhang, L., Buczynski, G., Rebstein, P., Weeks, G., Spiegelman, G., *et al.* (1999) The small Mr Ras-like GTPase Rap1 and the phospholipase C pathway act to regulate phagocytosis in Dictyostelium discoideum. *Mol. Biol. Cell*, **10**, 393.
34. Massol, P., Montcourrier, P., Guillemot, J. C., and Chavrier, P. (1998) Fc receptor-mediated phagocytosis requires CDC42 and Rac1. *EMBO J.*, **17**, 6219.
35. Castellano, F., Montcourrier, P., Guillemot, J. C., Gouin, E., Machesky, L., Cossart, P., *et al.* (1999) Inducible recruitment of Cdc42 or WASP to a cell-surface receptor triggers actin polymerization and filopodium formation. *Curr. Biol.*, **9**, 351.
36. Carballo, E., Pitterle, D. M., Stumpo, D. J., Sperling, R. T., and Blackshear, P. J. (1999) Phagocytic and macropinocytic activity in MARCKS-deficient macrophages and fibroblasts. *Am. J. Physiol.*, **277**, C163.
37. Cox, D., Wessels, D., Soll, D. R., Hartwig, J., and Condeelis, J. (1996) Re-expression of ABP-120 rescues cytoskeletal, motility, and phagocytosis defects of ABP-120- Dictyostelium mutants. *Mol. Biol. Cell*, **7**, 803.
38. Rivero, F., Furukawa, R., Noegel, A. A., and Fechheimer, M. (1996) Dictyostelium discoideum cells lacking the 34,000-dalton actin-binding protein can grow, locomote, and develop, but exhibit defects in regulation of cell structure and movement—a case of partial redundancy. *J. Cell Biol.*, **135**, 965.
39. Prassler, J., Stocker, S., Marriott, G., Heidecker, M., Kellermann, J., and Gerisch, G. (1996) Interaction of a Dictyostelium member of the plastin/fimbrin family with actin filaments and actin-myosin complexes. *Mol. Biol. Cell*, **8**, 83.
40. Skoudy, A., Nhieu, G. T., Mantis, N., Arpin, M., Mounier, J., Gounon, P., *et al.* (1999) A functional role for ezrin during Shigella flexneri entry into epithelial cells. *J. Cell Sci.*, **112**, 2059.
41. Bretscher, A., Reczek, D., and Berryman, M. (1997) Ezrin: a protein requiring conformational activation to link microfilaments to the plasma membrane in the assembly of cell surface structures. *J. Cell Sci.*, **110**, 3011.
42. Mackay, D. J., Esch, F., Furthmayr, H., and Hall, A. (1997) Rho- and rac-dependent assembly of focal adhesion complexes and actin filaments in permeabilized fibroblasts: an essential role for ezrin/radixin/moesin proteins. *J. Cell Biol.*, **138**, 927.
43. Niewöhner, J., Weber, I., Maniak, M., Müller-Taubenberger, A., and Gerisch, G. (1997)

Talin-null cells of Dictyostelium are strongly defective in adhesion to particle and substrate surfaces and slightly impaired in cytokinesis. *J. Cell Biol.*, **138**, 349.
44. Hitt, A. L., Hartwig, J. H., and Luna, E. J. (1994) Ponticulin is the major high-affinity link between the plasma-membrane and the cortical actin network in Dictyostelium. *J. Cell Biol.*, **126**, 1433.
45. Machesky, L. M. and Way, M. (1998) Cell motility—actin branches out. *Nature*, **394**, 125.
46. Machesky, L. M., Reeves, E., Wientjes, F., Mattheyse, F. J., Grogan, A., Totty, N. F., *et al.* (1997) Mammalian actin-related protein 2/3 complex localizes to regions of lamellipodial protrusion and is composed of evolutionarily conserved proteins. *Biochem. J.*, **328**, 105.
47. Machesky, L. M., Mullins, R. D., Higgs, H. N., Kaiser, D. A., Blanchoin, L., May, R. C., *et al.* (1999) Scar, a WASp-related protein, activates nucleation of actin filaments by the Arp2/3 complex. *Proc. Natl. Acad. Sci. USA*, **96**, 3739.
48. Grogan, A., Reeves, E., Keep, N., Wientjes, F., Trotty, N., Burlingame, A. L., *et al.* (1997) Cytosolic phox proteins interact with and regulate the assembly of coronin in neutrophils. *J. Cell Sci.*, **110**, 3071.
49. Saito, T., Lamy, F., Roger, P. P., Lecocq, R., and Dumont, J. E. (1994) Characterization and identification as cofilin and destrin of two thyrotropin- and phorbol ester-regulated phosphoproteins in thyroid cells. *Exp. Cell Res.*, **212**, 49.
50. Suzuki, K., Yamaguchi, T., Tanaka, T., Kawanishi, T., Nishimaki-Mogami, T., Yamamoto, K., *et al.* (1995) Activation induces dephosphorylation of cofilin and its translocation to plasma membranes in neutrophil-like differentiated HL-60 cells. *J. Biol. Chem.*, **270**, 19551.
51. Lavoie, J. N., Hickey, E., Weber, L. A., and Landry, J. (1993) Modulation of actin microfilament dynamics and fluid phase pinocytosis by phosphorylation of heat shock protein 27. *J. Biol. Chem.*, **268**, 24210.
52. Machesky, L. M., Atkinson, S. J., Ampe, C., Vandekerckhove, J., and Pollard, T. D. (1994) Purification of a cortical complex containing 2 unconventional actins from Acanthamoeba by affinity-chromatography on profilin-agarose. *J. Cell Biol.*, **127**, 107.
53. Bear, J. E., Rawls, J. F., and Saxe, C. L. (1998) Scar, a wasp-related protein, isolated as a suppressor of receptor defects in late dictyostelium development. *J. Cell Biol.*, **142**, 1325.
54. de Hostos, E. L., Bradtke, B., Lottspeich, F., Guggenheim, R., and Gerisch, G. (1991) Coronin, an actin binding protein of Dictyostelium discoideum localized to cell surface projections, has sequence similarities to G protein β subunits. *EMBO J.*, **10**, 4097.
55. de Hostos, E. L., Rehfueß, C., Bradtke, B., Waddell, D. R., Albrecht, R., Murphy, J., *et al.* (1993) Dictyostelium mutants lacking the cytoskeletal protein coronin are defective in cytokinesis and cell motility. *J. Cell Biol.*, **120**, 163.
56. Aizawa, H., Fukui, Y., and Yahara, I. (1997) Live dynamics of Dictyostelium cofilin suggests a role in remodeling actin latticework into bundles. *J. Cell Sci.*, **110**, 2333.
57. Aizawa, H., Sutoh, K., Tsubuki, S., Kawashima, S., Ishii, A., and Yahara, I. (1995) Identification, characterization, and intracellular-distribution of cofilin in Dictyostelium discoideum. *J. Biol. Chem.*, **270**, 10923.
58. Aizawa, H., Katadae, M., Maruya, M., Sameshima, M., Murakami-Murofushi, K., and Yahara, I. (1999) Hyperosmotic stress-induced reorganization of actin bundles in Dictyostelium cells over-expressing cofilin. *Genes Cells*, **4**, 311.
59. Konzok, A., Weber, I., Simmeth, E., Hacker, U., Maniak, M., and Müller-Taubenberger, A. (1999) DAip1, a Dictyostelium homologue of the yeast actin-interacting protein 1, is involved in endocytosis, cytokinesis, and motility. *J. Cell Biol.*, **146**, 453.
60. Mermall, V., Post, P., and Mooseker, M. S. (1998) Unconventional myosins in cell movement, membrane traffic, and signal transduction. *Science*, **279**, 527.

61. Buss, F., Kendrick-Jones, J., Lionne, C., Knight, A. E., Cote, G. P., and Paul Luzio, J. (1998) The localization of myosin VI at the golgi complex and leading edge of fibroblasts and its phosphorylation and recruitment into membrane ruffles of A431 cells after growth factor stimulation. *J. Cell Biol.*, **143**, 1535.
62. Fukui, Y., Lynch, T. J., Brzeska, H., and Korn, E. D. (1989) Myosin I is located at the leading edges of locomoting Dictyostelium amoebae. *Nature*, **341**, 328.
63. Ostap, E. M. and Pollard, T. D. (1996) Overlapping functions of myosin-I isoforms. *J. Cell Biol.*, **133**, 221.
64. Uyeda, T. and Titus, M. (1997) The myosins in Dictyostelium. In *Dictyostelium—a model system for cell and developmental biology* (ed. Y. Maeda, K. Inouye, and I. Takeuchi), p. 43. Universal Academy Press, Tokyo.
65. Swanson, J. A., Johnson, M. T., Beningo, K., Post, P., Mooseker, M., and Araki, N. (1999) A contractile activity that closes phagosomes in macrophages. *J. Cell Sci.*, **112**, 307.
66. Aubry, L., Klein, G., and Satre, M. (1997) Cytoskeletal dependence and modulation of endocytosis in Dictyostelium discoideum amoebae. In *Dictyostelium—a model system for cell and developmental biology* (ed. Y. Maeda, K. Inouye, and I. Takeuchi), p. 65. Universal Academy Press, Tokyo.
67. Best, A., Ahmed, S., Kozma, R., and Lim, L. (1996) The Ras-related GTPase Rac1 binds tubulin. *J. Biol. Chem.*, **271**, 3756.
68. Waterman-Storer, C., Worthylake, R., Liu, B., Burridge, K., and Salmon, E. (1999) Microtubule growth activates rac1 to promote lamellipodial protrusion in fibroblasts. *Nat. Cell Biol.*, **1**, 45.
69. Hoffstein, S. T., Friedman, R. S., and Weissmann, G. (1982) Degranulation, membrane addition, and shape change during chemotactic factor-induced aggregation of human neutrophils. *J. Cell Biol.*, **95**, 234.
70. Garrigues, J., Anderson, J., Hellström, K. E., and Hellström, I. (1994) Anti-tumor antibody BR96 blocks cell migration and binds to a lysosomal membrane glycoprotein on cell surface microspikes and ruffled membranes. *J. Cell Biol.*, **125**, 129.
71. Bretscher, M. S. and Aguado-Velasco, C. (1998) EGF induces recycling membrane to form ruffles. *Curr. Biol.*, **8**, 721.
72. Merrifield, C., Moss, S., Ballestrem, C., Imhof, B., Giese, G., Wunderlich, I., *et al.* (1999) Endocytic vesicles move at the tips of actin tails in cultured mast cells. *Nat. Cell Biol.*, **1**, 72.
73. Swanson, J. A. and Watts, C. (1995) Macropinocytosis. *Trends Cell Biol.*, **5**, 424.
74. Berthiaume, E. P., Medina, C., and Swanson, J. A. (1995) Molecular size-fractionation during endocytosis in macrophages. *J. Cell Biol.*, **129**, 989.
75. Desjardins, M., Huber, L. A., Parton, R. G., and Griffiths, G. (1994) Biogenesis of phagolysosomes proceeds through a sequential series of interactions with the endocytic apparatus. *J. Cell Biol.*, **124**, 677.
76. Racoosin, E. L. and Swanson, J. A. (1993) Macropinosome maturation and fusion with tubular lysosomes in macrophages. *J. Cell Biol.*, **121**, 1011.
77. Hewlett, L. J., Prescott, A. R., and Watts, C. (1994) The coated pit and macropinocytic pathways serve distinct endosome populations. *J. Cell Biol.*, **124**, 689.
78. Veithen, A., Amyere, M., Van Der Smissen, P., Cupers, P., and Courtoy, P. J. (1998) Regulation of macropinocytosis in v-Src-transformed fibroblasts: cyclic AMP selectively promotes regurgitation of macropinosomes. *J. Cell Sci.*, **111**, 2329.
79. Maniak, M. (1999) Green fluorescent protein in the visualisation of particle uptake and fluid-phase endocytosis. *Methods Enzymol.*, **302**, 43.

80. Souza, G. M., Metha, D. P., Lammertz, M., Rodriuez-Paris, J., Wu, R., Cardelli, J. A., *et al.* (1997) Dictyostelium lysosomal proteins with different sugar modifications sort to functionally distinct compartments. *J. Cell Sci.*, **110**, 2239.
81. Aubry, L., Klein, G., Martiel, J. L., and Satre, M. (1993) Kinetics of pH evolution in Dictyostelium discoideum. *J. Cell Sci.*, **105**, 861.
82. Padh, H., Ha, J., Lavasa, M., and Steck, T. L. (1993) A post-lysosomal compartment in Dictyostelium discoideum. *J. Biol. Chem.*, **268**, 6742.
83. Jenne, N., Rauchenberger, R., Hacker, U., Kast, T., and Maniak, M. (1998) Targeted gene disruption reveals a role for vacuolin B in the late endocytic pathway and exocytosis. *J. Cell Sci.*, **111**, 61.
84. Sallusto, F., Cella, M., Danieli, C., and Lanzavecchia, A. (1995) Dendritic cells use macropinocytosis and the mannose receptor to concentrate macromolecules in the major histocompatibility complex class II compartment: downregulation by cytokines and bacterial products. *J. Exp. Med.*, **182**, 389.
85. Norbury, C. C., Chambers, B. J., Prescott, A. R., Ljunggren, H. G., and Watts, C. (1997) Constitutive macropinocytosis allows TAP-dependent major histocompatibility complex class I presentation of exogenous soluble antigen by bone marrow-derived dendritic cells. *Eur. J. Immunol.*, **27**, 280.
86. Brossart, P. and Bevan, M. J. (1997) Presentation of exogenous protein antigens on major histocompatibility complex class I molecules by dendritic cells: pathway of presentation and regulation by cytokines. *Blood*, **90**, 1594.
87. Norbury, C. C., Hewlett, L. J., Prescott, A. R., Shastri, N., and Watts, C. (1995) Class I MHC presentation of exogenous soluble antigen via macropinocytosis in bone marrow macrophages. *Immunity*, **3**, 783.
88. Nesbitt, S. A. and Horton, M. A. (1997) Trafficking of matrix collagens through bone-resorbing osteoclasts. *Science*, **276**, 266.
89. Salo, J., Lehenkari, P., Mulari, M., Metsikkö, K., and Väänänen, H. K. (1997) Removal of osteoclast bone resorption products by transcytosis. *Science*, **276**, 270.
90. Pierce, A. and Lindskog, S. (1988) Coated pits and vesicles in the osteoclast. *J. Submicrosc. Cytol. Pathol.*, **20**, 161.
91. Akamine, A., Tsukuba, T., Kimura, R., Maeda, K., Tanaka, Y., Kato, K., *et al.* (1993) Increased synthesis and specific localization of a major lysosomal membrane sialoglycoprotein (LGP107) at the ruffled border membrane of active osteoclasts. *Histochemistry*, **100**, 101.
92. Palokangas, H., Mulari, M., and Väänänen, H. K. (1997) Endocytic pathway from the basal plasma membrane to the ruffled border membrane in bone-resorbing osteoclasts. *J. Cell Sci.*, **110**, 1767.
93. Holm, P., Hansen, S., Sandvig, K., and van Deurs, B. (1993) Endocytosis of desmosomal plaques depends on intact actin filaments and leads to a non-degradative compartment. *Eur. J. Cell Biol.*, **62**, 362.
94. Eriksson, A., Siegbahn, A., Westermark, B., Heldin, C. H., and Claesson-Welsh, L. (1992) PDGF alpha- and beta-receptors activate unique and common signal transduction pathways. *EMBO J.*, **11**, 543.
95. Altun-Gultekin, Z. F. and Wagner, J. A. (1996) Src, ras, and rac mediate the migratory response elicited by NGF and PMA in PC12 cells. *J. Neurosci. Res.*, **44**, 308.
96. Wyckoff, J. B., Insel, L., Khazaie, K., Lichtner, R. B., Condeelis, J. S., and Segall, J. E. (1998) Suppression of ruffling by the EGF receptor in chemotactic cells. *Exp. Cell Res.*, **242**, 100.
97. Bailly, M., Condeelis, J. S., and Segall, J. E. (1998) Chemoattractant-induced lamellipod extension. *Microsc. Res. Tech.*, **43**, 433.

98. Fabbri, M., Fumagalli, L., Bossi, G., Bianchi, E., Bender, J. R., and Pardi, R. (1999) A tyrosine-based sorting signal in the beta2 integrin cytoplasmic domain mediates its recycling to the plasma membrane and is required for ligand-supported migration. *EMBO J.*, **18**, 4915.
99. Fournier, A. E., Nakamura, F., Kawamoto, S., Goshima, Y., Kalb, R. G., and Strittmatter, S. M. (2000) Semaphorin 3A enhances endocytosis at sites of receptor-f-actin colocalization during growth clone collapse. *J. Cell Biol.*, **149**, 411.
100. Murphy, C., Saffrich, R., Grummt, M., Gournier, H., Rybin, V., Rubino, M., *et al.* (1996) Endosome dynamics regulated by a Rho protein. *Nature*, **384**, 427.

5 | Molecular mechanisms of membrane fusion in the endocytic pathway

HARALD STENMARK and MARINO ZERIAL

1. Introduction

The endocytic pathway of mammalian cells consists of distinct membrane-bound compartments that receive internalized molecules from the plasma membrane and recycle them back to the surface (early endosomes and recycling endosomes) or sort them to degradative compartments (late endosomes and lysosomes) (see Fig. 1). Endocytic membrane traffic between these organelles is a dynamic process that involves membrane budding and fusion reactions as well as the movement of endocytic vesicles and endosomes. The complexity of the pathway and the extensive membrane exchange between compartments necessitate a tight regulation of membrane fusion events, so that the overall organization of the endomembrane system remains intact during membrane turnover. Since the mechanisms of formation of endocytic vesicles at the plasma membrane are covered elsewhere (see Chapters 1–4), in this review we will first give a brief overview of how endocytic membrane docking and fusion are being studied, and then discuss our current understanding of the molecular machinery underlying this process, with special emphasis on what has been learnt from the fusion between early endosomes in mammalian cells.

2. Experimental systems to study endocytic membrane docking and fusion

Microscopy of living cells has revealed the plasticity and dynamics of endocytic organelles, which undergo continuous fusion and fission reactions (1–4). Our current view of the underlying molecular mechanisms is the result of the convergence of several independent experimental approaches. Like with studies of secretion, yeast genetics has proven to be a powerful tool to identify the various constituents of the molecular apparatus involved (5, 6). Genetic screens, such as those for *vps* (vacuolar

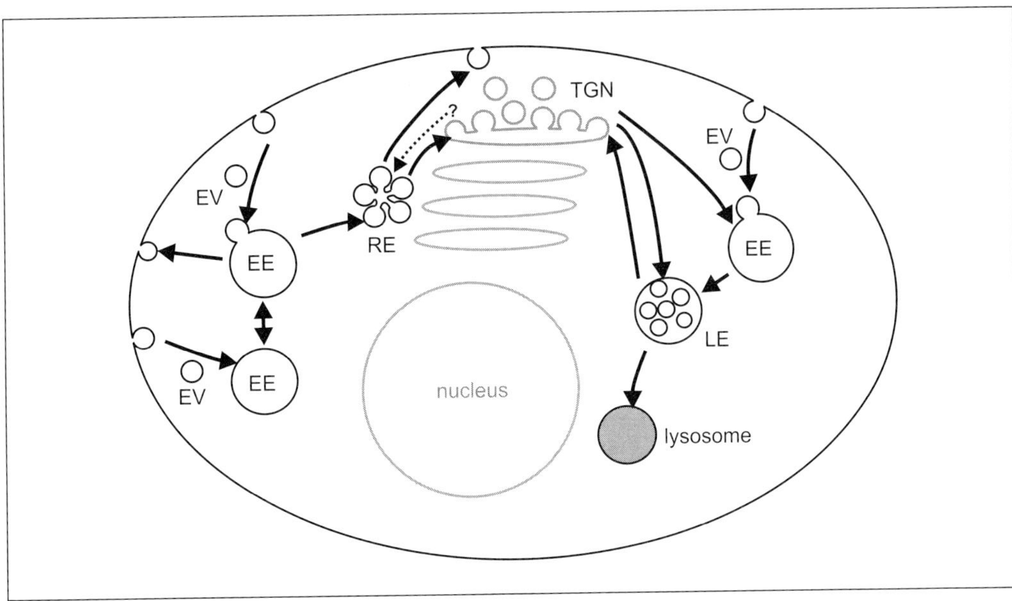

Fig. 1 Endocytic membrane traffic. In the left half of the schematized cell, endocytosis, homotypic early endosome fusion, and recycling (which can either take place from early endosomes or from recycling endosomes), are illustrated. In the right half, transport from early endosomes to late endosomes and lysosomes, and trafficking between the TGN and endosomes, is illustrated. Abbreviations: EE, early endosome; EV, endocytic vesicle; LE, late endosome; RE, recycling endosome; TGN, *trans*-Golgi network.

protein sorting), *end* (endocytosis defective), and *pep* (peptidase deficient) mutants have yielded several candidate molecules for the regulation of endocytic membrane fusion (see Chapter 10). From yeast genetics alone it is difficult to pin-point the precise function of a given gene product. However, the phenotypic analysis of yeast mutants has often been informative as to the possible role of a protein in membrane docking and fusion. This is the case with proteins whose functional inactivation leads to endocytic transport defects as well as the accumulation of transport intermediates (as verified by EM). Vps class D proteins are good examples (see Section 7). In mammalian cells, the expression of dominant inhibitory or stimulatory mutants of small GTPases has yielded important information, as exemplified by Rab5 (Section 4).

Because genetic studies and expression of dominant interfering mutants can only provide indirect evidence for a role in membrane fusion, more direct approaches have been developed. One powerful assay in mammalian cells reconstitutes the homotypic fusion between early endosomes (7, 8). In this assay, two early endosome fractions are loaded with distinct markers that can form a complex if they meet (e.g. antigen–antibody, biotin–avidin), and are mixed in the presence of cytosol and ATP. If fusion occurs, the two markers will react to form complexes, and these can subsequently be retrieved (for instance, on antibody-coated magnetic beads) and quantified. In a similar manner, the heterotypic fusion between endocytic vesicles and early endosomes can be studied (9, 10). Likewise, several related assays have been employed to reconstitute the fusion between endocytic carrier vesicles (ECVs) and late endosomes (11), and between late endosomes and lysosomes (12). While endosome fusion has been difficult to reconstitute in yeast, another homotypic fusion event, that of vacuoles (the yeast equivalent of lysosomes), has successfully been reconstituted. The experimental set-up is equivalent to that of homotypic early endosome fusion, except that the vacuole fusion assay measures a biochemical reaction that takes place when vacuoles from two appropriately genetically engineered yeast strains fuse. This assay was originally established to measure vacuole inheritance from mother to daughter cells (13), but it has proven useful for studies of the general molecular mechanisms involved in membrane fusion.

The introduction of the green fluorescent protein (GFP) in cell biology has recently provided a novel tool for the studies of membrane fusion. Now it is possible to follow membrane fusion in living cells by video microscopy, upon the expression of GFP fusion proteins that are targeted to endocytic organelles. Such studies have yielded information about the dynamics of endosome fusion (2, 3) as well as endosome motility (4).

3. Rab GTPases and SNAREs in membrane docking and fusion

The main players in membrane docking and fusion have been identified through a combination of biochemical, genetic, and molecular biological approaches. A hexameric ATPase, NSF (*N*-ethyl maleimide-sensitive factor), is required for membrane traffic in yeast and mammalian cells (14). The ATPase activity of NSF is regulated by another essential protein, SNAP (soluble NSF attachment protein) (15), and the NSF/SNAP complex binds to SNAP receptors known as SNAREs (16). SNAREs are divided into two subfamilies. One subfamily of SNAREs was originally found enriched on vesicles (v-SNARES), whereas the other subfamily was associated with target membranes (t-SNAREs). As will be further discussed below, such a strict segregation of SNAREs between vesicles and organelles does not exist and this terminology is perhaps misleading. Therefore, the alternative names of R- and Q-SNAREs (after conserved arginine and glutamine residues that distinguish the two subfamilies) have been proposed (17). Nevertheless, for simplicity, we will use here the original terminology. Results from the *in vitro* vacuole fusion assay (see Section 2) have revealed that fusion requires the formation of *trans*-SNARE complexes, that is, complexes between v-SNAREs on the one membrane and t-SNAREs on the other one (18). X-ray crystallography and NMR have unravelled the core structure of the SNARE complex involved in the fusion of synaptic vesicles with the presynaptic plasma membrane (19, 20). These studies reveal that the two t-SNAREs, SNAP-25 and syntaxin1, form a tight coiled-coil complex with the v-SNARE VAMP-2. Interestingly, the four α-helices involved in the complex formation (two from SNAP-25 and one each from syntaxin1 and VAMP-2) are all oriented in a **parallel** manner. This suggests that their zippering-like complex formation may release sufficient energy to drive the fusion of

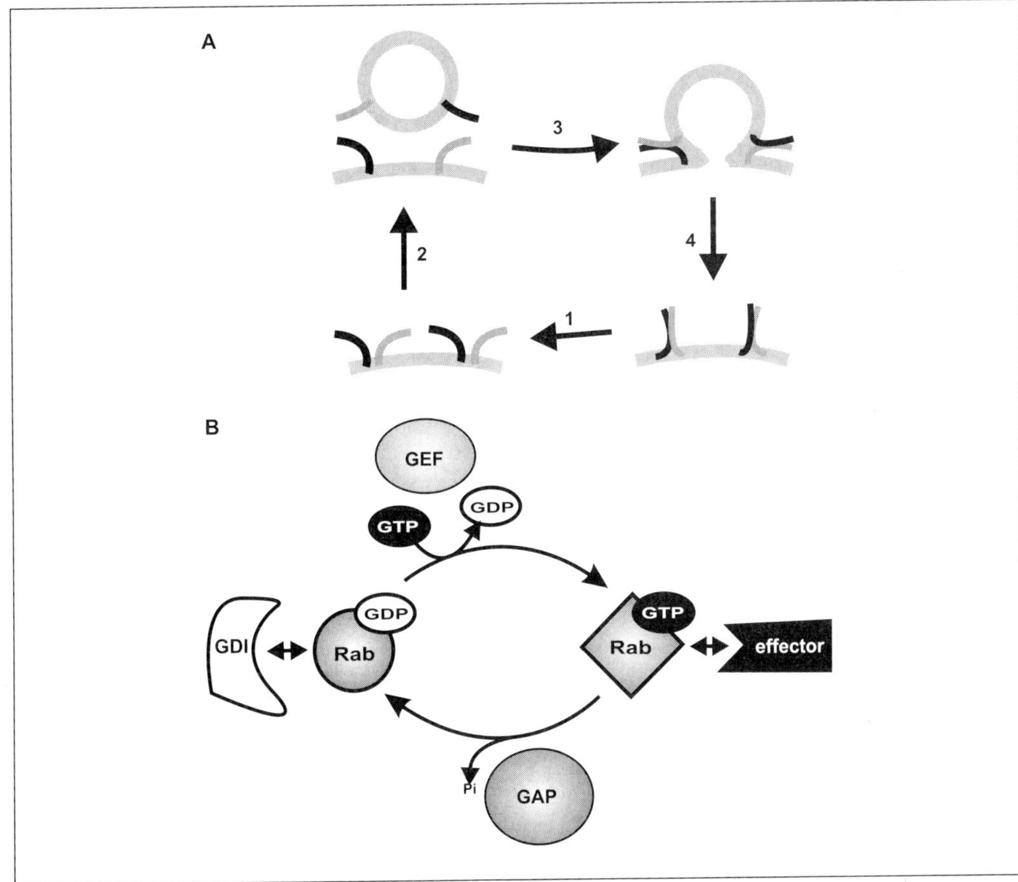

Fig. 2 SNARE and Rab proteins in membrane docking and fusion. (A) *trans*-SNARE complexes in membrane docking/fusion. 1 Pre-existing SNARE complexes are dissociated ('primed') through the activity of NSF/SNAP. 2 Primed SNAREs on a vesicle interact with primed SNAREs on a target membrane. 3 The SNAREs form tight complexes through parallel coiled-coil interactions, and the energy released may power membrane fusion. 4 After fusion, the SNARE complexes reside in the target membrane. The 't-SNAREs' are indicated in black and 'v-SNAREs' in grey. (B) the Rab GTPase cycle. The Rab protein switches between two different conformations depending on whether GDP or GTP is bound. GDP/GTP exchange is catalysed by a GDP/GTP exchange factor (GEF), whereas a GTPase activating protein (GAP) stimulates the GTPase activity of the Rab protein. The GDP-bound form of the Rab protein is recognized by GDP dissociation inhibitor (GDI), which regulates its membrane association (not illustrated here). The GTP-bound form of the Rab protein interacts with effector molecules.

the two opposing bilayers (Fig. 2A). Indeed, *trans*-SNARE complex formation is sufficient to cause the fusion of lipid/detergent micelles *in vitro,* albeit with low efficiency (21–23). NSF has similarity to heat shock ATPases involved in protein folding, and the role of NSF/SNAP may be to disassemble pre-existing SNARE complexes, in order for SNAREs to be reused for a new round of vesicular transport.

The original SNARE hypothesis implicates SNAREs in vesicle targeting as well as in fusion (16). However, while recent work has indeed confirmed the role of SNAREs in the fusion of lipid bilayers, their role as the sole molecular determinants in vesicle targeting and fusion seems implausible. First, although distinct SNAREs may show different intracellular localizations, SNAREs are generally found on multiple organelles. For instance, syntaxin1 is found on synaptic vesicles, endosomes, and on the plasma membrane, and syntaxin6 is found both in the TGN and on early endosomes (24, 25). Taking into account that SNAREs are integral membrane proteins and consequently need to be recycled by membrane trafficking, this comes as no surprise. Secondly, v- and t-SNAREs form complexes in a rather unspecific manner *in vitro* (26). Thirdly, the function of NSF in post-mitotic Golgi vesicle docking appears to be independent of its ability to prime SNAREs (27). Fourthly, a wealth of yeast genetic data indicate that SNAREs do not determine fusion specificity (28). It thus seems unlikely that these molecules by themselves would play the leading role in vesicle targeting. There is indeed an additional layer of regulation provided by another group of proteins, the Rab GTPases (29, 30).

Rab proteins belong to a family of about 40 members, and just like SNAREs, different Rab GTPases are localized to distinct compartments and regulate distinct trafficking steps. In contrast to SNAREs, Rab GTPases are reversibly attached to membranes, thus enabling their efficient recycling from acceptor to donor membranes. This is accomplished through the association of lipophilic geranylgeranyl groups, attached to C-terminal cysteine residues, with the lipid bilayer. Geranylgeranylated Rab GTPases are presented to, and removed from membranes by, an essential protein, Rab GDP dissociation inhibitor (GDI) (31, 32). Like most other GTPases, Rab GTPases switch their conformation depending on whether GDP or GTP is bound, and this determines their ability to interact with effector molecules (which specifically occurs with the GTP-bound form) (see Fig. 2B). Recent studies indicate that one Rab GTPase may have numerous effectors, some of which may mediate vesicle 'tethering' prior to SNARE complex formation (see Section 4). The latter is in agreement with both yeast genetic studies and biochemical studies in vacuole or endosome fusion assays (Section 2), which indicate that Rab GTPases act prior to SNAREs in membrane docking and fusion. It thus appears that Rab GTPases and their effectors cause transport vesicles to tether to their correct target membranes, whereas SNAREs then take over to trigger the actual membrane fusion process. However, as discussed in Section 6, it is still possible that not only SNAREs, but also Rab effectors, may participate in the fusion reaction.

4. Rab5 and its effectors in early-endosome fusion

Several Rab GTPases and SNARE proteins are thought to function in endocytic membrane traffic (33, 34). Among the Rab GTPases, Rab5 has been most studied, and much of our knowledge about Rab function in general derives from studies of this GTPase. Rab5 is found on the cytosolic side of early endosomes, endocytic vesicles, and the plasma membrane (35, 36). Expression of a GTPase-deficient mutant ($Rab5^{Q79L}$) causes an increased rate of endocytosis and the formation of giant early endosomes. Conversely, the expression of a mutant with preferential affinity for GDP ($Rab5^{S34N}$)

inhibits endocytosis and causes the formation of very small early-endocytic profiles (37). Rab5 is essential for homotypic early-endosome fusion *in vitro* (8), and such fusion is stimulated by $Rab5^{Q79L}$ and inhibited by $Rab5^{S34N}$ (37). Moreover, a mutant ($Rab5^{D136N}$) that binds xanthine instead of guanine nucleotides stimulates endosome fusion in the presence of the non-hydrolysable XTPγS, and inhibits fusion in the presence of XDP (38). These results indicate that Rab5, in its GTP-bound form, stimulates early-endosome fusion. With $Rab5^{D136N}$, XTP hydrolysis occurs even when endosome fusion is inhibited, indicating that nucleoside triphosphate hydrolysis is not required for membrane fusion but rather exerts a regulatory function on this process.

The search for Rab5 effectors, first with the yeast two-hybrid system and more recently through affinity chromatography, has yielded a surprisingly high number of Rab5:GTP-interacting molecules. In fact, no less than 22 cytosolic proteins could be eluted from a Rab5:GTPγS affinity column (39). Even though not all these proteins may bind directly to Rab5:GTP, this illustrates that Rab proteins, contrary to early belief, regulate multiple effectors and control more than one biochemical event. The latter has been demonstrated in the case of Rab5, which, in addition to its role in endocytic membrane fusion, also plays a role in the formation of endocytic vesicles (10), and in the motility of endocytic vesicles along microtubules (4).

Which then are the effectors of Rab5 in endocytic membrane fusion? The first one to be identified was Rabaptin-5 (40), a dimeric coiled-coil protein that is found in a complex with Rabex-5, a GDP/GTP exchange factor (GEF) for Rab5 (10). Rabaptin-5 is recruited to early endosomes in a Rab5- and GTP-dependent manner, and its immunodepletion from cytosol strongly inhibits early endosome fusion *in vitro* (40). Upon such immunodepletion, endosome fusion can be restored by the addition of the Rabaptin-5/Rabex5 complex, but not by Rabaptin-5 alone. The presence of a GEF in a complex with a Rab5 effector suggests that this protein is needed in order to generate sufficient amounts of Rab5:GTP on the endosome membrane (10). The association of Rabex5 with a Rab5 effector suggests the possibility that a Rab5 signal may be locally amplified at the membrane: The binding of Rabaptin-5 to Rab5:GTP may bring Rabex-5 in to a position to convert a neighbouring Rab5:GDP into Rab5:GTP. This Rab5:GTP may then recruit another Rabaptin-5/Rabex5 complex, and so on, resulting in a patch of Rab5:GTP in the membrane (Fig. 3). This, together with biochemical studies indicating that Rab5 effectors are present in large oligomers on the membrane of early endosomes (3), may explain the 'hot spots' of Rab5/effector complexes observed on endosomes (2, 3).

While the C-terminus of Rabaptin-5 contains a Rab5-binding domain, a Rab4-binding domain has been detected at its N-terminus (41). Like Rab5, Rab4 is found on early endosomes, but this GTPase appears to control membrane recycling rather than incoming endocytic traffic (42, 43). This raises the possibility that Rabaptin-5 may serve to co-ordinate endocytic membrane transport with recycling. Interestingly, cleavage of Rabaptin-5 by caspase-3, thus separating the Rab4- and Rab5-binding domains, seems to provide the molecular explanation for the inhibition of endosome fusion observed during programmed cell death (apoptosis) (44, 45).

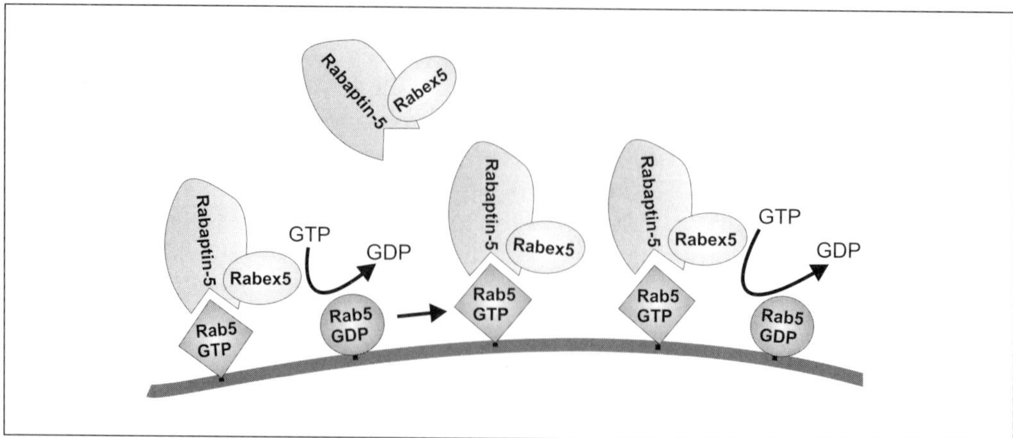

Fig. 3 Recruitment of Rabaptin-5/Rabex5 to endosomes by Rab5. Interaction of Rabaptin-5 with Rab5:GTP brings Rabex5 in position to promote GDP/GTP exchange of a neighbouring Rab5:GDP. The resulting Rab5:GTP may then recruit another Rabaptin-5/Rabex5 complex, which may activate another Rab5:GDP, and so on. This results in a local amplification of Rab5 activation.

Even though the Rabaptin-5/Rabex5 complex is essential for both homotypic early-endosome fusion and for the fusion between endocytic vesicles and early endosomes (10), it is not sufficient for these transport reactions. Endocytic membrane fusion requires several other Rab5 effectors (39). Strikingly, another Rab5 effector, EEA1 (early endosome antigen 1) can alone support membrane fusion under special experimental conditions. In fact, if sufficient amounts of EEA1 are added to early endosomes, their fusion is stimulated even in the absence of cytosol (32). EEA1 thus appears to be a core component of the membrane docking and fusion machinery, whereas the main role of Rabaptin-5/Rabex5 might be to maintain a high local level of endosomal Rab5:GTP, so that EEA1 may be efficiently recruited to endosomes. Surprisingly, however, in the presence of excess EEA1, endosome fusion can occur even when the endosomes have been treated with GDI in order to remove Rab5. This suggests that Rab5-mediated recruitment is dispensable at high EEA1 concentrations, and that EEA1 may interact with additional membrane molecules (39). Indeed, EEA1 has been found to interact both with a membrane lipid (see Section 5) and with endosomal SNARE molecules (see Section 6). Rab5 is symmetrically distributed between endocytic vesicles and early endosomes (46), and its presence on both membranes is required both in heterotypic (endocytic vesicle to endosome) and homotypic (endosome to endosome) fusion (47, 48). In contrast, EEA1 is only found on early endosomes, and the finding that it nevertheless is required for heterotypic membrane fusion makes EEA1 a good candidate for conferring directionality to vesicular traffic from the plasma membrane to early endosomes (46). The multiplicity of Rab5 effectors, however, suggests that EEA1 is only one component of the early endosome transport machinery and other molecules are likely to co-operate with EEA1 and the Rabaptin-5 complex in endocytic membrane docking and fusion.

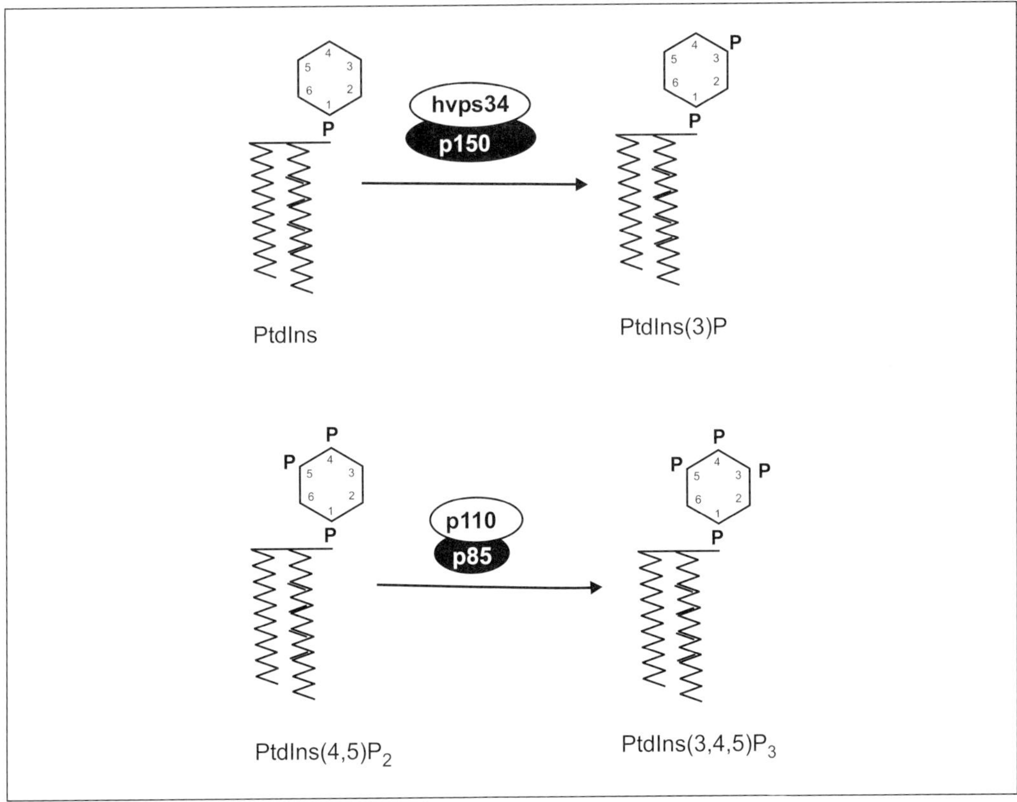

Fig. 4 Generation of 3′-phosphoinositides by PI 3-kinases.

5. PI 3-kinase and endocytic membrane fusion

Phosphatidylinositol 3-kinases (PI 3-kinases) phosphorylate phosphatitylinositol (PtdIns) and its derivatives PtdIns-4-phosphate (PtdIns(4)P) and PtdIns-4,5-bisphosphate (PtdIns(4,5)P_2) to yield the 3′-phosphoinositides PtdIns(3)P, PtdIns(3,4)P_2, and PtdIns(3,4,5)P_3, respectively (see Fig. 4) (49, 50). The finding that PI 3-kinase activity is essential for vacuolar protein sorting in yeast (see later) (51), prompted several studies of the involvement of this kinase in mammalian membrane trafficking. These studies revealed that, at least in some cell types, PI 3-kinase is essential for endocytosis as well as for endocytic recycling and late endocytic trafficking (52–57). The studies of how wortmannin, a specific PI 3-kinase inhibitor, blocks endosome fusion (58–60) have contributed to shed light on one of the mechanisms involved: The PI 3-kinase product, PtdIns(3)P binds to proteins containing FYVE zinc finger domains (61), including EEA1 (34, 62, 63). EEA1 contains Rab5-binding domains at its N- and C-termini, and the C-terminal Rab5-binding domain is adjacent to a FYVE finger. The C-terminus of EEA1 (comprising the Rab5-binding domain and the FYVE finger) is necessary and sufficient for the targeting of EEA1 to early endosomes (64), and both Rab5-binding (through the Rab5-binding domain) and PtdIns(3)P-binding (through the FYVE finger) are required for the efficient recruitment of EEA1 to endosomes *in vivo* (65). This ensures that EEA1 is only recruited to membranes that contain both Rab5:GTP and PtdIns(3)P and may explain how EEA1 becomes specifically localized to early endosomes.

The above model implies that early endosomes are enriched in PtdIns(3)P, and this has recently been verified by the use of a PtdIns(3)P-specific probe (H. Stenmark and R. Parton, unpublished). What directs the localized production of PtdIns(3)P on early endosome membranes? A clue has emerged through the finding that hvps34, the PI 3-kinase that appears to be the principle enzyme responsible for PtdIns(3)P production in mammalian cells (66), is an effector of Rab5 (67). It binds to Rab5:GTP together with its regulatory subunit, p150, and this binding probably ensures the 3′-phosphorylation of PtdIns on early endosomes. Rab5 thus plays a dual role in the recruitment of EEA1 to membranes: first, through a direct binding to EEA1, and secondly, through the stimulation of PtdIns(3)P production on endosome membranes (see Fig. 5).

It is intriguing that a second PI 3-kinase, p110-β, has also been identified as a Rab5 effector (67). This PI 3-kinase is thought to mainly phosphorylate PtdIns(4,5)P_2 *in vivo*, thus yielding PtdIns(3,4,5)P_3, and, unlike hvps34, whose activity is constitutive, p110-β is activated via the binding of various agonists to their receptors (66). Unlike hvps34, this PI 3-kinase neither seems to function at the level of Rab5-dependent early endosome fusion, nor in endosome motility along microtubules and it is not yet known if p110-β and PtdIns(3,4,5)P_3 play a role in the other Rab5-regulated functions, such as vesicle formation.

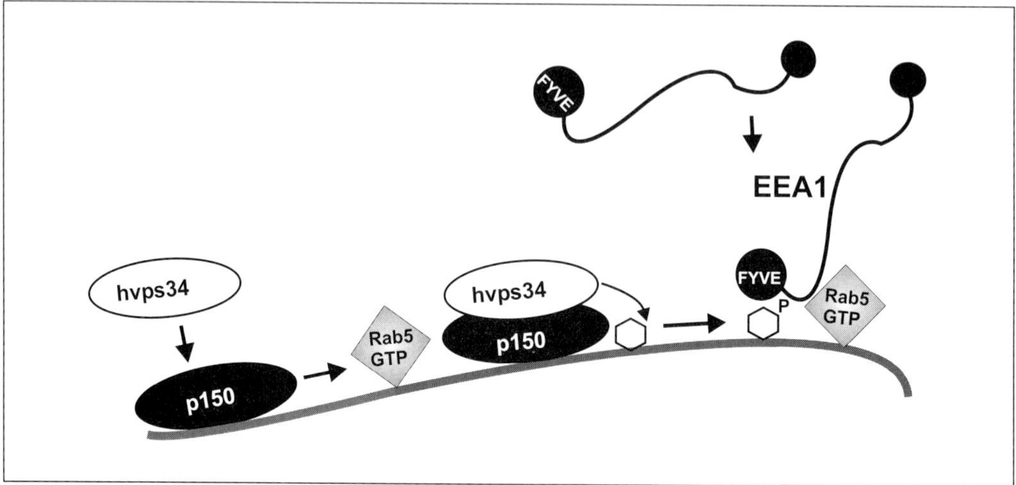

Fig. 5 Recruitment of EEA1 to endosome membranes via Rab5 and PtdIns(3)P. Rab5 serves a dual role in EEA1 recruitment. First, to recruit locally the PI 3-kinase hvps34 (by interacting with its membrane adaptor p150), thus locally stimulating the phosphorylation of PtdIns into PtdIns(3)P. Secondly, by interacting with the Rab5 binding domain of EEA1. This interaction is of low affinity, and efficient EEA1 recruitment relies on the additional binding of the FYVE domain to PtdIns(3)P.

6. EEA1 and SNARE complex formation

As mentioned in Section 4, early-endosome fusion *in vitro* can proceed in the apparent absence of Rab5, provided that excess EEA1 is present. Excess EEA1 even obviates the need for PI 3-kinase activity (39), suggesting that both Rab5:GTP and PtdIns(3)P play synergistic rather than obligatory roles in the recruitment of EEA1 to endosomes. This further suggests that other EEA1-binding molecules exist on early endosome membranes. Indeed, EEA1 has been found to interact with the t-SNAREs syntaxin6 and syntaxin13 (3, 68), which are both present on early endosomes (syntaxin6 is abundant in the *trans*-Golgi network (TGN) as well). Rab GTPases and their effectors are thought to operate prior to SNARE complex formation in membrane fusion (see Section 3), and the EEA1–syntaxin interaction may provide insight into the molecular dynamics of this process.

While the functional role of the EEA1–syntaxin6 interaction is not known (the potential regulation of TGN to endosome traffic by EEA1 has not been investigated), several results link EEA1 to syntaxin13 function (3). Recombinant syntaxin13 that lacks the transmembrane domain inhibits early-endosome fusion *in vitro*, and the same is the case with anti-syntaxin13 antibodies. Furthermore, a synthetic FYVE peptide that inhibits the interaction of syntaxin13 with EEA1 inhibits early-endosome fusion with similar concentration dependence. Finally, EEA1, NSF, Rabaptin-5, Rabex5,

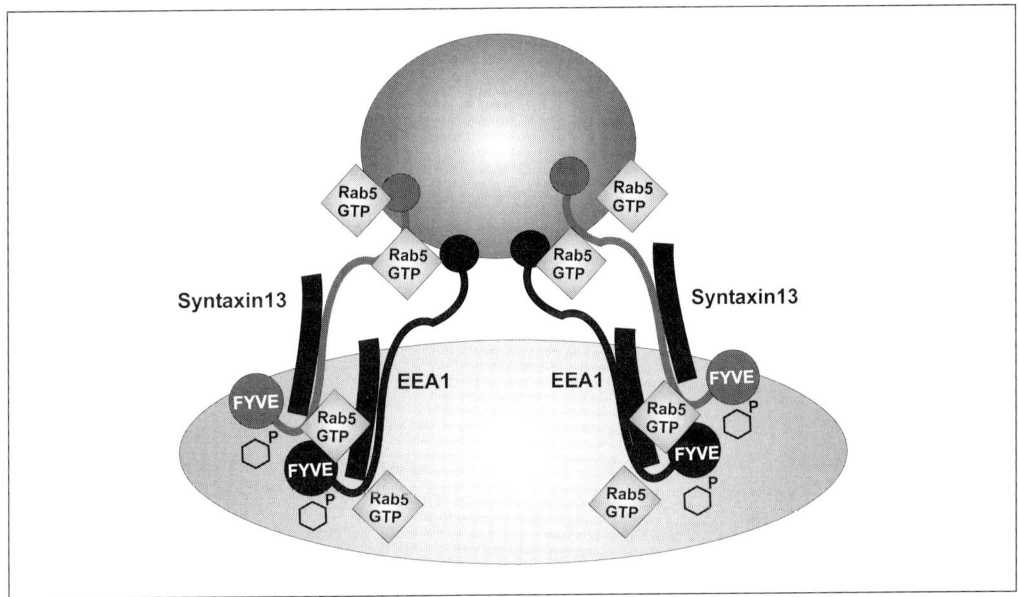

Fig. 6 The function of EEA1 in endocytic membrane docking and fusion. EEA1 may tether two Rab5-positive structures, such as an endocytic vesicle and an early endosome, by virtue of its two Rab5-binding sites. On the membranes, EEA1 assembles into oligomeric complexes that contain Rabaptin-5/Rabex5 (not shown), NSF (not shown), and syntaxin13. This ensures a spatially controlled SNARE priming. Membrane fusion following this tethering step conceivably involves the complex formation between syntaxin13 and a v-SNARE (not shown), as well as rearrangements of EEA1 oligomers.

and syntaxin13 (but not a number of other t-SNAREs) can be found together in high molecular weight complexes that are stabilized in the presence of a C-terminal EEA1 construct that appears to block endosome fusion at a post-docking stage. Interestingly, these membrane-associated complexes are regulated by the ATPase activity of NSF, suggesting that NSF may serve as a chaperone for Rab effector complexes as well as for SNAREs. The association of NSF with the high molecular weight EEA1 complexes suggests that EEA1 may recruit NSF to prime SNAREs locally (Fig. 6). The exact nature and function of the high molecular weight EEA1 complexes that are formed transiently during endosome fusion is not known. Conceivably, this system would make it possible that the SNARES are active only in the site where membrane docking occurs and ensure that promiscuous SNARE complex formation would not lead to uncontrolled membrane fusion. Moreover, since it has been postulated that viral proteins form fusion pores through the formation of oligomeric complexes, in an analogous manner EEA1 might co-operate with syntaxin13 to form a fusion pore (3). Clearly, further work is required in order to establish whether EEA1 functions only in vesicle tethering or whether it also participates directly in membrane fusion. Such studies will ultimately require the development of assays that reconstitute endocytic docking and fusion using purified components (21).

7. Conservation of the endocytic membrane fusion machinery

The general membrane trafficking regulators NSF, SNAP, and Rab GDI are all structurally and functionally conserved between mammals and yeast (69). The same is the case with Rab5 (70), whose yeast homologue, Vps21p/Ypt51p, is involved in vacuolar protein sorting and endocytic membrane traffic (71–73). Interestingly, Vps21p is found in the class D subclass of Vps proteins, which include the Rabex5 homologue, Vps9p (74), the PI 3-kinase, Vps34p (51), and the possible EEA1 homologue, Vac1p/Vps19p (71). Members of this subclass are thought to regulate the docking of Golgi-derived or endocytic vesicles with endosomes. Moreover, the FYVE finger of Vac1p/Vps19p is essential for its function, and this protein interacts directly with Vps21p (75, 76). Vac1p/Vps19p can also be found in a complex with the possible syntaxin13 homologue, Pep12p (77). It thus appears that endocytic membrane fusion in yeast and mammals employs similar molecular machinery. However, so far, early endocytic membrane fusion has not been studied in detail in yeast, and there may still be important differences between yeast and mammalian endosome fusion. For instance, no Rabaptin-5 homologue has been found in the yeast *Saccharomyces cerevisiae* and, interestingly, this yeast also lacks a Rab4 homologue. Furthermore, while Vps21p is involved in Golgi to endosome traffic, it remains to be established whether Rab5 regulates such traffic. Given that syntaxin6, which interacts with EEA1, has been implicated in *trans*-Golgi network to endosome trafficking (25), the latter issue deserves further experimentation.

The fact that the machinery of early-endocytic membrane fusion appears to be fairly well conserved from yeast to humans raises the question: How general are the mechanisms involved? Studies of homotypic vacuole fusion in yeast (see Section 2) have provided important clues. First, vacuole membrane docking requires a Rab GTPase, Ypt7, whose presence on both membranes is essential for fusion (78, 79). Secondly, homologues of NSF and SNAP are required at a pre-fusion step (80). Thirdly, *trans*-SNARE-pairing is required for fusion (18, 81). Similarly, Rab7 and SNARE proteins have been found to be essential for late endocytic trafficking in mammalian cells (82, 83). This conservation is by no means limited to the endocytic pathway. It is well established that exocytic membrane fusion and intra-Golgi traffic also require the participation of both Rab and SNARE proteins as well as a complex of cytosolic proteins, many of which contain coiled-coil domains (84–86). Thus, general principles emerge from the studies of intracellular transport in various steps (biosynthetic and endocytic pathways) and experimental systems, namely that distinct trafficking routes are based on common molecular mechanisms for membrane docking and fusion.

8. Conclusion

Current evidence indicates that membrane fusion in the early endocytic pathway is regulated by a protein complex whose recruitment to specific membrane domains is ensured by the GTPase Rab5 and the 3'-phosphoinositide, PtdIns(3)P. EEA1, which is central in this complex, mediates membrane tethering and possibly plays a role in membrane fusion in concert with syntaxin13 and NSF/SNAP. This conserved molecular machinery confers a spatial and temporal control of early endocytic membrane docking and fusion.

The molecular machinery that causes early-endocytic membrane fusion consists of general components, such as NSF and SNAP, as well as unique endocytic components such as Rab5, syntaxin13, and EEA1. While the specific recruitment of EEA1 to early endosomes is ensured by its binding to Rab5:GTP and PtdIns(3)P, prior to its interaction with syntaxin13, it is not known how Rab5 and syntaxin13 become targeted to early endosomes. The identification of the molecules involved remains a major task. It also remains to be established which v-SNARE(s) interact(s) with syntaxin13 on early endosomes.

Even though we know most of the key molecules in early-endocytic membrane fusion, the exact mechanism of the fusion process remains enigmatic. Further progress will depend on the successful reconstitution of endocytic membrane fusion *in vitro*, starting from purified components.

Acknowledgements

The authors were supported by the Norwegian Cancer Society (H. S.), the Research Council of Norway (H. S.), the Novo-Nordisk Foundation (H. S.), the Max Planck Gesellschaft (M. Z.), and an EU TMR grant [ERB-CT96-0020] (H. S. and M. Z.).

References

1. Hopkins, C. R., Gibson, A., Shipman, M., and Miller, K. (1990) Movement of internalized ligand-receptor complexes along a continuous endosomal reticulum [see comments]. *Nature*, **346**, 335.
2. Roberts, R. L., Barbieri, M. A., Pryse, K. M., Chua, M., and Stahl, P. D. (1999) Endosome fusion in living cells overexpressing GFP-rab5. *J. Cell Sci.*, **112**, 3667.
3. McBride, H. M., Rybin, V., Murphy, C., Giner, A., Teasdale, R., and Zerial, M. (1999) Oligomeric complexes link Rab5 effectors with NSF and drive membrane fusion via interactions between EEA1 and syntaxin 13. *Cell*, **98**, 377.
4. Nielsen, E., Severin, F., Backer, J. M., Hyman, A. A., and Zerial, M. (1999) Rab5 regulates motility of early endosomes on microtubules. *Nat. Cell Biol.*, **1**, 376.
5. Horazdovsky, B. F., DeWald, D. B., and Emr, S. D. (1995) Protein transport to the yeast vacuole. *Curr. Opin. Cell Biol.*, **7**, 544.
6. Geli, M. I. and Riezman, H. (1998) Endocytic internalization in yeast and animal cells: similar and different. *J. Cell Sci.*, **111**, 1031.
7. Mayorga, L. S., Diaz, R., and Stahl, P. D. (1989) Regulatory role for GTP-binding proteins in endocytosis. *Science*, **244**, 1475.
8. Gorvel, J. P., Chavrier, P., Zerial, M., and Gruenberg, J. (1991) rab5 controls early endosome fusion *in vitro*. *Cell*, **64**, 915.
9. Woodman, P. G. and Warren, G. (1991) Isolation of functional, coated, endocytic vesicles. *J. Cell Biol.*, **112**, 1133.
10. Horiuchi, H., Lippé, R., McBride, H. M., Rubino, M., Woodman, P., Stenmark, H., *et al.* (1997) A novel Rab5 GDP/GTP exchange factor complexed to Rabaptin-5 links nucleotide exchange to effector recruitment and function. *Cell*, **90**, 1149.
11. Aniento, F., Emans, N., Griffiths, G., and Gruenberg, J. (1994) Cytoplasmic dynein-dependent vesicular transport form early to late endosomes. *J. Cell Biol.*, **124**, 1373.
12. Mullock, B. M., Perez, J. H., Kuwana, T., Gray, S. R., and Luzio, J. P. (1994) Lysosomes can fuse with a late endosomal compartment in a cell-free system from rat liver. *J. Cell Biol.*, **126**, 1173.
13. Conradt, B., Haas, A., and Wickner, W. (1994) Determination of four biochemically distinct, sequential stages during vacuole inheritance *in vitro*. *J. Cell Biol.*, **126**, 99.
14. Whiteheart, S. W., Rossnagel, K., Buhrow, S. A., Brunner, M., Jaenicke, R., and Rothman, J. E. (1994) N-ethylmaleimide-sensitive fusion protein: a trimeric ATPase whose hydrolysis of ATP is required for membrane fusion. *J. Cell Biol.*, **126**, 945.
15. Wilson, D. W., Whiteheart, S. W., Wiedmann, M., Brunner, M., and Rothman, J. E. (1992) A multisubunit particle implicated in membrane fusion. *J. Cell Biol.*, **117**, 531.
16. Söllner, T., Whiteheart, S. W., Brunner, M., Erdjument-Bromage, H., Geromanos, S., Tempst, P., *et al.* (1993) SNAP receptors implicated in vesicle targeting and fusion. *Nature*, **362**, 318.
17. Fasshauer, D., Sutton, R. B., Brunger, A. T., and Jahn, R. (1998) Conserved structural features of the synaptic fusion complex: SNARE proteins reclassified as Q- and R-SNAREs. *Proc. Natl. Acad. Sci. USA*, **95**, 15781.
18. Ungermann, C., Sato, K., and Wickner, W. (1998) Defining the functions of *trans*-SNARE pairs. *Nature*, **396**, 543.
19. Poirier, M. A., Xiao, W. Z., Macosko, J. C., Chan, C., Shin, Y. K., and Bennett, M. K. (1998) The synaptic SNARE complex is a parallel four-stranded helical bundle. *Nat. Struct. Biol.*, **5**, 765.

20. Sutton, R. B., Fasshauer, D., Jahn, R., and Brunger, A. T. (1998) Crystal structure of a SNARE complex involved in synaptic exocytosis at 2.4 Å resolution. *Nature*, **395**, 347.
21. Weber, T., Zemelman, B. V., McNew, J. A., Westermann, B., Gmachl, M., Parlati, F., *et al.* (1998) SNAREpins: Minimal machinery for membrane fusion. *Cell*, **92**, 759.
22. Parlati, F., Weber, T., McNew, J. A., Westermann, B., Söllner, T. H., and Rothman, J. E. (1999) Rapid and efficient fusion of phospholipid vesicles by the a-helical core of a SNARE complex in the absence of an N-terminal regulatory domain. *Proc. Natl. Acad. Sci. USA*, **96**, 12565.
23. Nickel, W., Weber, T., McNew, J. A., Parlati, F., Söllner, T. H., and Rothman, J. E. (1999) Content mixing and membrane integrity during membrane fusion driven by pairing of isolated v-SNAREs and t-SNAREs. *Proc. Natl. Acad. Sci. USA*, **96**, 12571.
24. Walch-Solimena, C., Blasi, J., Edelmann, L., Chapman, E. R., von Mollard, G. F., and Jahn, R. (1995) The t-SNAREs syntaxin 1 and SNAP-25 are present on organelles that participate in synaptic vesicle recycling. *J. Cell Biol.*, **128**, 637.
25. Davanger, S., Bock, J. B., Klumperman, J., and Scheller, R. H. (1997) Syntaxin 6 functions in trans-Golgi network vesicle trafficking. *Mol. Biol. Cell*, **8**, 1261.
26. Yang, B., Gonzalez, L., Jr., Prekeris, R., Steegmaier, M., Advani, R. J., and Scheller, R. H. (1999) SNARE interactions are not selective Implications for membrane fusion specificity. *J. Biol. Chem.*, **274**, 5649.
27. Muller, J. M., Rabouille, C., Newman, R., Shorter, J., Freemont, P., Schiavo, G., *et al.* (1999) An NSF function distinct from ATPase-dependent SNARE disassembly is essential for Golgi membrane fusion. *Nat. Cell Biol.*, **1**, 335.
28. Pelham, H. R. B. (1999) SNAREs and the secretory pathway Lessons from yeast. *Exp. Cell Res.*, **247**, 1.
29. Novick, P. and Zerial, M. (1997) The diversity of Rab proteins in vesicle transport. *Curr. Opin. Cell Biol.*, **9**, 496.
30. Schimmöller, F., Simon, I., and Pfeffer, S. R. (1998) Rab GTPases, directors of vesicle docking. *J. Biol. Chem.*, **273**, 22161.
31. Pfeffer, S. R., Dirac-Svejstrup, A. B., and Soldati, T. (1995) Rab GDP dissociation inhibitor: putting rab GTPases in the right place. *J. Biol. Chem.*, **270**, 17057.
32. Garrett, M. D., Zahner, J. E., Cheney, C. M., and Novick, P. J. (1994) *GDI1* encodes a GDP dissociation inhibitor that plays an essential role in the yeast secretory pathway. *EMBO J.*, **13**, 1718.
33. Olkkonen, V. M. and Stenmark, H. (1997) Role of rab GTPases in membrane traffic. *Int. Rev. Cytol.*, **176**, 1.
34. Patki, V., Lawe, D. C., Corvera, S., Virbasius, J. V., and Chawla, A. (1998) A functional PtdIns(3)P-binding motif. *Nature*, **394**, 433.
35. Chavrier, P., Parton, R. G., Hauri, H. P., Simons, K., and Zerial, M. (1990) Localization of low molecular weight GTP binding proteins to exocytic and endocytic compartments. *Cell*, **62**, 317.
36. Bucci, C., Parton, R. G., Mather, I. H., Stunnenberg, H., Simons, K., Hoflack, B., *et al.* (1992) The small GTPase rab5 functions as a regulatory factor in the early endocytic pathway. *Cell*, **70**, 715.
37. Stenmark, H., Parton, R. G., Steele-Mortimer, O., Lütcke, A., Gruenberg, J., and Zerial, M. (1994) Inhibition of rab5 GTPase activity stimulates membrane fusion in endocytosis. *EMBO J.*, **13**, 1287.
38. Rybin, V., Ullrich, O., Rubino, M., Alexandrov, K., Simon, I., Seabra, M. C., *et al.* (1996) GTPase activity of Rab5 acts as a timer for endocytic membrane fusion. *Nature*, **383**, 266.

39. Christoforidis, S., McBride, H. M., Burgoyne, R. D., and Zerial, M. (1999) The Rab5 effector EEA1 is a core component of endosome docking. *Nature*, **397**, 621.
40. Stenmark, H., Vitale, G., Ullrich, O., and Zerial, M. (1995) Rabaptin-5 is a direct effector of the small GTPase Rab5 in endocytic membrane fusion. *Cell*, **83**, 423.
41. Vitale, G., Rybin, V., Christoforidis, S., Thornqvist, P. Ö., McCaffrey, M., Stenmark, H., *et al.* (1998) Distinct Rab-binding domains mediate the interaction of rabaptin-5 with GTP-bound rab4 and rab5. *EMBO J.*, **17**, 1941.
42. van der Sluijs, P., Hull, M., Zahraoui, A., Tavitian, A., Goud, B., and Mellman, I. (1991) The small GTP-binding protein rab4 is associated with early endosomes. *Proc. Natl. Acad. Sci. USA*, **88**, 6313.
43. van der Sluijs, P., Hull, M., Webster, P., Male, P., Goud, B., and Mellman, I. (1992) The small GTP-binding protein rab4 controls an early sorting event on the endocytic pathway. *Cell*, **70**, 729.
44. Cosulich, S. C., Horiuchi, H., Zerial, M., Clarke, P. R., and Woodman, P. G. (1997) Cleavage of Rabaptin-5 blocks endosome fusion during apoptosis. *EMBO J.*, **16**, 6182.
45. Swanton, E., Bishop, N., and Woodman, P. (1999) Human Rabaptin-5 is selectively cleaved by caspase-3 during apoptosis. *J. Biol. Chem.*, **274**, 37583.
46. Fykse, E. M. (1998) Depolarization of cerebellar granule cells increases phosphorylation of rabphilin-3A. *J. Neurochem.*, **71**, 1661.
47. Barbieri, M. A., Hoffenberg, S., Roberts, R., Mukhopadhyay, A., Pomrehn, A., Dickey, B. F., *et al.* (1998) Evidence for a symmetrical requirement for Rab5-GTP in *in vitro* endosome-endosome fusion. *J. Biol. Chem.*, **273**, 25850.
48. Rubino, M., Miaczynska, M., Lippé, R., and Zerial, M. (2000) Selective membrane recruitment of EEA1 suggests a role in directional transport of clathrin-coated vesicles to early endosomes. *J. Biol. Chem.*, **275**, 3745.
49. Leevers, S. J., Vanhaesebroeck, B., and Waterfield, M. D. (1999) Signalling through phosphoinositide 3-kinases: the lipids take centre stage. *Curr. Opin. Cell Biol.*, **11**, 219.
50. Rameh, L. E. and Cantley, L. C. (1999) The role of phosphoinositide 3-kinase lipid products in cell function. *J. Biol. Chem.*, **274**, 8347.
51. Schu, P. V., Takegawa, K., Fry, M. J., Stack, J. H., Waterfield, M. D., and Emr, S. D. (1993) Phosphatidylinositol 3-kinase encoded by yeast VPS34 gene essential for protein sorting. *Science*, **260**, 88.
52. Brown, W. J., DeWald, D. B., Emr, S. D., Plutner, H., and Balch, W. E. (1995) Role for phosphatidylinositol 3-kinase in the sorting and transport of newly synthesized lysosomal enzymes in mammalian cells. *J. Cell Biol.*, **130**, 781.
53. Davidson, H. W. (1995) Wortmannin causes mistargeting of procathepsin D. Evidence for the involvement of a phosphatidylinositol 3-kinase in vesicular transport to lysosomes. *J. Cell Biol.*, **130**, 797.
54. Clague, M. J., Thorpe, C., and Jones, A. T. (1995) Phosphatidylinositol 3-kinase regulation of fluid phase endocytosis. *FEBS Lett.*, **367**, 272.
55. Joly, M., Kazlauskas, A., and Corvera, S. (1995) Phosphatidylinositol 3-kinase activity is required at a postendocytic step in platelet-derived growth factor receptor trafficking. *J. Biol. Chem.*, **270**, 13225.
56. Reaves, B. J., Bright, N. A., Mullock, B. M., and Luzio, J. P. (1996) The effect of wortmannin on the localisation of lysosomal type I integral membrane glycoproteins suggests a role for phosphoinositide 3-kinase activity in regulating membrane traffic late in the endocytic pathway. *J. Cell Sci.*, **109**, 749.
57. Fernandez-Borja, M., Wubbolts, R., Calafat, J., Janssen, H., Divecha, N., Dusseljee, S., *et al.*

(1999) Multivesicular body morphogenesis requires phosphatidylinositol 3-kinase activity. *Curr. Biol.*, **14**, 55.
58. Li, G., D'Souza-Schorey, C., Barbieri, M. A., Roberts, R. L., Klippel, A., Williams, L. T., *et al.* (1995) Evidence for phosphatidylinositol 3-kinase as a regulator of endocytosis via activation of Rab5. *Proc. Natl. Acad. Sci. USA*, **92**, 10207.
59. Jones, A. T. and Clague, M. J. (1995) Phosphatidylinositol 3-kinase activity is required for early endosome fusion. *Biochem. J.*, **311**, 31.
60. Spiro, D. J., Boll, W., Kirchhausen, T., and Wessling-Resnick, M. (1996) Wortmannin alters the transferrin receptor endocytic pathway *in vivo* and *in vitro*. *Mol. Biol. Cell*, **7**, 355.
61. Stenmark, H. and Aasland, R. (1999) FYVE-finger proteins—effectors of an inositol lipid. *J. Cell Sci.*, **112**, 4175.
62. Gaullier, J.-M., Simonsen, A., D'Arrigo, A., Bremnes, B., Aasland, R., and Stenmark, H. (1998) FYVE fingers bind PtdIns(3)P. *Nature*, **394**, 432.
63. Burd, C. G. and Emr, S. D. (1998) Phosphatidylinositol(3)-phosphate signaling mediated by specific binding to RING FYVE domains. *Mol. Cell*, **2**, 157.
64. Stenmark, H., Aasland, R., Toh, B. H., and D'Arrigo, A. (1996) Endosomal localization of the autoantigen EEA1 is mediated by a zinc-binding FYVE finger. *J. Biol. Chem.*, **271**, 24048.
65. Simonsen, A., Lippé, R., Christoforidis, S., Gaullier, J.-M., Brech, A., Callaghan, J., *et al.* (1998) EEA1 links PI(3)K function to Rab5 regulation of endosome fusion. *Nature*, **394**, 494.
66. Vanhaesebroeck, B., Leevers, S. J., Panayotou, G., and Waterfield, M. D. (1997) Phosphoinositide 3-kinases: a conserved family of signal transducers. *Trends Biochem. Sci.*, **22**, 267.
67. Christoforidis, S., Miaczynska, M., Ashman, K., Wilm, M., Zhao, L., Yip, S.-C., *et al.* (1999) Phosphatidylinositol-3-OH kinases are Rab5 effectors. *Nat. Cell Biol.*, **1**, 249.
68. Simonsen, A., Gaullier, J.-M., D'Arrigo, A., and Stenmark, H. (1999) The Rab5 effector EEA1 interacts directly with syntaxin-6. *J. Biol. Chem.*, **274**, 28857.
69. Rothman, J. E. and Wieland, F. T. (1996) Protein sorting by transport vesicles. *Science*, **272**, 227.
70. Singer-Krüger, B., Stenmark, H., and Zerial, M. (1995) Yeast Ypt51p and mammalian Rab5: counterparts with similar function in the early endocytic pathway. *J. Cell Sci.*, **108**, 3509.
71. Horazdovsky, B. F., Busch, G. R., and Emr, S. D. (1994) VPS21 encodes a rab5-like GTP binding protein that is required for the sorting of yeast vacuolar proteins. *EMBO J.*, **13**, 1297.
72. Singer-Krüger, B., Stenmark, H., Düsterhöft, A., Philippsen, P., Yoo, J. S., Gallwitz, D., *et al.* (1994) Role of three rab5-like GTPases, Ypt51p, Ypt52p, and Ypt53p, in the endocytic and vacuolar protein sorting pathways of yeast. *J. Cell Biol.*, **125**, 283.
73. Gerrard, S. R., Bryant, N. J., and Stevens, T. H. (2000) *VPS21* controls entry of endocytosed and biosynthetic proteins into the yeast prevacuolar compartment. *Mol. Biol. Cell*, **11**, 613.
74. Hama, H., Tall, G. G., and Horazdovsky, B. (1999) Vps9p is a guanine nucleotide exchange factor involved in vesicle-mediated vacuolar protein transport. *J. Biol. Chem.*, **274**, 15284.
75. Peterson, M. R., Burd, C. G., and Emr, S. D. (1999) Vac1p coordinates rab and phosphatidylinositol 3-kinase signaling in Vps45p-dependent vesicle docking/fusion at the endosome. *Curr. Biol.*, **9**, 159.
76. Tall, G. G., Hama, H., DeWald, D. B., and Horazdovsky, B. F. (1999) The phosphatidylinositol 3-phosphate binding protein Vac1p interacts with a Rab GTPase and a Sec1p homologue to facilitate vesicle-mediated vacuolar protein sorting. *Mol. Biol. Cell*, **10**, 1873.

77. Burd, C. G., Peterson, M., Cowles, C. R., and Emr, S. D. (1997) A novel Sec18p/NSF-dependent complex required for Golgi-to-endosome transport in yeast. *Mol. Biol. Cell*, **8**, 1089.
78. Haas, A., Scheglmann, D., Lazar, T., Gallwitz, D., and Wickner, W. (1995) The GTPase Ypt7p of *Saccharomyces cerevisiae* is required on both partner vacuoles for the homotypic fusion step of vacuole inheritance. *EMBO J.*, **14**, 5258.
79. Mayer, A. and Wickner, W. (1997) Docking of yeast vacuoles is catalyzed by the Ras-like GTPase Ypt7p after symmetric priming by Sec18p (NSF). *J. Cell Biol.*, **136**, 307.
80. Mayer, A., Wickner, W., and Haas, A. (1996) Sec18p (NSF)-driven release of Sec17p (α-SNAP) can precede docking and fusion of yeast vacuoles. *Cell*, **85**, 83.
81. Nichols, B. J., Ungermann, C., Pelham, H. R. B., Wickner, W. T., and Haas, A. (1997) Homotypic vacuolar fusion mediated by t- and v-SNAREs. *Nature*, **387**, 199.
82. Robinson, L. J., Aniento, F., and Gruenberg, J. (1997) NSF is required for transport from early to late endosomes. *J. Cell Sci.*, **110**, 2079.
83. Mullock, B. M., Bright, N. A., Fearon, G. W., Gray, S. R., and Luzio, J. P. (1998) Fusion of lysosomes with late endosomes produces a hybrid organelle of intermediate density and is NSF dependent. *J. Cell Biol.*, **140**, 591.
84. Jahn, R. and Südhof, T. C. (1999) Membrane fusion and exocytosis. *Annu. Rev. Biochem.*, **68**, 863.
85. TerBush, D. R., Maurice, T., Roth, D., and Novick, P. (1996) The Exocyst is a multiprotein complex required for exocytosis in *Saccharomyces cerevisiae*. *EMBO J.*, **15**, 6483.
86. Hsu, S. C., Hazuka, C. D., Foletti, D. L., and Scheller, R. H. (1999) Targeting vesicles to specific sites on the plasma membrane: the role of the sec6/8 complex. *Trends Cell Biol.*, **9**, 150.

6 Lysosomes and other late compartments of the endocytic pathway

MATTHEW N. J. SEAMAN and J. PAUL LUZIO

1. Introduction

By definition, late compartments of the endocytic pathway are those compartments which endocytosed tracers enter after early endosomes, i.e. they are time defined. In mammalian cells these late endocytic compartments are usually regarded as comprising late endosomes (which have the morphological appearance of multi-vesicular bodies) and lysosomes. They also include specialized late endocytic compartments in differentiated mammalian cells, such as the MHC class II compartment essential for antigen presentation in B lymphocytes, macrophages, and dendritic cells (1, 2) and organelles grouped as secretory lysosomes found in melanocytes, platelets, cytotoxic T lymphocytes, NK, and other cell types (3, 4).

Whilst it has been relatively easy to identify and characterize late endosomes and lysosomes by subcellular fractionation and morphological techniques, establishing cell-free assays to examine interactions between them has been more difficult. This is because many endocytosed macromolecules are rapidly digested once they reach the degradative environment in these compartments. Nevertheless, content mixing and/or exchange of membrane proteins has been observed between late endosomes (5), between lysosomes (6–8), and between late endosomes and lysosomes (9). These data show that the organelles interact with each other and are in dynamic equilibrium both *in vivo* and *in vitro*. They also receive membrane proteins and lumenal content from early endosomes in the endocytic pathway, from the TGN (*trans*-Golgi network) in the biosynthetic pathway, and from autophagosomes in the autophagocytic pathway (Fig. 1). In each of these pathways there has been remarkable recent progress in identifying proteins required at specific vesicle formation and fusion steps. Some of these have been identified as a result of the establishment of mammalian cell-free systems reconstituting steps in the pathway. Another method of finding proteins involved has been the study of *vps* (vacuolar protein sorting) and *vam* (vacuolar morphology) mutants in the budding yeast *Saccharomyces cerevisiae*. In this organism

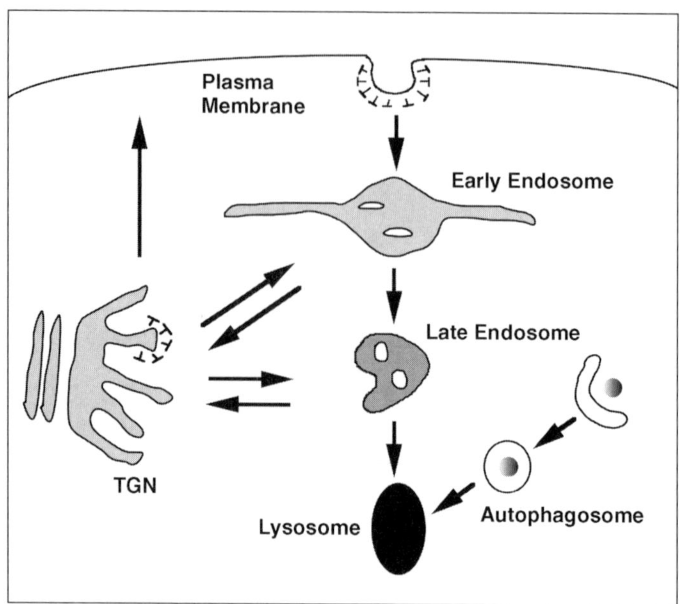

Fig. 1 A conventional model of autophagic, biosynthetic, and endocytic routes of delivery to lysosomes. Arrows indicate traffic routes between compartments. The lysosome is represented as the terminal degradation compartment of the endocytic and autophagic pathways.

the vacuole is functionally equivalent to the lysosome, both being acidic compartments containing a variety of hydrolytic enzymes involved in the degradation of macromolecules (10, 11). Recently, *Drosophila* genetics and the study of genes involved in diseases of the secretory lysosome in mammals have also proved useful in identifying novel proteins required for the function of the endocytic pathway. In this chapter we examine present knowledge of the structure, function, and biogenesis of late endosomes and lysosomes in mammalian cells. We also discuss the contribution of genetic screens to identifying proteins involved in vacuolar protein sorting and autophagy in yeast and review recent data from studies of *Drosophila* eye colour, mouse coat colour, and human disease that have shed light on the molecular mechanisms of membrane traffic in the late endocytic pathway.

2. The heterogeneity of lysosomes

The discovery of lysosomes almost fifty years ago (12) was an early success of subcellular fractionation using ultracentrifugation (for overviews of the history of the discovery see refs 13, 14). The discovery relied on the association of a number of hydrolases with an organelle of defined centrifugal properties and the observation that these enzymes demonstrated structure-linked latency dependent on the integrity of the surrounding membrane. From the earliest days it was recognized that lysosomes showed heterogeneous morphology, containing dense deposits and membrane whorls (reviewed in ref. 15). In mammalian cells they characteristically appear as organelles of ~ 0.5–2 μm diameter, often with electron dense cores. They can make up 0.5–5% of the cell volume and are concentrated near the microtubule organizing centre (16). Lysosomes are now defined as intracellular membrane-bound organelles containing many hydrolytic enzymes which are optimally active at acid pH and are distinguished from endosomes because they lack the two mannose-6-phosphate receptors (MPRs) and recycling cell surface receptors. Lysosomes are usually regarded as the terminal degradation compartment of the endocytic pathway (17) (Fig. 1).

3. The endocytic pathway to lysosomes

The identification of endosomal compartments on the major endocytic pathway from the clathrin-coated pit at the plasma membrane to the lysosome was achieved by a combination of microscopy, subcellular fractionation, and kinetic analysis of the passage of endocytosed molecules through different compartments, with many research groups making important contributions. Since the early 1980s it has been clear that, like lysosomes, endosomal compartments in mammalian cells are morphologically heterogeneous and constitute a pleiomorphic smooth membrane system consisting of tubular and vesicular elements with many of the latter containing intra-organelle vesicles (18–20). Despite this complexity, a generalized route map for the passage of endocytosed material to lysosomes has been established. The timing of delivery to early endosomes and then to late endosomes varies between cell types and between cells in culture and in living tissue (21). Early endosomes are loaded within 1–5 minutes of uptake, are tubular with varicosities, and many are peripherally located within the cell close to the plasma membrane. Late endosomes become loaded 4–30 minutes after uptake, are more spherical, and have the appearance of multivesicular bodies. They are mostly juxtanuclear being concentrated near the microtubule organizing centre. Early and late endosomes are characterized by different lumenal pH (22), different protein composition (23), and association with different small GTPases of the Rab family, with Rab5 on early endosomes and Rab7 on late endosomes (24).

The early endosome is the major sorting compartment of the endocytic pathway where many ligands dissociate from their receptors in the acid pH of the lumen and from which many of the receptors recycle to the cell surface (25, 26). It is also the site of sorting into transcytotic pathways in polarized cells (27, 28), to the TGN (29, 30), and to lysosomes. The membrane traffic pathway from early to late endosomes was the subject of some dispute a few years ago with the proposal of maturation and vesicular transport models. In essence the maturation model envisages the continuous removal and addition of material to the maturing early endosome until it becomes, by definition, a late endosome (31–33). In contrast the vesicular transport model proposes the formation of an endocytic carrier vesicle (ECV) that buds from the early endosome and fuses with the late endosome thereby delivering lumenal content and membrane proteins (34, 35). The existence of such ECVs was inferred from studies of cell-free content mixing between endosomal compartments and from morphological analysis of the endocytic pathway. Although the maturation and vesicular transport models of delivery from early to late endosomes appear at first sight to be contradictory, there is in fact great similarity, since both models provide for an intermediate between early and late endosomes, differing mainly in whether this is, in essence, what is left after removal of components of early endosomes that are to be recycled or delivered elsewhere or a newly formed ECV (36). The identification of a specific cytosolic coat recruited to form the ECV would help to show that this was not simply the product of maturation. At present the best candidate for such a coat is an isoform of the COP-I coat responsible for retrograde traffic from the Golgi complex to the ER (reviewed in ref. 35). The endosomal form appears to have a different subunit composition and its role is controversial since it is difficult to rule out the possibility that the effects on endocytic traffic of inhibiting COP-I function are secondary to a primary effect on the secretory pathway.

4. The delivery of endocytosed material from late endosomes to lysosomes

A cell-free system to study content mixing between late endosomes and lysosomes derived from rat hepatocytes has been established with the lysosomes containing a slowly digested ligand (9, 21). The demonstration of content mixing between the organelles showed that maturation of late endosomes to lysosomes could not be a sufficient explanation of endocytic delivery to lysosomes nor of lysosome biogenesis. It is relevant to note that experiments in which late endosomes have been incubated with cytosol and ATP, whilst showing changes in density, have never shown sufficient increases in density to be consistent with the formation of lysosomes (21, 31). Although content mixing in cell-free assays between donor and acceptor organelles on membrane traffic pathways is often explained by vesicular transport, no evidence has been obtained for such a process occurring between late endosomes and lysosomes.

Two hypotheses have been proposed to account for content mixing between late endosomes and lysosomes. The first is termed 'kiss-and-run' and proposes that late endosomes and lysosomes are undergoing continuous repeated transient fusion and fission (6). This hypothesis explains data from experiments in which small soluble markers appeared to move more readily between lysosomes and endosomes/phagosomes than larger soluble markers, and from experiments in which horse radish peroxidase (HRP) was delivered to lysosomes faster than lysosomal membrane glycoproteins (lgps). The second or 'hybrid organelle' hypothesis, proposes direct fusion between late endosomes and lysosomes with subsequent recovery of lysosomes for reuse. It may be regarded as an extension of 'kiss-and-run', where some kisses result in fusion (37, 38). Griffiths (37) pointed out that, morphologically, lysosomes resemble electron dense secretory granules. He speculated that their function is to store mature lysosomal enzymes and proposed that they fuse periodically with late endosomes in a regulated fashion to form a compartment, 'the cell stomach', where degradation occurs. He also proposed that lysosomes subsequently bud off this compartment and are reused as required.

In the rat liver cell-free system evidence has been obtained for direct fusion between late endosomes and lysosomes. In this system, the immunoprecipitable product of content mixing betwen late endosomes and lysosomes can be recovered in a membrane-bound organelle intermediate in density between late endosomes and lysosomes (9). This late endosome–lysosome hybrid organelle contains markers derived from both lysosomes (cathepsin L) and late endosomes (endocytosed avidin-labelled asialofetuin and the cation independent MPR), has a mean diameter (0.96 μm) larger than either rat liver lysosomes (0.38 μm) or late endosomes (0.34 μm), and is less electron dense than lysosomes (9). The hybrid organelle has the characteristics of Griffiths' 'cell stomach'.

Further evidence for the hybrid organelle hypothesis has come from the ability to isolate directly from rat liver an organelle fraction with the appropriate centrifugation and morphological characteristics (9) and also from EM experiments on cultured cells. One study on cultured HEp-2 cells by Futter *et al.* (39) showed that transfer of complexes of EGF and its receptor to a late endocytic degradative compartment involved maturation of late endosomal MVBs (multivesicular bodies) which fused directly with MPR-negative lysosomes. These lysosomes were pre-labelled by loading with HRP in a 15 minute pulse, 4 hour chase protocol. Activation of the lysosomal HRP with diaminobenzidine and H_2O_2 before EGF uptake prevented fusion of the lysosomes with the MVBs, but not maturation of the latter, nor docking of the MVBs with lysosomes. Experiments with cultured NRK cells are consistent with the existence of the hybrid organelle *in vivo* and have provided additional information about its properties (40). In one experiment electron dense lysosomes were pre-loaded with BSA–5 nm gold using a 4 hour pulse, 20 hour chase protocol and the mixing of this lysosomal marker with subsequently endocytosed BSA–15 nm gold was examined. Mixing of the two labels was first observed 15 minutes after uptake of the BSA–15 nm gold in a MPR-positive compartment with the expected characteristics of the hybrid organelle. Flocculation of the BSA–15 nm gold occurred in these organelles consistent with proteolysis of the BSA which did not occur in earlier endosomal compartments. Only after uptake of BSA–15 nm gold for 30 minutes was it observed in electron dense lysosomes. At later times increasing amounts of BSA–15 nm gold were observed in electron dense lysosomes. Interestingly, the amount of both labels in the hybrid compartment reached a plateau of ~ 15% within 1 hour of uptake of BSA–15 nm gold suggesting that at any one time ~ 15% of lysosomes are fused with late endosomes to form the hybrid compartment in NRK cells and that lysosomes can be re-formed from this compartment. In a further experiment using cultured NRK cells endocytosed sucrose was observed to accumulate in swollen late endosomes which fused with electron dense lysosomes such that after 24 hours all of the lysosomes had been consumed. Subsequent endocytosis of invertase resulted in the re-formation of electron dense lysosomes with almost 100% recovery after 8 hours (40).

5. The mechanism of fusion of late endosomes and lysosomes

Formation of hybrid organelles in the rat liver cell-free system is ATP-, cytosol-, and temperature-dependent (9) and requires the presence of NSF (*N*-ethyl maleimide-sensitive factor), SNAPs (soluble NSF attachment proteins), and a small GTPase of the Rab family (9). Together, these characteristics are consistent with the action of the common cytosolic fusion machinery used at many other sites on the secretory and endocytic pathways with it functioning according to the tenets of the SNARE (SNAP receptor) hypothesis that proposed specific pairing between v- (vesicle) and t- (target organelle) SNAREs (41). Site-specific components of the fusion machinery, including specific SNAREs and the Rab, required for late endosome–lysosome fusion have not yet been identified.

In contrast, there has been much greater progress in understanding homotypic fusion events at the early endosome and this has been illuminating with respect to events later in the endocytic pathway. The fusion of endocytic vesicles with early endosomes and early endosomes with each other is mediated by a process involving Rab5 and the t-SNARE syntaxin13 (see Chapter 5). Recent work on these early events in the endocytic pathway has suggested that SNARE complex formation is preceded by the recruitment of tethering proteins into an oligomeric assembly on the endosome membrane (42). Tethering has been defined as involving links that extend over distances >25 nm from a given membrane surface, and docking as holding membranes within a bilayer's distance, <5–10 nm of one another (43). Tether recruitment for early endosome fusion is dependent on the presence of Rab5 and also of PtdIns 3-phosphate (phosphatidylinositol-3-phosphate) patches formed as the result of PtdIns 3-kinase (phosphatidylinositol-3-kinase) activity and one protein thought to act as a tether is EEA1 (44–46). Oligomeric assembly of the tethers is modulated by the ATPase activity of NSF, and syntaxin13 is transiently incorporated into the large oligomer via direct interactions with one of the tether proteins. It has been proposed that the oligomeric tether complex and NSF mediate the local activation of syntaxin13 upon membrane tethering, leading to membrane fusion (42). To date no specific tethering proteins have been identified for late endosome–lysosome fusion. Nevertheless, the evidence for their existence is strong. Fine striations have been observed between adjacent endosomes and lysosomes in several morphological studies on cultured cells (39, 40, 47). The observation that PtdIns 3-kinase inhibitors block late endosome–lysosome fusion (9) is also relevant in view of the evidence suggesting a role for a lipid product of PtdIns 3-kinase activity in tether recruitment onto early endosomes.

Whilst there is now good evidence that syntaxin13 is a t-SNARE functioning early in the mammalian endocytic pathway, the roles of specific SNAREs at other stages of this pathway are poorly understood. Localization studies have identified VAMP-3 (cellubrevin), VAMP-5, VAMP-7, VAMP-8 (endobrevin), syntaxin7, and syntaxin8 as other SNAREs in the endocytic pathway but localization is rarely to a single organelle and in some cases extends to compartments of the secretory pathway. Functional data on most of the SNAREs is presently lacking. The ability of VAMP-3 to form complexes with syntaxin13 suggests a role in the early endosome (48). Data from permeabilized cell assays in which antibodies to specific SNAREs have been found to inhibit degradation of endocytosed EGF have implicated both VAMP-7 and syntaxin8 in the pathway from early endosomes to lysosomes, most likely in trafficking from early to late endosomes rather than late endosomes to lysosomes (49, 50). Syntaxin7 has also been implicated in traffic late in the pathway because overexpression of the cytoplasmic domain inhibits delivery of endocytosed markers to late endocytic compartments (51). Very little is known about the role of SNAP-25 homologues in the endocytic pathway, other than the observation that in biochemical assays some are able to bind to SNAREs localized in the pathway (52, 53).

Data from a liposome fusion assay suggests that SNAREs may be membrane fusion catalysts (54–56). Fusion may also require downstream events after SNARE complex formation. In yeast vacuole fusion, discussed further below, Ca^{2+} release from the vacuole lumen is required in a post-docking phase of fusion (57, 58) and this may also be the case for homotypic fusion between mammalian early endosomes (59) and heterotypic fusion of late endosomes and lysosomes (60). By analogy with early endosome fusion it is possible to propose a model for late endosome–lysosome fusion with tethering, docking, and fusion steps (Fig. 2).

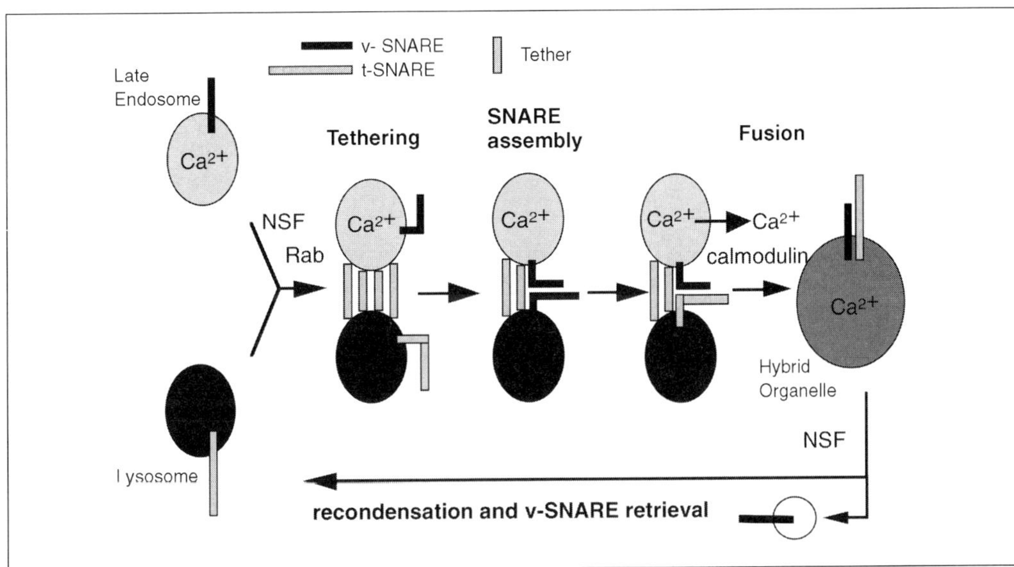

Fig. 2 The mechanism of late endosome–lysosome fusion. The diagram shows the possible stages of fusion of late endosomes with lysosomes and re-formation of lysosomes from the subsequent hybrid organelles. Tethering by unknown proteins may precede SNARE assembly and fusion. Ca^{2+} release from the organelle lumens (for simplicity, shown here only from the endosome) may be necessary for fusion. Re-formation of lysosomes from the hybrid organelles should require a process to recycle the v-SNARE for subsequent rounds of fusion and the condensation of lumenal content to produce dense core lysosomes.

6. Lysosome recovery from hybrid organelles

Since the fusion of late endosomes with lysosomes to form hybrid structures results in consumption of the starting organelles, it is apparent that a recovery system involving membrane retrieval and a lysosomal content recondensation process occurs *in vivo* (9, 38, 40). Condensation of content must occur since lysosomes are characterized both by their greater density compared to endosomes when analysed by isopycnic centrifugation, in a variety of density gradient media, and by the frequent electron dense appearance of lumenal content when analysed by EM. Another mammalian cell organelle in which condensation of content has been well described is the regulated dense core secretory granule. In this case it is clear that protein aggregation, under conditions of low intraorganelle pH and high Ca^{2+} (61, 62), plays an important role in the formation of the mature secretory granule. There is also evidence for clathrin/AP-1 adaptor coats forming on immature secretory granules, possibly for removal of lysosomal enzymes, as these organelles mature. Thus, it seems possible that the re-formation of lysosomes from late endosomes/lysosome hybrid organelles occurs by an analagous process to the formation of regulated secretory granules from the TGN. This hypothesis is supported by:

(a) Data from a cell-free assay measuring re-formation of lysosomes from hybrid organelles which is inhibited by chelation of lumenal Ca^{2+} or addition of the vacuolar ATPase inhibitor bafilomycin A1 (60).

(b) Experiments showing that bafilomycin A1, prevents transport of endocytosed HRP from endosomes to lysosomes in cultured hepatoma cells (63).

Recruitment of clathrin and AP-2 adaptors to late endocytic compartments including lysosomes has also been demonstrated (64–66) and could contribute to the process of removing endosomal membrane proteins that are not present in lysosomes.

7. The biosynthetic route to the yeast vacuole

In contrast to the paucity of information about specific molecules required for membrane traffic on the pathways to the lysosome in mammalian cells, many gene products necessary for membrane traffic to the yeast vacuole have been identified. In earlier studies, genetic screens in the budding yeast *Saccharomyces cerevisiae* had been sucessfully applied to identify mutants that were defective in secretion. These pioneering studies showed that classical yeast genetics could be used to uncover the mechanisms responsible for membrane trafficking. However before screens for mutants defective in sorting to the vacuole could be undertaken two further observations were crucial. These observations were made from experiments upon the trafficking of CPY (carboxypeptidase Y) to the vacuole. Cloning of the CPY gene allowed its expression to be regulated by plasmid copy number. When CPY was expressed from a high copy plasmid it was found that much of the CPY produced was secreted from the yeast (67). In contrast when CPY was expressed from a single copy plasmid, the CPY was matured in the vacuole. Therefore delivery of CPY to the vacuole is gene dosage-dependent suggesting that the mechanism for sorting CPY away from secreted proteins is saturable. These studies were extended by site-directed mutagenesis of CPY to identify a specific tetrapeptide sequence within the pro-peptide of CPY that directs it away from secreted proteins to the vacuole (68–70). The pro-peptide of CPY was found to be both necessary and sufficient to direct either CPY or a reporter protein such as invertase to the vacuole (71). Clearly this distinguishes yeast from mammalian cells where a carbohydrate tag provides the means to recognize and sort many lysosomal enzymes in the TGN. These observations paved the way for screens to identify mutants defective in trafficking of CPY to the vacuole. Two laboratories independently screened for mutants that secrete CPY. One screen identified mutants using an artifical substrate for CPY that when cleaved by secreted CPY would provide the sole source of leucine for cells that were leucine auxotrophs (72). The other screen ingeniously utilized the CPY–invertase fusion protein that was normally targeted to the vacuole. A yeast strain was engineered that was defective for the wild-type invertase gene *SUC2*. This strain was unable to grow on plates where sucrose was the only fermentable carbon source. A selection scheme was used whereby spontaneous mutants that were defective in the vacuole targeting of the CPY–invertase protein would grow on sucrose containing plates spread with antimycin A, to block normal respiration and induce fermentation (71). Both screens were successful in isolating many different complementation groups. Genetic analysis of the two sets of mutant revealed significant overlap. The two sets of vacuole tafficking mutants were combined into one set comprising more than forty genes and named *vps* (vacuole protein sorting mutants; see Table 1) (73, 74). Recent findings have indentifed a parallel pathway in yeast that is responsible for the sorting and delivery of the membrane protein alkaline phosphatase (ALP) to the yeast vacuole. Mutants that specifically block this pathway, but do not affect the transport of CPY, have been identified and are discussed below.

The *vps* mutants displayed a range of different phenotypes with respect to growth, CPY sorting, and vacuolar morphology and given the large number of mutants obtained, it seemed likely that trafficking to the vacuole is a multistep process. One crucial aspect of the *vps* mutants is that they do not affect the early secretory pathway hence CPY is secreted in its Golgi-modifed form. Initial classification of the *vps* mutants was made on the basis of vacuolar morphology, with three different phenotypes being scored; A, normal morphology; B, fragmented vacuole; C, no recognizable vacuole (73). Further analysis indicated that a subset of mutants were accumulating a prevacuolar compartment which also functioned as an endosomal intermediate (74). There were other subtle variations of morphology within the original three groups prompting the designation of three more groups. In terms of vacuolar morphology, the six groups can be summarized as follows:

A Normal vacuolar morphology.
B/B* Fragmented vacuole.
C No recognizable vacuole.
D Slightly enlarged single vacuole.
E Enlarged prevacuole juxta a normal vacuole.
F One vacuole surrounded by smaller framented vacuoles.

Table 1 Vacuolar protein sorting genes

VPS	Class	Function/comments	Also known as
1	F	Dynamin-like GTPase may function at multiple steps.	
3	D	Overexpression results in vacuole fragmentation.	*PEP6*
4	E	AAA-type ATPase required for proper endosomal maintenance.	
5	B	Sorting-nexin-1 homologue. Component of retromer complex.	*PEP10*
6	D	Endosomal t-SNARE.	*PEP12*
8	D	Has Zn binding domain.	
9	D	Exchange factor for Vps21p. Homologue to mammalian Rabex.	
10	A	Type I transmembrane receptor for CPY and other hydrolases.	*PEP1*
11	C	Has Zn binding domain. In a complex with Vps18p.	*PEP5*
13	A	Large protein (~ 300 kDa) facilitates Vps10p, Kex2p localization.	*SOI1*
15	D	Ser/thr kinase in complex with Vps34p PtdIns 3-kinase.	
16	C	Part of complex that also contains Vps11p, Vps18p, and Vps33p.	*VAM9*
17	B	Component of retromer complex.	
18	C	Homologue of *Drosophila* dor protein. In complex with Vps11p.	*VAM8/PEP3*
19	D	Homologue to mammalian EEA1. Required for fusion at endosome.	*VAC1/PEP7*
21	D	Small GTPase of Rab family. Homologous to mammalian Rab5.	
23	E	Homologue of mammalian TSG101.	*STP22*
24	E	Required for proper endosome maintenance.	
26	F	Component of retromer complex.	*PEP8*
27	E	Homologue of mammalian Hrs protein. Functions at endosome.	
28	E	Required for proper endosome maintenance.	
29	A	Component of retromer complex.	*PEP11*
30	A	Homologue of mammalian Beclin. Also functions in autophagy.	*APG6*
32	E	Required for proper endosome maintenance.	*SNF7*
33	C	Member of Sec1 family. Required for fusion at vacuole.	
34	D	PtdIns kinase. Phosphorylates phosphatidyl inositol at 3 position.	
35	A	Component of retromer complex.	
36	E	Required for proper endosome maintenance.	*VAC3*
38	A/E	As yet uncharacterized.	
39	B*	In a complex with Vps41p. Part of AP3 coat?	*VAM6*
41	B*	In a complex with Vps39p. Binds yeast Apl5p component of AP3 coat.	*CVT8*
43	B*	Part of vacuolar t-SNARE machinery with VAM3.	*VAM7*
45	D	Member of Sec1 family. Required for fusion at endosome.	
VAM3	B*	Vacuolar t-SNARE. Also required for autophagy.	
YPT7	B*	Rab GTPase. Required for fusion at vacuole.	*VAM4*

NB: B and B* *vps* mutants, respectively, display different degrees of vacuolar fragmentation from almost complete to very severe.

These six groups not only indicate the vacuolar morphology phenotype but also have proved to be remarkably accurate in predicting which *VPS* gene products are acting together at a single step in transport to the vacuole. Cloning of the respective *VPS* genes by complementation of the CPY sorting defect has opened the door to functional studies and elucidation of the molecular mechanisms of vacuolar protein sorting. Trafficking to the vacuole can be broken down into a series of discrete steps each of which is mediated by a specific set of Vps proteins (see Fig. 3). These steps are:

1. Sorting of CPY at the late-Golgi into a vesicle.
2. Transport to an endosomal intermediate.
3. Fusion of vesicle with endosome.
4. Separation of CPY from receptor—recycling of receptor.
5. Fusion of endosome with vacuole to deliver contents.

The key observations that the sorting of CPY at the Golgi is saturable and propeptide dependent strongly suggested that there was a CPY sorting receptor responsible for binding CPY in the late-Golgi to divert it away from secretory vesicles

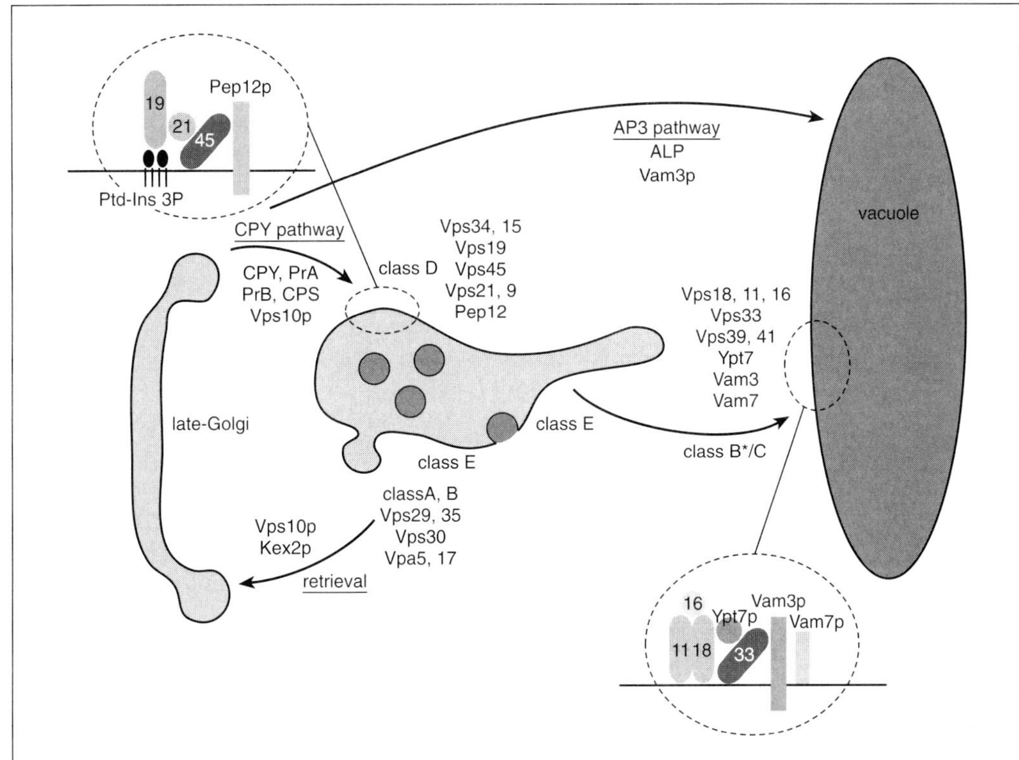

Fig. 3 Schematic diagram of vacuolar protein sorting in *S. cerevisiae*. Transport of proteins to the vacuole is a multistep process that diverges at the late-Golgi resulting in two pathways; the CPY pathway and the ALP pathway. Sorting of soluble hydrolases such as CPY at the late-Golgi is receptor-mediated requiring the Vps10p protein. Transport of ligand–receptor complexes to the endosome is followed by docking and fusion which is regulated by a complex of proteins that contains the t-SNARE Pep12p, the Sec1p homologue, Vps45p, and a putative tethering factor, Vps19p. Vps19p binds PtdIns 3-phosphate and is regulated by the small GTPase Vps21p. Ligand and receptor dissociate and the receptor is recycled back to the Golgi for further rounds of sorting. Retrieval of Vps10p is directed by a membrane-associated complex dubbed retromer that contains five Vps proteins, Vps35p, Vps29p, Vps26p, Vps5p, and Vps17p. Class E Vps proteins modulate endosomal function and are important in the formation of intra-lumenal vesicles within the endosome. Fusion of the endosome (and also ALP containing vesicles) with the vacuole is SNARE-mediated requiring the function of the Vam3 protein along with Vam7p. Vps33p is a Sec1p homologue and a complex of Vps11p, Vps18p, and Vps16p may provide tethering activity. The small GTPase Ypt7p regulates the fusion process.

destined for the plasma membrane. This was indeed the case and the CPY receptor was identified as the product of the *VPS10* gene, a class A *VPS* gene (75). Vps10p is a type I transmembrane receptor with a large luminal domain and smaller cytoplasmic tail. Vps10p recognizes the Golgi-modified form of CPY via the QRPL tetrapeptide sequence in the pro-domain. Curiously it was found that deletion of *VPS10* resulted in a selective sorting defect in which CPY was secreted but other soluble hydrolases such as proteinase A (PrA) were delivered to the vacuole and correctly matured. This observation was to prove useful in understanding the trafficking of Vps10p.

If the sorting of CPY is receptor-mediated then there is presumably a coat protein(s) that concentrates the receptor ligand complex into vesicles for delivery to the endosome. To date the best candidate for this coat protein is clathrin. Temperature-sensitive alleles of the clathrin heavy chain (CHC) mis-sort and secrete CPY after a shift to the non-permissive temperature (76, 77). However, after prolonged incubation at the non-permissive temperature, some clathrin ts alleles recover and the chc^{ts} cells correctly sort CPY again. Furthermore, clathrin heavy chain deletion mutants have no CPY sorting defect (78). It has therefore been suggested that clathrin mutants have the ability to adapt to the loss of clathrin, possibly by up-regulating another pathway. Clathrin therefore may be not strictly necessary for CPY sorting at the late-Golgi, but it appears that the product of the *VPS1* gene is necessary (79). *VPS1* encodes a large GTPase of the dynamin family (80). *VPS1* mutants have somewhat pleiotropic phenotypes which cause a CPY sorting defect, mis-localization of the Golgi protein Kex2p (81), and also mis-sorting of alkaline phosphatase to the plasma membrane (82). This argues for Vps1p acting at multiple sites possibly by facilitating vesicle budding.

Whilst the mechanisms required to sort CPY at the Golgi remain to be fully understood, the picture of what happens at the endosome is much clearer. At the endosomal membrane a complex of proteins encoded by class D genes function together in the docking and fusion of vesicles containing Vps10p and CPY (83). A key component of the docking machinery is Pep12p, a syntaxin-like SNARE molecule (84). Pep12p inactivation results in a rapid and severe CPY sorting defect (83). Associated with Pep12p is Vps45p, a homologue of Sec1p (85, 86). Vps21p, a Rab5-like small GTPase is believed to regulate the docking and fusion possibly through interactions with Vps19p/Vac1p (87–89). Vps9p promotes GDP for GTP exchange on Vps21p and is essential for CPY sorting (90, 91). Both *vps45* and *vps21* mutants accumulate small vesicles which are believed to be transport intermediates (85, 87). Vps19p/Vac1p is a functional homologue of the mammalian protein EEA1 and shares a conserved domain known as a FYVE domain. FYVE domains co-ordinate two molecules of Zn^{2+} and have been shown to bind specifically to PtdIns 3-phosphate (92, 93). Vps34p is a PtdIns 3-kinase and is associated with Vps15p, a serine/threonine kinase (94–97). Inactivation of Vps34p also results in a rapid and severe CPY sorting defect similar to *pep12* mutants (97). As Pep12p is believed to fulfil a syntaxin-like role in vesicle docking, there must also be a VAMP-like SNARE present on the Golgi-derived vesicles. The best candidate for this molecule is Vti1p which may act as a v-SNARE at multiple sites but does display genetic interactions with

PEP12 to suggest a functional interaction (98, 99). Lastly, the ubiquitous fusion machinery components, NSF and α-SNAP are also required for docking and fusion of CPY-containing vesicles with the endosomal membrane (83).

The last stage in the transport of CPY to the vacuole is delivery from the endosome to the vacuole. The proteins that mediate this step fall mostly in the class C group of Vps proteins including the gene products of *VPS18, VPS11, VPS16,* and *VPS33.* These proteins function together in the fusion of endosomes with vacuoles and mutations in these proteins result in a phenotype with no recognizable vacuolar structures (73). Another key component of the docking/fusion machinery is the SNARE protein Vam3p. Like Pep12p, Vam3p is a syntaxin-like SNARE, but it is localized to the vacuolar membrane instead. Associated with Vam3p are Vam7p and Vps33p (100, 101). Vam7p has some homology to Sec9p and the mammalian protein SNAP25 and like SNAP25 may function as part of the t-SNARE in vesicle docking/fusion. The cognate v-SNARE required for endosome/vacuole fusion is likely to be either Vti1p which has been demonstrated to have genetic interactions with Vam3p or the v-SNARE Nyv1p which may perform this function (99). *In vitro* vacuole–vacuole fusion assays have shown a requirement for Nyv1p (102, 103) but there is no CPY sorting defect in *nyv*1 mutants which argues against a role in endosome–vacuole fusion (99). Vps33p is a Sec1p homologue. This protein has also been shown to associate with a complex containing Vps18p, Vps11p, and Vps16p (104). Both Vps18p and Vps11p have Zn^{2+} co-ordinating domains but these domains do not conform to FYVE domain consensus and are not believed to bind PtdIns 3-phosphate. Their role in docking/fusion of endosomes with vacuoles is as yet undefined but biochemical studies have indicated that Vps18p and Vps11p may form large membrane-associated oligomeric complexes which could function to tether the two membranes together prior to fusion (104). Regulating these protein–protein interactions necessary for docking/fusion is a small GTPase of the Rab family, Ypt7p (105). Again, NSF and α-SNAP are also required for the sucessful docking/fusion of endosomes with vacuoles.

From a mechanistic view point there are clearly many similarities between the docking and fusion of Golgi-derived vesicles with the endosome and the docking/fusion of endosomes with vacuoles. Both steps are SNARE-mediated requiring NSF and α-SNAP function and both also involve Sec1p homologues and Rab GTPase proteins. The processes are so similar in fact that Vam3p and Pep12p can functionally substitute for one another when expressed at high copy number (100). This feature was used in a novel screen in which *VAM3* was mutagenized so that low copy number *vam*3 could functionally substitute for *PEP12.* These mutant Vam3 proteins were subsequently shown to be altered in their trafficking such that they were no longer trafficking to the vacuole by the AP3 pathway but instead were being transported via the CPY pathway to the prevacuolar endosome (106).

Two key stages of transport to the vacuole, namely docking/fusion at the endosome and vacuole are now fairly well understood, but what happens in between? There must be further sorting at the endosome to ensure that the CPY receptor, Vps10p is recycled back to the late-Golgi. Class E *vps* mutants have been shown to accumulate an exaggerated endosomal structure in which markers of the late-Golgi, including Vps10p, become concentrated (74, 107, 108). Class E mutants also exhibit reduced delivery of endocytosed proteins to the vacuole indicating that traffic out of the endosome is impaired in both directions, ie back to the late-Golgi or onwards to the vacuole (74, 108, 109) . How or why class E mutants affect endosomal function is not yet clear but the large collection of proteins encoded by class E genes are definitely of paramount importance to sorting and trafficking within the endocytic system. This is illustrated most strikingly by recent studies on the processing of a membrane-bound vacuolar hydrolase, CPS (carboxy-peptidase S). CPS is a type II membrane protein which upon arrival in the vacuole is cleaved to produce the mature enzyme. Experiments to localize CPS surprisingly indicated that it was not present in membrane fractions enriched in vacuolar membranes containing markers proteins such as ALP and Vam3p. GFP-tagging of CPS revealed the reason why. CPS is localized to the lumen of the vacuole in a wild-type cells (110). However in certain class E mutants, CPS was found to be present in the vacuolar membrane indicating that sorting of CPS into the lumen occurs at the endosome and results in the incorporation of CPS into lumenal vesicles inside the endosome (110). This topological shift is indentical to the process that mediates the down-regulation of EGF receptors in mammalian cells by inclusion of activated EGF receptors into multivesicular bodies (39).

Recently it has been found that CPS is not the only protein to undergo this sorting event. The α-factor mating pheromone receptor Ste2p is also sorted into the lumen of endosomes for delivery to the vacuolar lumen in a class E Vps protein-dependent fashion (111). The product of the *STP22/VPS23* gene, was shown to be necessary for this sorting. The mammalian homologue of this gene is TSG101, a gene which has been implicated in tumour susceptibility and which functions in traffic to late endosomes (112). Clearly these data could have profound implications for understanding the mechanisms by which signal transducing receptor proteins are down-regulated by incorporation into intra-lumenal vesicles in the endocytic system. Regulation of these events may occur by lipid modification. A PtdIns 3-phosphate 5-kinase, the product of the *FAB1* gene in yeast is also required for CPS localization to intraluminal vesicles (110). How the PtdIns 3,5-bisphosphate functions in sorting of proteins such as CPS into internal endosomal vesicles in not clear but it seems likely that the lipid moitety may regulate the localization or activity of a collection of proteins, many of which may be encoded by class E *VPS* genes. The substrate for Fab1p is PtdIns 3-phosphate, the product of the Vps34p PtdIns 3-kinase. As mentioned previously, the PtdIns 3-phosphate produced by the Vps34p kinase is bound by FYVE domain containing proteins, one of which is Vps19p/Vac1p. Vps19p is however, not the only Vps protein with a FYVE domain, Vps27p also has a FYVE domain and binds PtdIns 3-phosphate (92). *VPS27* is a class E vacuolar protein sorting gene (113) which displays significant homology to the mammalian gene Hrs1. Vps34p should therefore be seen as one of the key mediators of trafficking within the endocytic system. Consistent with this is the finding that yeast *vps34* mutants also exhibit a defect in endocytosis (114). There are also data supporting functions for its mammalian homologue in the endocytic (115, 116) and autophagic (117) pathways.

Biosynthetic and endocytic traffic to the vacuole converges at the endosome and it seems likely that a process of maturation occurs prior to endosome fusion with the vacuole. Mutants such as *vam3* which block fusion at the vacuole accumulate large structures which resemble endosomes and hence small vesicular carriers are unlikely to mediate transport from endosomes to the vacuole. Part of the endosomal maturation process will be to retrieve Vps10p (and other proteins) back to the late-Golgi so that further rounds of CPY sorting can occur. This retrieval has recently been shown to require the function of at least six *VPS* genes which fall in classes A, B, and F. Initial studies revealed that three class A *VPS* mutants, *vps29*, *vps30*, and *vps35* all displayed a phenotype indentical to *vps10* mutants. They mis-sorted and secreted CPY whilst PrA was delivered to the vacuole and matured correctly (118). This cargo-specific defect hinted that the gene products of *VPS29*, *VPS30*, and *VPS35* may somehow regulate Vps10p function. Localization of Vps10p in these mutants revealed the cause of the CPY sorting defect, Vps10p was mis-localized to the vacuolar membrane. Epistasis experiments with class E mutants or *pep12* mutants placed Vps35p function at the endosome consistent with a role in Vps10p retrieval from the endosome to the Golgi (118).

Subsequent studies showed that Vps35p and Vps29p interact together with three other proteins, the products with the Vps5p, Vps17p, and Vps26p genes (119). Vps5p and Vps17p have been shown to stably associate and are important for Vps10p localization (120), however, *vps5* and *vps17* are class B mutants (121) and have somewhat different phenotypes to the class A mutants *vps35* and *vps29*. Class B mutants mis-sort significant amounts of PrA and have fragmented vacuoles (73, 74). *vps26* is a class F mutant which has a phenotype intermediate to that of the class A and class B mutants (74). Vps5p is homologous to mammalian sorting nexin-1 (SNX1) (120) which has been shown to associate with the cytoplasmic tail of transmembrane receptors such as EGF receptor (122, 123). Epitope tagged Vps5p was localized to the endosomal membrane, again consistent with a role in the retrieval of Vps10p. The protein complex containing Vps35p, Vps29p, Vps26p, Vps5p, and Vps17p was designated the 'retromer' complex due to its role in endosome to Golgi retrieval (119). Why would there be phenotypic differences between the *vps35* and *vps5* mutants? One possibility is that within the complex, Vps35p and Vps5p have different functions. Vps35p is a peripheral membrane protein that co-localizes with Vps10p, even in a *vps29* mutant (118). Furthermore, genetic evidence has revealed alleles of *vps35* that affect the localization of another late-Golgi protein dipeptidylamino peptidase I (DPAPA) but not Vps10p localization (82). This strongly argues for Vps35p playing a cargo-selective role in retrieval. Vps5p on the other hand has been shown to self-assemble *in vitro* into large homogeneous oligomers which hints at Vps5p having a more structural role in retrieval (119). These characteristics, i.e. cargo selection and vesicle formation by self-assembly are the properties normally associated with vesicle-coat proteins. It seems likely therefore that phenotypic differences between *vps35* and *vps5* mutants can be explained by the different roles each protein plays within the complex. In the absence of Vps35p function some retrieval may occur although it will be inefficient due to a lack of positive cargo selection. Without Vps5p (or Vps17p) function there would be no retrieval and hence a more severe defect would be observed.

The identification of a protein complex that mediates retrieval from the endosome to the Golgi puts an important piece into the vacuole protein sorting pathway jigsaw but there are a few pieces still missing. For example, it is not known whether a small GTPase is required for retromer complex function. If this complex is indeed a vesicle coat then a role for a small ARF-like GTPase in regulating membrane association would be predicted. There are also questions to be addressed regarding what fusion machinery in required for endosome to Golgi retrieval. It is presumably different from the fusion machinery required for endosomal delivery of cargo such as CPY from the Golgi. If proteins can be retrieved from the endosome to the Golgi is it possible that retrieval can also occur from the vacuole? It would clearly be advantageous if membrane proteins such as the endosomal v-SNARE which mediates fusion with the vacuole could be recycled back to the endosome (or the Golgi) for further rounds of endosome–vacuole fusion. Recent studies have suggested that not only is this retrieval pathway possible but that the endosomal v-SNARE Vti1p does in fact recycle between the vacuole and Golgi via the prevacuolar endosome. Bryant *et al.*, (124) used a reporter protein of ALP carrying the DPAPA retrieval sequence to demonstrate that upon arrival of the DPAPA-ALP chimera in the vacuole (which was monitored by proteolytic clipping of ALP) the molecule could be retrieved from the vacuole to the prevacuolar compartment. This retrieval was dependent upon the product of the *VAC7* gene which has been implicated in vacuolar inheritance (125). Furthermore, in a *vac7* mutant, Vti1p was also mis-localized to the vacuolar membrane indicating that it normally recycles out of the vacuole back to the prevacuolar compartment. This finding throws into question the notion that the yeast vacuole is a terminal compartment and hints that like mammalian lysosomes, the yeast vacuole may be quite a dynamic organelle.

8. Delivery of newly synthesized proteins to mammalian lysosomes

In mammalian cells the bulk of newly synthesized lysosomal hydrolases are delivered to the lysosome after binding to the MPRs in the TGN. The main trafficking route of MPRs appears to involve recruitment into AP-1 containing clathrin-coated pits at the TGN, vesicle traffic and fusion with early endosomes, followed by traffic to late endosomes, before return to the TGN (126–129). At first sight the suggestion that lysosomes fuse with late endosomes to create a hybrid organelle containing an efficient proteolytic environment is difficult to reconcile with the need to recycle MPRs from late endosomes to the TGN. However, in HEp-2 cells where a detailed kinetic study has been carried out, it was found that the accumulation of the cation-independent MPR does not occur in the late endosome when the ligand dissociates, but that much of the MPR is retrieved before the late endosomes fuse with lysosomes (129), i.e. as in yeast there appears to be endosomal maturation before fusion with the lysosome

(39). Thus, only a small proportion of the MPR will be exposed to the proteolytic environment of the hybrid as the bulk of newly synthesized hydrolases are delivered to the hybrid and subsequently to the re-formed lysosome.

The delivery of newly synthesized lysosomal membrane glycoproteins (lgps, LAMPs, and LIMPs) to lysosomes is less well understood than the delivery of mannose 6-phosphate tagged hydrolases. Except for lysosomal acid phosphatase which is delivered first from the TGN to the cell surface and then via endocytosis to lysosomes (see Chapter 7), lgps are thought to traffic mostly via an intracellular route from the TGN. The two different routes are reflected in the half-times of delivery of newly synthesized lgps from the TGN to lysosomes which for lysosomal acid phosphatase is ~ 6 hours (as a result of multiple rounds of recycling between early endosomes and the cell surface before delivery to lysosomes) and for other lgps 0.5–2 hours (130–132). Two cytosolic tail motifs have been identified in lgps as mediating lysosomal targeting. One of these, the GYXXØ motif (where X is any amino acid and Ø is a bulky hydrophobic amino acid) contains the YXXØ motif known to be important for interaction with the μ2 subunit of the AP-2 adaptor and endocytosis via clathrin-coated pits. The other is a di-leucine motif. One possible intracellular route is for the lgps to exit the TGN in AP-1 associated clathrin-coated vesicles (CCVs) and traffic with the MPR to endosomal compartments. There is some evidence to support this hypothesis. First, the GYQTI motif (present in LAMP-1) interacts with μ1 subunit of AP-1 in the yeast two-hybrid system (133). Secondly, LAMP-1 has been visualized with MPR in CCVs budding from the TGN (134). However, a subsequent study analysing TGN-derived vesicles found that LAMP-1 and LAMP-2 were in separate vesicles from those containing MPR and AP-1 (135).

9. The AP-3 route to lysosomes and lysosome-like organelles

Recently, a novel heterotetrameric complex with homology to the adaptor complexes, AP-1 and AP-2 has been identified, and named AP-3 (136–139). The first components of the AP-3 coat identified were two μ subunits, a brain-specific isoform μ3B, originally named p47B, and a ubiquitous isoform μ3A, originally named p47A (140). A brain-specific homologue β3B, of the AP-2 β subunit, was also identified and initially named β-NAP (141). Later a ubiquitous homologue named β3A was identified by screening an EST database (137, 139). The other two subunits named σ3 and δ (the homologue of the α subunit of AP-2) were also identified by screening an EST database (137, 138). Formation of a functional complex by these proteins was shown by demonstrating that the various subunits co-immunoprecipitate (137). On the basis that the proteins form a functional complex they were renamed β3A, μ3A (p47A), σ3, and δ for the ubiquitous form and β3B (β-NAP) and μ3B (p47B) for the brain-specific isoforms (137). The function and localization of the AP-3 complex are still poorly understood. Using antibodies to the β3B subunit, EM studies have demonstrated that AP-3 is associated with the TGN, but it does not co-localize with AP-1 (136). AP-

3 is thus a candidate for a non-AP-1 coat that might bud from the TGN and transport lysosomal glycoproteins to the lysosome. However, AP-3 has also been localized to the tubular portion of early endosomes (142). Whether AP-3 is clathrin associated is still a matter of debate. Immunolocalization by both confocal microscopy and EM revealed that there was very little co-localization between AP-3 and clathrin (136, 137) and the brain-specific isoform, β3B was found on a different population of vesicles to CCVs (141). However, Dell'Angelica *et al.* (142) found that AP-3 co-localized with clathrin on sorting endosomes and found that a fusion protein of glutathione S-transferase with part of β3A could bind clathrin heavy chain in an affinity column.

A homologous AP-3 complex in yeast is encoded by the genes *APL5*, *APL6*, *APM3*, and *APS3*. Characterization of various *vps* mutants but specifically the class D mutants *vps45* and *pep12* mutants indicated that ALP was being transported to the vacuole by a different pathway to that of CPY (83, 143). Further evidence for a separate ALP pathway came from experiments in which the cytoplasmic tail of CPS was swapped with the tail of ALP. Attaching the tail of CPS to ALP rendered the chimera *PEP12* dependent in its transport to the vacuole, whilst attaching the tail of ALP to CPS had the opposite effect (144). These observations led to a screen for mutants defective in ALP trafficking to the vacuole. The mutants identified turned out to encode subunits of the yeast AP3 complex (145). Deletion of any of the AP-3 complex subunits resulted in a severe defect in trafficking of not only ALP to the vacuole but also the vacuolar t-SNARE Vam3p (145, 146). Mutational studies on *VAM3* later showed that its sorting into the AP-3 pathway was dependent upon a specific di-leucine-type signal which when mutated resulted in Vam3p being mis-sorted into the CPY pathway (106). A similar di-leucine signal appears to mediate the AP-3-dependent sorting of ALP. Interestingly, AP-3 mutants did not totally block ALP transport to the vacuole but instead resulted in a pronounced kinetic delay. It was subsequently shown that in the absence of AP-3 function, ALP (and also presumably Vam3p) could reach the vacuole by the CPY pathway in a *PEP12* dependent fashion (145). Whilst biosynthetic traffic from the yeast Golgi appears to diverge, with CPY and ALP taking different routes to the vacuole, both pathways utilize the same machinery for the fusion of vesicle with the vacuole. For instance, arrival of both CPY and ALP at the vacuole requires the function of Ypt7p and the class C *VPS* genes along with Vam3p and Vam7p (100, 101). Possibly acting together with AP-3 in the sorting/transport of ALP is the Vps41 protein. This protein has recently been implicated in the formation of AP-3 vesicles and has been shown to bind to the carboxy terminus of the Apl5p protein (147). Vps41p may play a dual role in membrane traffic since it (and Vps39p) has also been implicated homotypic vacuole fusion (148, 149). Since clathrin has been shown not to be involved in ALP trafficking, yeast AP-3 may utilize Vps41p (possibly with Vps39p) to promote vesicle formation. Intriguingly, Vps41p and its homologues in *Caenorhabditis elegans*, *Drosophila melanogaster*, and various plant species contain a clathrin heavy chain homology region which is thought to be involved in oligomerization and formation of the clathrin lattice, raising the possibility that in an AP-3 vesicle coat Vps41p carries out the functions performed by clathrin in AP-1 and AP-2 coats.

Mutations in the AP-3 subunits have been identified in *Drosophila melanogaster*, mice, and man. The *garnet* gene of *Drosophila* has been found to encode the *Drosophila* homologue of the δ subunit of AP-3 (137, 150, 151). *Drosophila* with mutations of this gene have decreased quantities of both brown and red pigment in eye pigment granules. Similarly, the *pearl* mutation in mice has been shown to result from a defect in the β subunit of AP-3 and the *mocha* mutation from a defect in the δ subunit of AP-3. Mice with either of these mutations have reduced skin colour, again suggesting a defect in the biogenesis of pigment granules/melanosomes. *Mocha* and *pearl* mice also exhibit prolonged bleeding times due to decreased numbers of platelet dense granules (152). Dell'Angelica *et al.* (153) screened 20 Hermansky-Pudlak patients by indirect immunofluorescent analysis of their fibroblasts. Two related patients were found to have a mutation in the β3A subunit.

Since the yeast vacuole, melanosome, and platelet dense granules are are all lysosome-related organelles, it has been inferred that AP-3 plays a role in intracellular targeting of newly synthesized lgps from the TGN to lysosomes. Fibroblasts transfected with antisense oligonucleotides to μ3A or fibroblasts from individuals lacking β3A have reduced quantities of functional AP-3 (153, 154). Both these cells show an increased trafficking of LAMP-1 via the cell surface detected as more efficient endocytosis of anti-LAMP-1 antibodies. The absence of functional AP-3 in cells does not grossly affect the steady-state localization of LAMP-1, LAMP-2, LIMP-II, or CD63 to the lysosome (153, 154).

Although a fourth adaptor complex, AP-4, with subunits homologous to those of AP-1 and AP-2 has been discovered its function is unknown (155, 156). It is expressed ubiquitously but at low abundance. Interactions of the μ4 subunit with lgp tails in a yeast two-hybrid system has led to speculation about a role in delivery to lysosomes but there is, as yet, no corroborative evidence for this.

10. Autophagy

Whilst, in mammalian cells, the endocytic pathway is the major route of delivery of macromolecules from the extracellular milieu to a site for digestion by lysosomal enzymes, it is not the only process delivering proteins for digestion. Four methods by which endogenous intracellular proteins are delivered to lysosomes for digestion have been described. These are crinophagy in which secretory granules or regions of the ER fuse with lysosomes (157), carrier-mediated transport of selected cytosolic proteins bearing the motif KFERQ across the lysosomal membrane (158), microautophagy in which small portions of cytoplasm are sequestered by lysosomes (159), and macroautophagy. This last process, which is responsible for the breakdown of a large number of cellular proteins, occurs in response to serum deprivation of cells and is particularly sensitive to levels of a regulatory group of amino acids which includes leucine, glutamine, proline, methionine, histidine, tryptophan, tyrosine, and phenylalanine (160). Autophagosomes appear rapidly, 10 minutes after withdrawal of serum (160, 161). The origin of autophagosomes is controversial, one hypothesis being that they are derived from ribososome-free regions of the rough ER as they lack endocytic and Golgi markers but contain proteins of the rough ER (161, 162). An alternative hypothesis suggests that they are formed by a specialized membrane structure called the phagophore which has a multilamellar morphology (163, 164) and contains complex oligosaccharides indicating that it originates from a post-Golgi source (165–167). Autophagosomes mature into autolysosomes after 15 minutes (168) by acquiring proton pumping ATPase(s), lgps, and acid hydrolases (169). This appears to be a multistep process with late autophagosomes containing lgps in their outer membranes but, unlike autolysosomes, lacking acid hydrolases (168, 170). It has been suggested that vesicles containing lgps but not acid hydolases fuse with the maturing autophagososome (171). There is morphological evidence to suggest that lysosomes, not endosomes, fuse with autophagosomes (169) and it appears that acquisition of a v-H^+ ATPase may be a prerequisite for fusion to lysosomes (172). Progress in understanding the molecular mechanism of microautophagy and macroautophagy in mammalian cells has been slow. In contrast there has recently been a great deal of progress in identifying gene products required for autophagy in yeast and this offers the future prospect of identifying mammalian homologues (173, 174).

Autophagy in yeast provides a survival mechanism that allows the cell to continue to synthesize essential proteins in the absence of external nutrients. Non-selective uptake of cytoplasmic proteins into the vacuole and their subsequent degradation releases amino acids that can be used for the synthesis of essential proteins. Mechanistically autophagy can be broken down to five discrete stages (Fig. 4):

(a) Starvation signalling. The cell must have a mechanism that is able to sense a lack of certain nutrients and, in response, induce autophagy.

(b) Autophagosome formation. This is the non-selective entrapment of cytoplasmic proteins by a double layered membrane which forms a sealed vesicle.

(c) Targeting to the vacuole. The autophagosome is targeted to and docks with the vacuole.

(d) Fusion and release of the autophagosome with the vacuole. The outer membrane of the autophagosome fuses with the vacuole releasing a single membrane-bound autophagic body into the lumen of the vacuole.

(e) Vacuolar lipases/proteases digest the autophagic body releasing lipid and amino acids that can leave the vacuole and be incorporated into newly synthesized protein.

Genetic screens in *S. cerevisiae* have been applied to identify the genes required for autophagy and recent analysis of these genes has greatly increased our understanding of this process (reviewed in ref. 174) Whilst autophagy is an induced response brought about by nutritional starvation, there is in yeast a constitutive pathway that targets certain cytoplasmic proteins to the vacuole (reviewed in ref. 175). Studies on the yeast API (amino-peptidase I enzyme) showed that it was matured in the vacuole like a conventional vacuolar hydrolase but did not require the secretory pathway for its delivery to the vacuole (176). This novel route taken by API to reach the vacuole was dubbed Cvt or 'cytoplasm to vacuole transport'. Subsequent experiments have

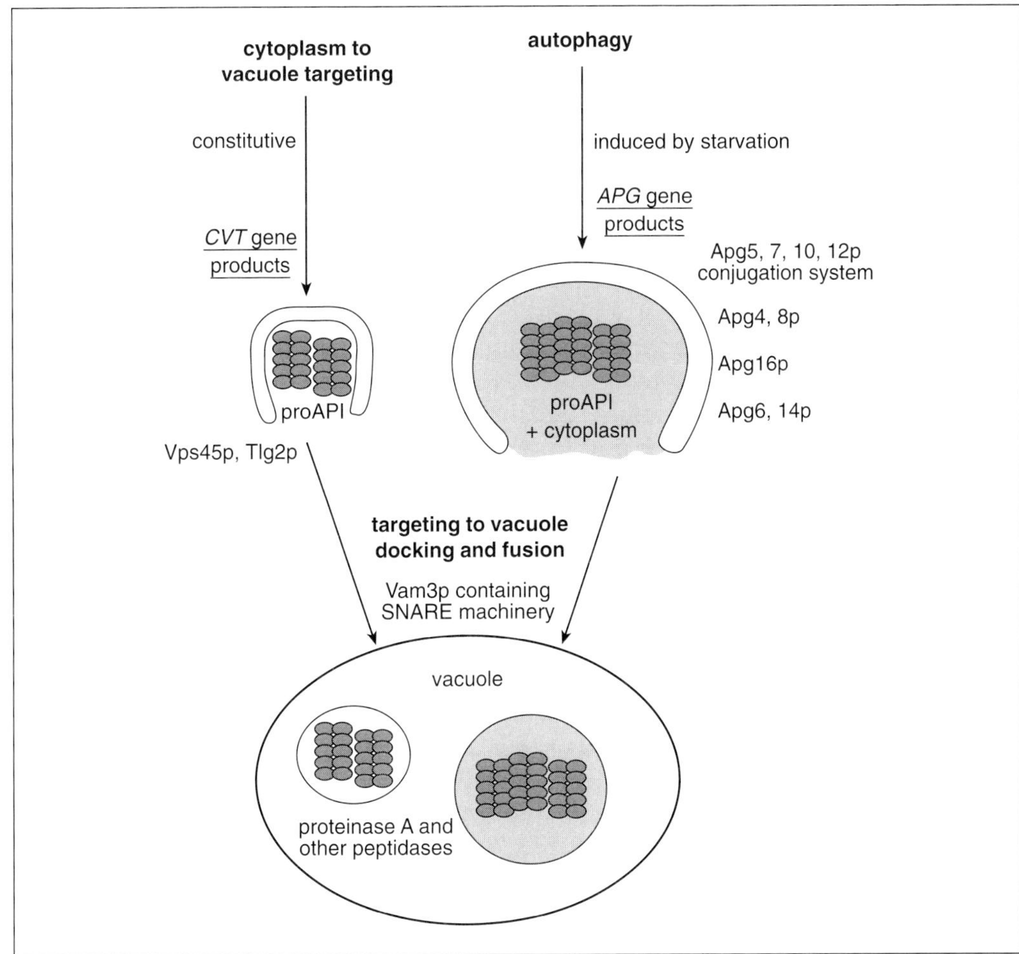

Fig. 4 Autophagy and the Cvt pathway. In the Cvt pathway, proAPI is enwrapped by a double membrane bilayer that forms around oligomeric clumps of proAPI in the cytoplasm. This process requires Vps45p and the t-SNARE Tlg2p. Autophagy is very similar but does not appear to require Vps45p and Tlg2p. Mechanistically the Cvt pathway is very similar to autophagy the key difference being that autophagy is induced and non-selective, whilst the Cvt pathway is constitutive and only a discrete set of proteins are taken up into the vacuole this way. Once the enwrapping process is complete, the Cvt vesicle or autophagosome will fuse with the vacuole. There is much overlap between the components required for autophagy and those functioning in the Cvt pathway. Crucially, both pathways converge at the vacuole and share similar docking and fusion machinery with the t-SNARE Vam3p at the heart of the process to deliver Cvt vesicles or autophagosomes to the vacuole. Delivery to the vacuole is followed by proteolytic digestion.

revealed much about the Cvt pathway and have shown that mechanistically it is very similar to autophagy (177, 178). The key difference is that whilst autophagy is induced and results in non-selective degradation of cytoplasmic proteins, the Cvt pathway is constitutive and to date the only two known proteins believed to be transported to the vacuole this way are α-mannosidase (179) and API. Transport of proAPI to the vacuole occurs by the entrapment of proAPI within a double membrane vesicle which forms within the cytoplasm (180). Biochemical and EM studies have shown that proAPI oligomerizes (181) to form an electron dense clump within the cytoplasm which is then enwrapped with a double membrane structure that appears to completely encircle the dense proAPI whilst also excluding cytoplasmic proteins. Subsequently the proAPI is targeted to the vacuole. Docking and fusion release the proAPI still surrounded by a single membrane bilayer into the vacuolar lumen. This series of events is remarkably similar to the process of autophagy. Indeed during the induced autophagic response, proAPI continues to be transported to the vacuole only this time in autophagic bodies (178). Electron micrographs have shown clumps of proAPI being surrounded by double membrane bilayers which are also enwrapping cytoplasmic proteins. A study of Cvt vesicles indicated that they have a mean diameter of approximately 150 nm but autophagic bodies are larger with a diameter of 400–900 nm (178).

Understanding the process of autophagy requires answers to many questions but perhaps the key problems to be solved are:

- What signalling occurs to induce the autophagic response?
- Where does the membrane required to enwrap the cytoplasmic proteins come from and how is this process driven?
- What regulates the targeting of the autophagosome to the vacuole?

To address these questions, screens have been undertaken in *S. cerevisiae* to uncover mutants that are defective in either autophagy (182, 183) or the Cvt pathway (184). Given the striking similarites in the apparent mechanisms of autophagy and the Cvt pathway it is perhaps not surprising to find very significant overlap between genes required for these processes (185).

Cloning of the genes that are defective in the autophagy mutants has revealed much about the mechanisms that are involved. For instance the product of the *APG*1 gene is a predicted serine/threonine kinase (186). Apg7p has homology to the ubiquiting activating enzyme E1 (187) and is part of a novel protein conjugation system that also involves Apg12p, Apg10p, and Apg5p (188, 189). Apg12p is a small protein tag that is conjugated onto Apg5p via firstly conjugation onto Apg7p and then conjugation to Apg10p. It is not known what role the Apg5p/Apg12p conjugate plays in autophagy or whether the Apg7p/Apg12p or Apg10p/Apg12p conjugates have any other function beyond being intermediates to forming Apg5p/Apg12p, but what is clear is that all four of these proteins are essential for autophagy. Recent studies have shown that the Apg5p/Apg12p conjugate requires the product of the *APG16* gene to function (190). Apg5p has been found to be necessary for sequestration of protein into the Cvt and autophagy pathways (191), hence it seems likely that this novel protein conjugation system is vital for autophagosome formation. Indeed no autophagosomes are observed in *apg5*, *apg7*, *apg10*, or *apg12* mutants.

As there is significant overlap between the genes required for the Cvt pathway and autophagy, it should be expected that most of the proteins required for autophagy would be expressed constitutively. This is indeed the case with the exception that the

APG8 gene is significantly induced upon switching to starvation conditions (192). Immunogold EM showed that Apg8p is localized to the membrane that enwraps the cytoplasmic proteins to form the autophagosome (192) and is delivered to the vacuole with the autophagosome, becoming localized to the vacuolar lumen. Whether Apg8 facilitates the recruitment of cytosolic proteins into autophagosomes remains unclear. Apg8p exists in a complex that also contains Apg4p (193) and Apg4p interacts with tubulin (193). In *apg8Δ* or *apg4Δ* cells autophagosomes accumulated in the cytoplasm due to a failure to target to and fuse with the vacuole. However, the role of microtubules in autophagosome formation remains controversial (192).

The products of the *APG6* and *APG14* genes also form a membrane-associated protein complex (194). Both Apg6p and Apg14p are required for autophagy but overexpression of Apg14p can partially supress an *apg6* mutant. Curiously, Apg6p is also required for vacuolar protein sorting. *APG6* is allelic to the *VPS30* gene (194). *VPS30* has been shown to be required for the proper localization of Vps10p and is believed to function in the retrieval of Vps10p from the endosome to the Golgi (118). Deletion of *APG14* had no effect upon vacuolar protein sorting suggesting that whilst Apg6p/Vps30p may have a dual role in both autophagy and vacuolar protein sorting, the *APG14* gene product functions solely in autophagy (194). A mammalian homologue of Apg6p/Vps30p has also recently been identified and has been named beclin because of its ability to interact with the pro-apoptosis protein Bcl-2 (195). Human beclin expressed in *apg6Δ/vps30Δ* yeast cells was able to complement the autophagy defect but not the vacuolar protein sorting phenotype (196).

Studies on autophagosome formation are currently at an early stage and in spite of the identification of genes whose products function in autophagosome biogenesis there is no clear picture yet of the molecular mechanisms that are required for autophagosome formation. However, later stages of autophagy in which the autophagosome fuses with the vacuole are now better understood. Analysis of the late acting *vps* mutants have revealed that many class C and class B* *VPS* genes function in the fusion of autophagosomes with the vacuole. The v-SNARE Vam3p is required for autophagy and in *vam3* mutants autophagosomes accumulate in the cytoplasm (100). There is a similar autophagy defect in *vps18* mutants (104) and it seems reasonable to expect most, if not all, late acting *VPS* genes required for the fusion of endosomes with the vacuole will also be required for autophagosome fusion with the vacuole. Indeed some of the late acting *VPS* genes have also been isolated as *cvt* mutants. This finding may be significant with respect to the origin of membrane for autophagosome formation. Given that there is a requirement for Vam3p in autophagosome/vacuole fusion it is reasonable to describe that fusion event as SNARE-mediated. This being the case, there should also be a t-SNARE on the autophagosome that pairs with Vam3p. The best candidate would be a SNARE that normally pairs with Vam3p for trafficking to the vacuole or vacuole/vacuole fusion. Candidates for this t-SNARE are Vti1p (99) and Nyv1p (103). Presently, however, there is no evidence to suggest that either Vti1p or Nyv1p are required for autophagosome fusion with the vacuole.

Another fusion step that appears to be SNARE mediated is the fusion of membranes that form around proAPI during packaging of proAPI into vesicles on the Cvt pathway. Recent evidence has indicated that the t-SNARE Tlg2p and the Sec1p homologue Vps45p function at the proAPI packaging/enwrapping stage in the Cvt pathway (197). Thus in a *tlg2* or a *vps45* mutant proAPI is not matured in the vacuole. Surprisingly, however, when autophagy is induced, proAPI is now matured in the vacuole indicating that Tlg2p and Vps45p are not essential in the autophagy pathway. Tlg2p has been localized to the endocytic and late-Golgi compartments of yeast (198, 199) and these data hint that endosomal or late-Golgi membranes provide the membrane for the formation of the Cvt vesicle. Autophagy may also utilize endosomal membranes in autophagosome formation as recent data has shown that the integral membrane protein Apg9p partially co-localizes with endosomal marker proteins (200). Identification of the SNARE molecules that perform the enwrapping step in autophagy or other markers of the autophagic membrane would be a major step towards identifying the membranes that are used in autophagy.

11. The impact of the genetics of multicellular organisms on undestanding lysosome biogenesis

Although yeast genetics has, to date, made the greatest impact on our knowledge of the proteins required for protein targeting to lysosomes, the genetics of multicellular organisms, in particular the fruitfly, mouse, and man is likely to provide equally important insights over the next few years. This is because there is now excellent genetic evidence showing that the biogenesis of pigment granules in the *Drosophila* eye and melanosomes in mouse and man require the same gene products as those needed for lysosome biogenesis. Eye colour mutants in *Drosophila* and mouse coat colour mutants have been known and studied throughout the twentieth century. Over 80 mutations affecting eye colour in *Drosophila* have been described and whilst some of the eye colour genes encode pigment synthesis enzymes and membrane transporters, a substantial group known as the 'granule group' function in the delivery of proteins to lysosomes and pigment granules (151). As discussed above, the first of these to be established with such a function was *garnet*, encoding the *Drosophila* homologue of the AP-3 δ (137). Since then it has been shown that the *light* gene is the *Drosophila* homologue of yeast *VPS41*, *carnation* the *Drosophila* homologue of *VPS33*, and *deep orange* the *Drosophila* homolgue of *VPS18* (151, 201). The protein products of *deep orange* and *carnation*, like those of *VPS18* and *VPS33* (104) function in large protein complexes required in traffic from late endosome-like to lysosome-like organelles. A further *Drosphila* mutant, *hook*, originally identified on the basis of a bristle phenotype encodes a 679 amino acid cytoplasmic protein that has a mammalian homologue and appears to function late in the endocytic pathway such that in mutant tissue there is a reduction in number of MVBs and endocytosed ligands are prematurely degraded in lysosomes (202, 203), suggesting that the wild-type hook protein acts to inhibit delivery to lysosomes.

Melanosomes may be grouped into a class of lysosome-like organelles known as secretory lysosomes, although this terminology may appear somewhat confusing in view of recent evidence that Ca^{2+} regulated fusion with the plasma membrane may be a property of lysosomes in all cell types (204). Secretory lysosomes are, in particular, found in cells of the haemopoietic lineage. The identification of genes associated with abnormalities in secretory lysosome function may also provide insights into the biogenesis and function of 'normal' lysosomes (for review see ref. 4). In human Chediak-Higashi syndrome (CHS) and the mouse model of the disease, the *beige* mouse (due to its coat colour), all lysosomes are abnormally enlarged but only cells with secretory lysosomes are functionally impaired. The CH gene encodes a cytosolic protein of ~ 400 kDa,designated LYST for lysosomal trafficking regulator protein. Overexpression of LYST in mutant fibroblasts resulted in abnormally small lysosomes suggesting a role for the protein in lysosome fission (205). LYST shows limited sequence homology with yeast Vps15p, the regulatory subunit of the yeast Vps34p PtdIns 3-kinase. Hermansky-Pudlak syndrome (HPS) was mentioned above since in a small subset of patients with this disease there is a deficiency of the AP-3 δ subunit (HPS2). Like CHS, HPS is an autosomal recessive disease characterized by abnormalities of pigmentation, blood clotting (resulting from platelet storage granule deficiency), and other systemic abnormalities associated with defective lysosomal function. The most common form of HPS has a mouse model, the *pale ear* mouse. Both *pale ear* and HPS1 genes have been cloned and HPS1 predicted to encode a 79 kDa protein (206). Interestingly, there are more than a dozen other coat colour mutants of mice that display HPS-like phenotypes all of which may provide further insights into proteins required for lysosome biogenesis (207). Recently, one of these, *gunmetal*, has been identified as a Rab geranylgeranyltransferase (208). It is also interesting to note recent data showing the interaction of pallidin, the protein encoded by the *pallid* gene, with syntaxin13, a t-SNARE implicated in early endosome fusion (209).

A final group of genetic diseases which may lead to greater insights into lysosome biogenesis and function are the mammalian lysosomal storage diseases. Further study of these may lead to a better understanding of the functional and morphological relationships of lysosomes, late endosomes, and the late endosome/lysosome hybrid organelle. In Niemann-Pick C disease, for example, it has recently been shown that cholesterol accumulation is principally in late endosomes rather than lysosomes (210) and at a different site to the NPC1 protein (211). The latter is predicted to be a polytopic membrane spanning protein with a sterol sensing domain that is presumably important in managing cholesterol homeostasis. It is not clear whether the NPC1 protein is in lysosomes, LAMP2 positive, MPR negative vesicles important in mediating retrograde traffic of lysosomal proteins to earlier parts of the endocytic pathway (212) and/or vesicles shuttling cholesterol from late endosomes/lysosomes to the TGN (213). The role played by the late endosome/lysosome hybrid organelle in sterol trafficking is not known. Other diseases where abnormalities of membrane traffic occur as a result of a change in the lumenal environment of late endosomes and lysosomes include sialic acid storage disease where the accumulation of acidic monosaccharides intereferes with the formation of lysosomes (214), mucolipidosis type IV disease where lipid efflux is impaired (215), and Gaucher's disease where the severe clinical phenotype is related to impaired intracellular transport of mutant acid β-glucosidase (216). What is also becoming clear from study of lysosomal storage disorders is that transcriptional regulation of the synthesis of lysosomal proteins including lgp's occurs and that lysosomal biogenesis in storage disorders is a regulated process, albeit poorly understood at present (217).

12. The effect of toxins on late endocytic organelles

Toxins have often been useful in dissecting membrane traffic pathways, with the clostridial neurotoxins which cleave SNAREs being particularly important in confirming the role of the SNARE complex in docking synaptic vesicles to the plasma membrane (218). *Helicobactor pylori,* the bacterium responsible for stomach ulcers produces a toxin VacA which causes the formation of vacuoles coated with Rab7 which are thought to be derived from late endosomes (219) and may be related to the late endosome/lysosome hybrid compartment (220). The target of this toxin is unknown. Possibly related vacuoles are formed after treatment of cultured cells with the fungal metabolite wortmannin, which is a potent inhibitor of some PtdIns 3-kinases including the mammalian homologue of Vps34p (221). Such toxins are likely to prove useful in the future in further understanding the biogenesis of late endocytic organelles. Similarly, the study of stable vacuoles in which parasites such as *Toxoplasma gondii* and *Mycobacterium tuberculosis* reside and replicate as well as the mechanisms of mobilization of lysosomes and formation of parasitophorous vacuoles containing *Trypanasoma cruzi* may also throw light on the formation and function of late endocytic organelles (see Chapter 11).

13. Conclusions

It is now clear that lysosomes and lysosome-like organelles, including the yeast vacuole, can no longer be regarded simply as the terminal degradative compartments of endocytic pathways. Lysosomes are, in fact, dynamic organelles which repeatedly fuse with late endosomes to form hybrid organelles in which digestion of endocytosed macromolecules takes place and from which lysosomes are re-formed. The outline of the molecular mechanisms of delivery of membrane proteins and lumenal contents to lysosomes by biosynthetic, autophagic, and endocytic pathways is becoming clear, in large part from studies of yeast mutants, but also, increasingly, from studies of mutations in other species and of cell-free systems. In the next few years we can look forward to a much fuller understanding of the molecular mechanisms of membrane traffic amongst late compartments of the endocytic pathway and better knowledge of how abnormalities in these pathways can contribute to human disease.

Acknowledgements

We thank the MRC and the Wellcome Trust for financial support for our experimental work which has led to some of the ideas discussed in this chapter.

References

1. Ferrari, G., Knight, A. M., Watts, C., and Pieters, J. (1997) Distinct intracellular compartments involved in invariant chain degradation and antigenic peptide loading of major histocompatibility complex (MHC) Class II molecules. *J. Cell Biol.*, **139**, 1433.
2. Kleijmeer, M. J., Morkowski, S., Griffith, J. M., Rudensky, A. Y., and Geuze, H. J. (1997) Major histocompatibility complex class II compartments in human and mouse B lymphoblasts represent conventional endocytic compartments. *J. Cell Biol.*, **139**, 639.
3. Page, L. J., Darmon, A. J., Uellner, R., and Griffiths, G. M. (1998) L is for lytic granules: lysosomes that kill. *Biochim. Biophys. Acta*, **1401**, 146.
4. Stinchcombe, J. C. and Griffiths, G. M. (1999) Regulated secretion from hemopoietic cells. *J. Cell Biol.*, **147**, 1.
5. Aniento, F., Emans, N., Griffiths, G., and Gruenberg, J. (1993) Cytoplasmic dynein-dependent vesicular transport from early to late endosomes. *J. Cell Biol.*, **123**, 1373.
6. Storrie, B. and Desjardins, M. (1996) The biogenesis of lysosomes: is it a kiss and run, continuous fusion and fission process? *BioEssays*, **18**, 895.
7. Bakker, A. C., Webster, P., Jacob, W. A., and Andrews, N. W. (1997) Homotypic fusion between aggregated lysosomes triggered by elevated $[Ca^{2+}]_i$ in fibroblasts. *J. Cell Sci.*, **110**, 2227.
8. Ward, D. M., Leslie, J. D., and Kaplan, J. (1997) Homotypic lysosome fusion in macrophages: analysis using an *in vitro* assay. *J. Cell Biol.*, **139**, 665.
9. Mullock, B. M., Bright, N. A., Fearon, C. W., Gray, S. R., and Luzio, J. P. (1998) Fusion of lysosomes with late endosomes produces a hybrid organelle of intermediate density and is NSF dependent. *J. Cell Biol.*, **140**, 591.
10. Klionsky, D. J., Herman, P. K., and Emr, S. D. (1990) The fungal vacuole: composition, function and biogenesis. *Microbiol. Rev.*, **54**, 266.
11. Klionsky, D. J. (1997) Protein transport from the cytoplasm to the vacuole. *J. Memb. Biol.*, **157**, 105.
12. De Duve, C., Pressman, B. C., Gianetto, R., Wattiaux, R., and Appelmans, F. (1955) Tissue fractionation studies. VI. Intracellular distribution patterns of enzymes in rat liver tissue. *Biochem. J.*, **60**, 604.
13. De Duve, C. (1983) Lysosomes revisited. *Eur. J. Biochem.*, **137**, 391.
14. Bowers, W. E. (1998) Christian de Duve and the discovery of lysosomes and peroxisomes. *Trends Cell Biol.*, **8**, 330.
15. Holtzmann, E. (1989). *Lysosomes*, p. 439. Plenum Press. New York, London.
16. Matteoni, R. and Kreis, T. E. (1987) Translocation and clustering of endosomes and lysosomes depends on microtubules. *J. Cell Biol.*, **105**, 1253.
17. Kornfeld, S. and Mellman, I. (1989) The biogenesis of lysosomes. *Annu. Rev. Cell Biol.*, **5**, 483.
18. Hopkins, C. R. (1983) The importance of the endosome in intracellular traffic. *Nature*, **304**, 684.
19. Hopkins, C. R., Gibson, A., Shipman, M., and Miller, K. (1990) Movement of internalized ligand-receptor complexes along a continuous endosomal reticulum. *Nature*, **346**, 335.
20. Trowbridge, I. S., Collawn, J. F., and Hopkins, C. R. (1993). Signal-dependent membrane protein trafficking in the endocytic pathway. *Annu. Rev. Cell Biol.*, **9**, 129.
21. Mullock, B. M., Perez, J. H., Kuwana, T., Gray, S. R., and Luzio, J. P. (1994) Lysosomes can fuse with a late endosomal compartment in a cell-free system from rat liver. *J. Cell Biol.*, **126**, 1173.

22. Schmid, S., Fuchs, R., Kielian, M., Helenius, A., and Mellman, I. (1989) Acidification of endosome subpopulations in wild-type Chinese hamster ovary cells and temperature-sensitive acidification-defective mutants. *J. Cell Biol.*, **108**, 1291.
23. Schmid, S. L., Fuchs, R., Male, P., and Mellman, I. (1988) Two distinct subpopulations of endosomes involved in membrane recycling and transport to lysosomes. *Cell*, **52**, 73.
24. Rodman, J. S. and Wandinger-Ness, A. (2000) Rab GTPases coordinate endocytosis. *J. Cell Sci.*, **113**, 183.
25. Mellman, I. (1996) Endocytosis and molecular sorting. *Annu. Rev. Cell Dev. Biol.*, **12**, 575.
26. Mukherjee, S., Ghosh, R. N., and Maxfield, F. R. (1997) Endocytosis. *Physiol. Rev.*, **77**, 759.
27. Knight, A., Hughson, E., Hopkins, C. R., and Cutler, D. F. (1995) Membrane protein trafficking through the common apical endosome compartment of polarised Caco-2 cells. *Mol. Biol. Cell*, **6**, 597.
28. Mostov, K. E. and Cardone, M. H. (1995) Regulation of protein traffic in polarized epithelial cells. *BioEssays*, **17**, 129.
29. Reaves, B. J., Banting, G., and Luzio, J. P. (1998) Lumenal and trans-membrane domains play a role in sorting type I membrane proteins on endocytic pathways. *Mol. Biol. Cell*, **9**, 1107.
30. Mallet, W. G. and Maxfield, F. R. (1999) Chimeric forms of furin and TGN38 are transported from the plasma membrane to the trans-Golgi network via distinct endosomal pathways. *J. Cell Biol.*, **146**, 345.
31. Roederer, M., Barry, J. R., Wilson, R. B., and Murphy, R. F. (1990) Endosomes can undergo an ATP-dependent density increase in the absence of dense lysosomes. *Eur. J. Cell Biol.*, **51**, 229.
32. Murphy, R. F. (1991) Maturation models for endosome and lysosome biogenesis. *Trends Cell Biol.*, **1**, 77.
33. Stoorvogel, W., Strous, G. J., Geuze, H. J., Oorschot, V., and Schwartz, A. L. (1991) Late endosomes derive from early endosomes by maturation. *Cell*, **65**, 417.
34. Griffiths, G. and Gruenberg, J. (1991) The arguments for pre-existing early and late endosomes. *Trends Cell Biol.*, **1**, 5.
35. Gu, F. and Gruenberg, J. (1999) Biogenesis of transport intermediates in the endocytic pathway. *FEBS Lett.*, **452**, 61.
36. Gruenberg, J. and Maxfield, F. R. (1995) Membrane transport in the endocytic pathway. *Curr. Opin. Cell Biol.*, **7**, 552.
37. Griffiths, G. (1996) On vesicles and membrane compartments. *Protoplasma*, **195**, 37.
38. Luzio, J. P., Rous, B. A., Bright, N. A., Pryor, P. R., Mullock, B. M., and Piper, R. C. (2000) Lysosome-endosome fusion and lysosome biogenesis. *J. Cell Sci.*, **113**, 1515.
39. Futter, C. E., Pearse, A., Hewlett, L. J., and Hopkins, C. R. (1996) Multivesicular endosomes containing internalized EGF-EGF receptor complexes mature and then fuse directly with lysosomes. *J. Cell Biol.*, **132**, 1011.
40. Bright, N. A., Reaves, B. J., Mullock, B. M., and Luzio, J. P. (1997) Dense core lysosomes can fuse with late endosomes and are re-formed from the resultant hybrid organelles. *J. Cell Sci.*, **110**, 2027.
41. Rothman, J. E. (1994) Mechanisms of intracellular protein transport. *Nature*, **372**, 55.
42. McBride, H. M., Rybin, V., Murphy, C., Giner, A., Teasdale, R., and Zerial, M. (1999) Oligomeric complexes link rab5 effectors with NSF and drive membrane fusion via interactions between EEA1 and syntaxin 13. *Cell*, **98**, 377.
43. Pfeffer, S. R. (1999) Transport-vesicle targeting: tethers before SNAREs. *Nat. Cell Biol.*, **1**, E17.

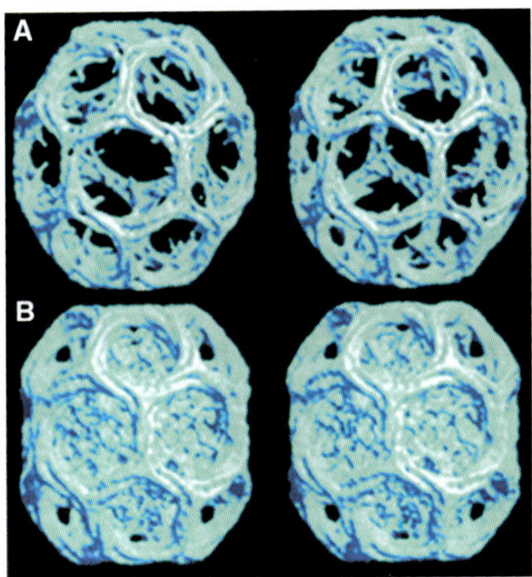

Plate 1 Clathrin baskets at 21 Å resolution. These images were generated by averaging electron micrograph images of (A) assembled clathrin and (B) clathrin co-assembled with purified AP complexes (15). The AP complexes are oriented inside the clathrin lattice, in contact with the clathrin N-terminal domains. The images are reproduced with permission from reference (15), copyright 1998, Oxford University Press.

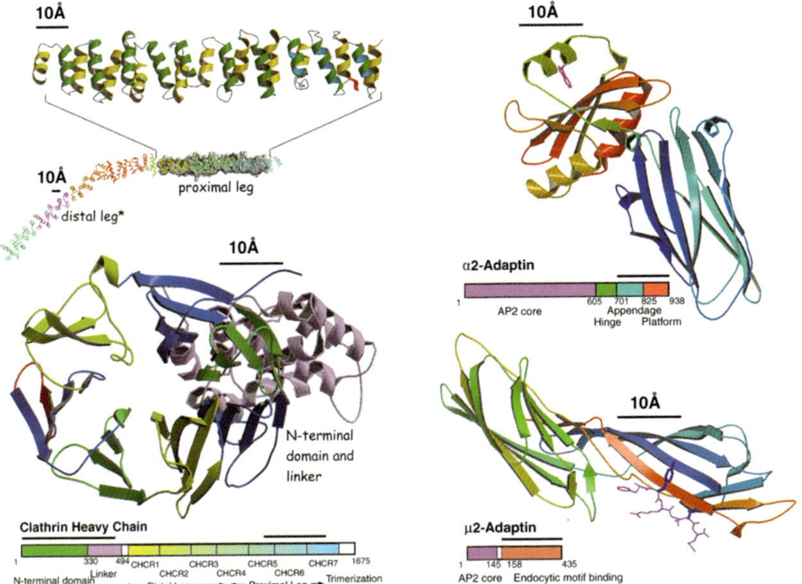

Plate 2 Crystallographic structures of portions of clathrin (*left*) and the AP-2 complex (*right*). The segment of each protein shown is indicated by the black line above a linear map of the protein sequence, with functional domains indicated. A scale bar is shown for each structure. These images were created from coordinates in the Protein Data Base for the review article by Wakeham *et al.* (93), which contains a more complete catalogue of structures of proteins involved in intracellular vesicle coat formation. The figures are reproduced with permission from reference (93), copyright 2000, Munksgaard International Publishers Ltd, Copenhagen, Denmark. We thank Diane E. Wakeham for preparation of this configuration. *Left* The top image is the proximal leg of the clathrin heavy chain (97), with the side chain of tyrosine 1477 shown in red. Tyrosine 1477 is near the predicted light chain binding region and is the target for phosphorylation by Src kinase during regulated endocytosis of epidermal growth factor (89). The solved structure reveals two repeated motifs (clathrin heavy chain repeats). These repeats are predicted to occur seven times in the clathrin triskelion leg and to account for the structure of the entire filamentous portion of the leg, as shown below the proximal leg structure. The alpha helical segments continue into a less ordered linker region, attached to the terminal domain structure shown in the bottom image (95). *Right* The top image is the structure of the appendage domain of the α subunit of the AP-2 complex, also called α2-adaptin (101, 102). The bottom image is the structure of the μ subunit of the AP-2 complex, also called μ2-adaptin, co-crystallized with a peptide containing an AP-2 binding motif. The peptide is shown in purple and the critical tyrosine side chain is highlighted in dark blue (101).

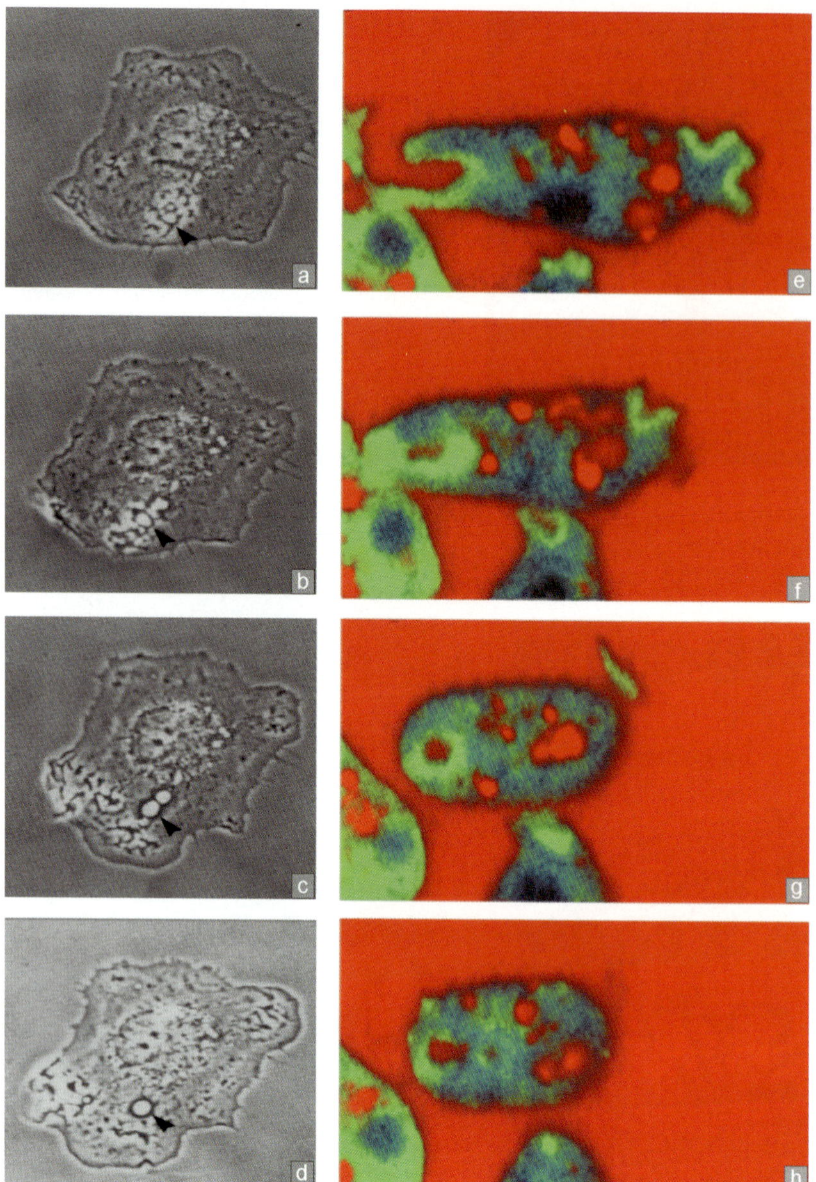

Plate 3 Time series of macropinocytosis. (a–d) Phase-contrast view of a macrophage. Frames are 65 μm wide and taken one minute apart. Reprinted from ref. 22. (a, b) The arrowhead indicates phase-dense ruffles. (c, d) The arrowhead points to newly formed macropinosome. (e–h) Single confocal section through a living *Dicytostelium* cell. Reprinted from ref. 5. The actin cytoskeleton is labelled with a GFP–coronin fusion protein (green). The medium contains a fluorescent tracer (red). Frames are 28 μm wide and were taken at 40 second intervals. (e, f) A membrane protrusion at the lower edge of the cell is driven by the actin cytoskeleton. (g, h) Uncoating of the newly formed macropinosome.

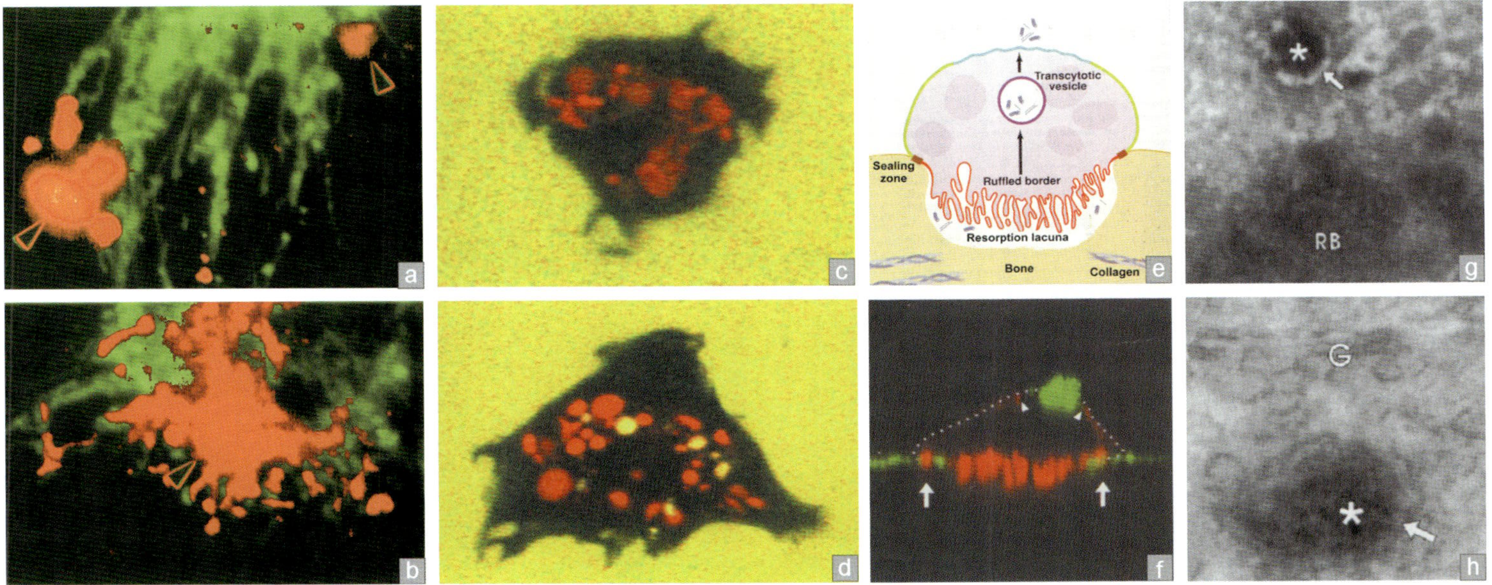

Plate 4 Fate of the macropinosome in professional phagocytes. (a, b) Macrophage with a pre-labelled tubular lysosome (green) receiving a pulse of red medium. Frame 22 μm wide. Modified after ref. 76. (a) Freshly formed macropinosome (red) in close proximity with a labelled lysosome. (b) Old macropinosome after merging with tubular lysosome. (c, d) Confocal section through living *Dictyostelium* cells in pH indicator medium. Neutral pH is indicated by yellow colour. Acidic endosomes appear in red. Frame 16 μm wide. Taken from ref. 83. (c) Acidic early phase of the endocytic pathway. (d) Cell after prolonged incubation. Some endosomes are neutralized indicating late stages of the endocytic pathway. The brightness allows to distinguish late stages (intense yellow colour) from nascent macropinosomes (dim yellow stain). (e–h) Bone resorption in the osteoclast. Reprinted with modifications from ref. 89. (e) Schematic drawing of an osteoclast attached to bone (yellow). Dissolved bone materials are internalized at the ruffled border and exocytosed at the dorsal side of the cell. (f) Confocal side view of an osteoclast labelled for transcytosed material (green) and filamentous actin (red). Arrows point to the sealing zone formed between plasma membrane and bone surface. Frame is 50 μm wide. (g, h) Transmission electron micrograph of thin sections through osteoclasts showing transcytotic vesicles (asterisk). (g) The membrane of the endocytic vesicle is not tightly associated with the internalized material suggesting that no sequential receptor–particle interactions (zippering) occur during uptake. RB, ruffled border. Frame is 3.3 μm wide. (h) Endocytic vesicles are large (above 2 μm in diameter) and contain amorphous rather than solid material, suggesting that dissolved bone material is internalized through macropinocytosis rather than through phagocytosis. G. Golgi apparatus. Frame is 5.4 μm wide.

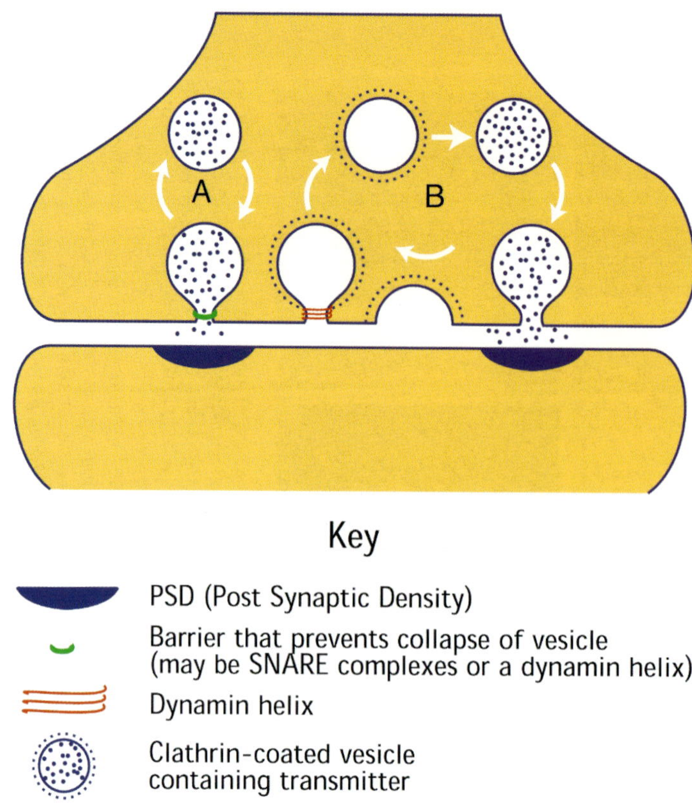

Plate 5 Mechanisms of synaptic vesicle recycling. (A) Transmitter release at the active zone may occur through the transient opening of a fusion pore (the 'kiss-and-run' hypothesis). (B) Transmitter release may also occur with the complete collapse of the vesicle into the plasma membrane followed by retrieval at sites to the edge of the active zone (the clathrin-mediated pathway).

44. Christoforidis, S., Miaczynska, M., Ashman, K., Wilm, M., Zhao, L., Yip, S-C., *et al.* (1999). Phosphatidylinositol-3-OH kinases are Rab5 effectors. *Nat. Cell Biol.*, **1**, 249.
45. Simonsen, A., Lippe, R., Christoforidis, S., Gaullier, J. M., Brech, A., Callaghan, J., *et al.* (1998) EEA1 links PI(3)K function to Rab5 regulation of endosome fusion. *Nature*, **394**, 494.
46. Stenmark, H. and Aasland, R. (1999) FYVE-finger proteins-effectors of an inositol lipid. *J. Cell Sci.*, **112**, 4175.
47. van Deurs, B., Holm, P. K., Kayser, L., and Sandvig, K. (1995). Delivery to lysosomes in the human carcinoma cell line HEp-2 involves an actin filament-facilitated fusion between mature endosomes and preexisting lysosomes. *Eur. J. Cell Biol.*, **66**, 309.
48. Prekeris, R., Klumperman, J., Chen, Y. A., and Scheller, R. H. (1998) Syntaxin 13 mediates cycling of plasma membrane proteins via tubulovesicular recycling endosomes. *J. Cell Biol.*, **143**, 957.
49. Advani, R. J., Yang, B., Prekeris, R., Lee, K. C., Klumperman, J., and Scheller, R. H. (1999) VAMP-7 mediates vesicular transport from endosomes to lysosomes. *J. Cell Biol.*, **146**, 765.
50. Prekeris, R., Yang, B., Oorschot, V., Klumperman, J., and Scheller, R. H. (1999) Differential roles of syntaxin 7 and syntaxin 8 in endosomal trafficking. *Mol. Biol. Cell*, **10**, 3891.
51. Nakamura, N., Yamamoto, A., Wada, Y., and Futai, M. (2000) Syntaxin 7 mediates endocytic trafficking to late endosomes. *J. Biol. Chem.*, **275**, 6523.
52. Steegmaier, M., Yang, B., Yoo, J-S., Huang, B., Shen, M., Yu, S., *et al.* (1998) Three novel proteins of the Syntaxin/SNAP-25 family. *J. Biol. Chem.*, **273**, 34171.
53. Valdez, A. C., Cabaniois, J-P., Brown, M. J., and Roche, P. A. (1999) Syntaxin 11 is associated with SNAP-23 on late endosomes and the trans-Golgi network. *J. Cell Sci.*, **112**, 845.
54. Weber, T., Zemelman, B. V., McNew, J. A., Westermann, B., Gmachl, M., Parlati, F., *et al.* (1998) SNAREpins: minimal machinery for membrane fusion. *Cell*, **92**, 759.
55. Nickel, W., Weber, T., McNew, J. A., Parlati, F., Sollner, T., and Rothman, J. E. (1999) Content mixing and membrane integrity during membrane fusion driven by pairing of isolated v-SNAREs and t-SNAREs. *Proc. Natl. Acad. Sci. USA*, **96**, 12571.
56. Parlati, F., Weber, T., McNew, J. A., Westermann, B., Sollner, T. H., and Rothman, J. E. (1999) Rapid and efficient fusion of phospholipid vesicles by the α-helical core of a SNARE complex in the absence of an N-terminal regulatory domain. *Proc. Natl. Acad. Sci. USA*, **96**, 12565.
57. Peters, C. and Mayer, A. (1998) Ca^{++}/calmodulin signals the completion of docking and triggers a late step of vacuole fusion. *Nature*, **396**, 575.
58. Peters, C., Andrews, P. D., Stark, M. J. R., Cesaro-Tadic, S., Glatz, A., Podtelejnikov, A., *et al.* (1999) Control of the terminal step of intracellular membrane fusion by protein phosphatase 1. *Science*, **285**, 1084.
59. Holroyd, C., Kistner, U., Annaert, W., and Jahn, R. (1999). Fusion of endosomes involved in synaptic vesicle recycling. *Mol. Biol. Cell*, **10**, 3035.
60. Pryor, P. R., Mullock, B. M., Bright, N. A., Gray, S. R., and Luzio, J. P. (2000) The role of intra-organellar Ca^{2+} in late endosome-lysosome heterotypic fusion and in the reformation of lysosomes from hybrid organelles. *J. Cell Biol.*, **149**, 1053.
61. Bauerfeind, R. and Huttner, W. B. (1993) Biogenesis of constitutive secretory vesicles, secretory granules and synaptic vesicles. *Curr. Opin. Cell Biol.*, **5**, 628.
62. Tooze, S. A. (1998) Biogenesis of secretory granule in the trans-Golgi network of neuroendocrine and endocrine cells. *Biochim. Biophys. Acta*, **1404**, 231.
63. van Weert, A. W. M., Dunn, K. W., Geuze, H. J., Maxfield, F. R., and Stoorvogel, W. (1995) Transport from late endosomes to lysosomes, but not sorting of integral membrane proteins in endosomes, depends on the vacuolar proton pump. *J. Cell Biol.*, **130**, 821.

64. Seaman, M. N. J., Ball, C. L., and Robinson, M. S. (1993) Targeting and mistargeting of plasma membrane adaptors in vitro. *J. Cell Biol.*, **123**, 1093.
65. Traub, L. M., Bannykh, S. I., Rodel, J. E., Aridor, M., Balch, W. E., and Kornfeld, S. (1996) AP-2-containing clathrin coats assemble on mature lysosomes. *J. Cell Biol.*, **135**, 1801.
66. Arneson, L. S., Kunz, J., Anderson, R. A., and Traub, L. M. (1999) Coupled inositide phosphorylation and phospholipase D activation initiates clathrin-coat assembly on lysosomes. *J. Biol. Chem.*, **274**, 17794.
67. Stevens, T. H., Rothman, J. H., Payne, G. S., and Schekman, R. (1986) Gene dosage-dependent secretion of yeast vacuolar carboxypeptidase Y. *J. Cell Biol.*, **102**, 1551.
68. Johnson, L. M., Bankaitis, V. A., and Emr, S. D. (1987) Distinct sequence determinants direct intracellular sorting and modification of a yeast vacuolar protease. *Cell*, **48**, 875.
69. Valls, L. A., Hunter, C. P., Rothman, J. H., and Stevens, T. H. (1987) Protein sorting in yeast: the localization determinant of yeast vacuolar carboxypeptidase Y resides in the propeptide. *Cell*, **48**, 887.
70. Valls, L. A., Winther, J. R., and Stevens, T. H. (1990) Yeast carboxypeptidase Y vacuolar targeting signal is defined by four propeptide amino acids. *J. Cell Biol.*, **111**, 361.
71. Bankaitis, V. A., Johnson, L. M., and Emr, S. D. (1986) Isolation of yeast mutants defective in protein targeting to the vacuole. *Proc. Natl. Acad. Sci. USA*, **83**, 9075.
72. Rothman, J. H. and Stevens, T. H. (1986) Protein sorting in yeast: mutants defective in vacuole biogenesis mislocalize vacuolar proteins into the late secretory pathway. *Cell*, **47**, 1041.
73. Robinson, J. S., Klionsky, D. J., Banta, L. M., and Emr, S. D. (1988) Protein sorting in *Saccharomyces cerevisiae*: isolation of mutants defective in the delivery and processing of multiple vacuolar hydrolases. *Mol. Cell. Biol.*, **8**, 4936.
74. Raymond, C. K., Howald-Stevenson, I., Vater, C. A., and Stevens, T. H. (1992) Morphological classification of the yeast vacuolar protein sorting mutants: evidence for a prevacuolar compartment in Class E *vps* mutants. *Mol. Biol. Cell*, **3**, 1389.
75. Marcusson, E. G., Horazdovsky, B. F., Cereghino, J. L., Gharakhanian, E., and Emr, S. D. (1994) The sorting receptor for yeast vacuolar carboxypeptidase Y is encoded by the *VPS10* gene. *Cell*, **77**, 579.
76. Seeger, M. and Payne, G. S. (1992) A role for clathrin in the sorting of vacuolar proteins in the Golgi complex of yeast. *EMBO J.*, **11**, 2811.
77. Chen, C. Y. and Graham, T. R. (1998) An arf1Delta synthetic lethal screen identifies a new clathrin heavy chain conditional allele that perturbs vacuolar protein transport in *Saccharomyces cerevisiae*. *Genetics*, **150**, 577.
78. Payne, G. S., Baker, D., van Tuinen, E., and Schekman, R. (1988) Protein transport to the vacuole and receptor-mediated endocytosis by clathrin heavy chain-deficient yeast. *J. Cell Biol.*, **106**, 1453.
79. Rothman, J. H., Raymond, C. K., Gilbert, T., O'Hara, P. J., and Stevens, T. H. (1990) A putative GTP binding protein homologous to Interferon-Inducible Mx proteins performs an essential function in yeast protein sorting. *Cell*, **61**, 1063.
80. Vater, C. A., Raymond, C. K., Ekena, K., Howald-Stevenson, I., and Stevens, T. H. (1992) The VPS1 protein, a homolog of dynamin required for vacuolar protein sorting in *Saccharomyces cerevisiae*, is a GTPase with two functionally separable domains. *J. Cell Biol.*, **119**, 773.
81. Wilsbach, K. and Payne, G. S. (1993) Vps1p, a member of the dynamin GTPase family, is necessary for Golgi membrane protein retention in *Saccharomyces cerevisiae*. *EMBO J.*, **12**, 3049.

82. Nothwehr, S. F., Bruinsma, P., and Strawn, L. A. (1999) Distinct domains within Vps35p mediate the retrieval of two different cargo proteins from the yeast prevacuolar/endosomal compartment. *Mol. Biol. Cell*, **10**, 875.
83. Burd, C. G., Peterson, M., Cowles, C. R., and Emr, S. D. (1997) A novel Sec18p/NSF-dependent complex required for Golgi-to-endosome transport in yeast. *Mol. Biol. Cell*, **8**, 1089.
84. Becherer, K. A., Rieder, S. E., Emr, S. D., and Jones, E. W. (1996) Novel syntaxin homologue, Pep12p, required for the sorting of lumenal hydrolases to the lysosome-like vacuole in yeast. *Mol. Biol. Cell.*, **7**, 579.
85. Cowles, C. R., Emr, S. D., and Horazdovsky, B. F. (1994) Mutations in the *VPS45* gene, a SEC1 homologue, result in vacuolar protein sorting defects and accumulation of membrane vesicles. *J. Cell Sci.*, **107**, 3449.
86. Piper, R. C., Whitters, E. A., and Stevens, T. H. (1994) Yeast Vps45p is a Sec1p-like protein required for the consumption of vacuole-targeted, post-Golgi transport vesicles. *Eur. J. Cell Biol.*, **65**, 305.
87. Horazdovsky, B. F., Busch, G. R., and Emr, S. D. (1994) *VPS21* encodes a rab5-like GTP binding protein that is required for the sorting ofyeast vacuolar proteins. *EMBO J.*, **13**, 1297.
88. Tall, G. G., Hama, H., DeWald, D. B., and Horazdovsky, B. F. (1999) The phosphatidylinositol 3-phosphate binding protein Vac1p interacts with a Rab GTPase and a Sec1p homologue to facilitate vesicle-mediated vacuolar protein sorting. *Mol. Biol. Cell*, **10**, 1873.
89. Peterson, M. R., Burd, C. G., and Emr, S. D. (1999) Vac1p coordinates Rab and phosphatidylinositol 3-kinase signaling in Vps45p-dependent vesicle docking/fusion at the endosome. *Curr. Biol.*, **9**, 159.
90. Hama, H., Tall, G. G., and Horazdovsky, B. F. (1999) Vps9p is a guanine nucleotide exchange factor involved in vesicle-mediated vacuolar protein transport. *J. Biol. Chem.*, **274**, 15284.
91. Burd, C. G., Mustol, P. A., Schu, P. V., and Emr, S. D. (1996) A yeast protein related to a mammalian Ras-binding protein, Vps9p, is required for localization of vacuolar proteins. *Mol. Cell. Biol.*, **16**, 2369.
92. Burd, C. G. and Emr, S. D. (1998) Phosphatidylinositol(3)-phosphate signaling mediated by specific binding to RING FYVE domains. *Mol. Cell*, **2**, 157.
93. Gaullier, J. M., Simonson, A., D'Arrigo, A., Bremnes, B., and Stenmark, H. (1998) FYVE fingers bind PtdIns(3)P. *Nature*, **394**, 432.
94. Herman, P. K., Stack, J. H., DeModena, J. A., and Emr, S. D. (1991) A novel protein kinase homolog essential for protein sorting to the yeast lysosome-like vacuole. *Cell*, **64**, 425.
95. Herman, P. K., Stack, J. H., and Emr, S. D. (1991) A genetic and structural analysis of the yeast Vps15 protein kinase: evidence for a direct role of Vps15p in vacuolar protein delivery. *EMBO J.*, **10**, 4049.
96. Schu, P. V., Takegawa, K., Fry, M. J., Stack, J. H., Waterfield, M. D., and Emr, S. D. (1993) Phosphatidylinositol 3-kinase encoded by yeast *VPS34* gene essential for protein sorting. *Science*, **260**, 88.
97. Stack, J. H., DeWald, D. B., Takegawa, K., and Emr, S. D. (1995) Vesicle-mediated protein transport: regulatory interactions between the Vps15 protein kinase and the Vps34 PtdIns 3-kinase essential for protein sorting to the vacuole in yeast. *J. Cell Biol.*, **129**, 321.
98. von Mollard, G. F. and Stevens, T. H. (1999) The *Saccharomyces cerevisiae* v-SNARE Vti1p is required for multiple membrane transport pathways to the vacuole. *Mol. Biol. Cell*, **10**, 1719.

99. von Mollard, G. F., Nothwehr, S. F., and Stevens, T. H. (1997) The yeast v-SNARE Vti1p mediates two vesicle transport pathways through interactions with the t-SNAREs Sed5p and Pep12p. *J. Cell Biol.*, **137**, 1511.
100. Darsow, T., Rieder, S. E., and Emr, S. D. (1997) A multispecificity syntaxin homologue, Vam3p, essential for autophagic and biosynthetic protein transport to the vacuole. *J. Cell Biol.*, **138**, 517.
101. Sato, T. K., Darsow, T., and Emr, S. D. (1998) Vam7p, a SNAP-25-like molecule, and Vam3p, a syntaxin homolog, function together in yeast vacuolar protein trafficking. *Mol. Cell. Biol.*, **18**, 5308.
102. Nichols, B. J., Ungermann, C., Pelham, H. R., Wickner, W. T., and Haas, A. (1997) Homotypic vacuolar fusion mediated by t- and v-SNAREs. *Nature*, **387**, 199.
103. Ungermann, C., von Mollard, G. F., Jensen, O. N., Margolis, N., Stevens, T. H., and Wickner, W. (1999) Three v-SNAREs and two t-SNAREs, presennt in a pentameric cis-SNARE complex on isolated vacuoles, are essential for homotypic fusion. (1999) *J. Cell Biol.*, **145**, 1435.
104. Rieder, S. E. and Emr, S. D. (1997) A novel RING finger protein complex essential for a late step in protein transport to the yeast vacuole. *Mol. Biol. Cell*, **8**, 2307.
105. Schimmoller, F. and Riezman, H. (1993) Involvement of Ypt7p, a small GTPase, in traffic from late endosome to the vacuole in yeast. *J. Cell Sci.*, **106**, 823.
106. Darsow, T., Burd, C. G., and Emr, S. D. Acidic di-leucine motif essential for AP-3-dependent sorting and restriction of the functional specificity of the Vam3p vacuolar t-SNARE. *J. Cell Biol.*, **142**, 913.
107. Cooper, A. A. and Stevens, T. H. (1996) Vps10p cycles between the late-Golgi and pre-vacuolar compartments in its function as the sorting receptor for multiple yeast vacuolar hydrolases. *J. Cell Biol.*, **133**, 529.
108. Rieder, S. E., Banta, L. M., Kohrer, K., McCaffery, J. M., and Emr, S. D. (1996) Multilamellar endosome-like compartment accumulates in the yeast vps28 vacuolar protein sorting mutant. *Mol. Biol. Cell*, **7**, 985.
109. Babst, M., Sato, T. K., Banta, L. M., and Emr, S. D. (1997) Endosomal transport function in yeast requires a novel AAA-type ATPase, Vps4p. *EMBO J.*, **16**, 1820.
110. Odorizzi, G., Babst, M., and Emr, S. D. (1998) Fab1p PtdIns(3)P 5-kinase function essential for protein sorting in the multivesicular body. *Cell*, **95**, 847.
111. Li, Y., Kane, T., Tipper, C., Spatrick, P., and Jenness, D. D. (1999) Yeast mutants affecting possible quality control of plasma membrane proteins. *Mol. Cell. Biol.*, **19**, 3588.
112. Babst, M., Odorizzi, G., Estepa, E. J., and Emr, S. D. (2000) Mammalian tumour susceptibility gene 101 (TSG101) and the yeast homologue Vps23p, both function in late endosomal trafficking. *Traffic*, **1**, 248.
113. Piper, R. C., Cooper, A. A., Yang, H., and Stevens, T. H. (1995) *VPS27* controls vacuolar and endocytic traffic through a prevacuolar compartment in *Saccharomyces cerevisiae*. *J. Cell Biol.*, **131**, 603.
114. Munn, A. L. and Riezman, H. (1994) Endocytosis is required for the growth of vacuolar H(+)-ATPase-defective yeast: identification of six new *END* genes. *J. Cell Biol.*, **127**, 373.
115. Siddhanta, U., McIlroy, J., Shah, A., Zhang, Y., and Backer, J. M. (1998) Distinct roles for the p110α and the hVPS34 phosphatidylinositol 3′-kinases in vesicular trafficking, regulation of the actin cytoskeleton and mitogenesis. *J. Cell Biol.*, **143**, 1647.
116. Nielsen, E., Severin, F., Backer, J. M., Hyman, A. A., and Zerial, M. (1999) Rab5 regulates motility of early endosomes on microtubules. *Nat. Cell Biol.*, **1**, 376.

117. Petiot, A., Ogier-Denis, E., Blommaart, E. F. C., Meijer, A. J., and Codogno, P. (2000) distinct classes of phosphatidylinositol 3′-kinase are involved in signaling pathways that control macroautophagy in HT-29 cells. *J. Biol. Chem.*, **275**, 992.
118. Seaman, M. N. J., Marcusson, E. G., Cereghino, J. L., and Emr, S. D. (1997) Endosome to Golgi retrieval of the vacuolar protein sorting receptor, Vps10p, requires the function of the *VPS29*, *VPS30*, and *VPS35* gene products. *J. Cell Biol.*, **137**, 79.
119. Seaman, M. N. J., McCaffery, J. M., and Emr, S. D. (1998) A membrane coat complex essential for endosome-to-Golgi retrograde transport in yeast. *J. Cell Biol.*, **142**, 665.
120. Horazdovsky, B. F., Davies, B. A., Seaman, M. N., McLaughlin, S. A., Yoon, S., and Emr, S. D. (1997) A sorting nexin-1 homologue, Vps5p, forms a complex with Vps17p and is required for recycling the vacuolar protein-sorting receptor. *Mol. Biol. Cell*, **8**, 1529.
121. Kohrer, K. and Emr, S. D. (1993) The yeast *VPS17* gene encodes a membrane-associated protein required for the sorting of soluble vacuolar hydrolases. *J. Biol. Chem.*, **268**, 559.
122. Kurten, R. C., Cadena, D. L., and Gill, G. N. (1996). Enhanced degradation of EGF receptors by a sorting nexin SNX1. *Science*, **272**, 1008.
123. Renfrew-Haft, C., Sierra, M., Haft, D. H., and Taylor, S. I. (1998). Identification of a family of sorting nexin molecules and characterisation of their association with receptors. *Mol. Cell. Biol.*, **18**, 7278.
124. Bryant, N. J., Piper, R. C., Weisman, L. S., and Stevens, T. H. (1998) Retrograde traffic out of the yeast vacuole to the TGN occurs via the prevacuolar/endosomal compartment. *J. Cell Biol.*, **142**, 651.
125. Bonangelino, C. J., Catlett, N. L., and Weisman, L. S. (1997) Vac7p, a novel vacuolar protein, is required for normal vacuole inheritance and morphology. *Mol. Cell. Biol.*, **17**, 6847.
126. Feng, Y., Press, B., and Wandinger-Ness, A. (1995) Rab 7: an important regulator of late endocytic membrane traffic. *J. Cell Biol.*, **131**, 1435.
127. Le Borgne, R. and Hoflack, B. (1998) Protein transport from the secretory to the endocytic pathway in mammalian cells. *Biochim. Biophys. Acta*, **1404**, 195.
128. Diaz, E. and Pfeffer, S. R. (1998) TIP47: a cargo selection device for mannose 6-phosphate receptor trafficking. *Cell*, **93**, 433.
129. Hirst, J., Futter, C. E., and Hopkins, C. R. (1998) The kinetics of mannose 6-phosphate receptor trafficking in the endocytic pathway in HEp-2 cells: the receptor enters and rapidly leaves multivesicular endosomes without accumulating in a prelysosomal compartment. *Mol. Biol. Cell*, **9**, 809.
130. Braun, M., Waheed, A., and Von Figura, K. (1989) Lysosomal acid phosphatase is transported to lysosomes via the cell surface. *EMBO J.*, **8**, 3633.
131. Green, S. A., Zimmer, K. P., Griffiths, G., and Mellman, I. (1987) Kinetics of intracellular transport and sorting of lysosomal membrane and plasma membrane proteins. *J. Cell Biol.*, **105**, 1227.
132. Barriocanal, J. G., Bonifacino, J. S., Yuan, L., and Sandoval, I. V. (1986) Biosynthesis, glycosylation, movement through the Golgi system, and transport to lysosomes by an N-linked carbohydrate-independent mechanism of three lysosomal integral membrane proteins. *J. Biol. Chem.*, **261**, 16755.
133. Ohno, H., Stewart, J., Fournier, M. C., Bosshart, H., Rhee, I., Miyatake, S., *et al.* (1995) Interaction of tyrosine-based sorting signals with clathrin-associated proteins. *Science*, **269**, 1872.
134. Honing, S., Griffith, J., Geuze, H. J., and Hunziker, W. (1996) The tyrosine-based lysosomal targeting signal in lamp-1 mediates sorting into Golgi-derived clathrin-coated vesicles. *EMBO J.*, **15**, 5230.

135. Karlsson, K. and Carlsson, S. R. (1998) Sorting of lysosomal membrane glycoproteins lamp-1 and lamp-2 into vesicles distinct from mannose 6-phosphate receptor/γ-adaptin vesicles at the trans-Golgi network. *J. Biol. Chem.*, **273**, 18966.
136. Simpson, F., Bright, N. A., West, M. A., Newman, L. S., Darnell, R. B., and Robinson, M. S. (1996) A novel adaptor-related protein complex. *J. Cell Biol.*, **133**, 749.
137. Simpson, F., Peden, A. A., Christopoulou, L., and Robinson, M. S. (1997) Characterization of the adaptor-related protein complex, AP-3. *J. Cell Biol.*, **137**, 835.
138. Dell'Angelica, E. C., Ohno, H., Ooi, C. E., Rabinovich, E., Roche, K. W., and Bonifacino, J. S. (1997) AP-3: an adaptor-like protein complex with ubiquitous expression. *EMBO J.*, **16**, 917.
139. Dell'Angelica, E. C., Ooi, C. E., and Bonifacino, J. S. (1997) β3A-adaptin, a subunit of the adaptor-like complex AP-3. *J. Biol. Chem.*, **272**, 15078.
140. Pevsner, J., Volknandt, W., Wong, B. R., and Scheller, R. H. (1994) Two rat homologs of clathrin-associated adaptor proteins. *Gene*, **146**, 279.
141. Newman, L. S., McKeever, M. O., Okano, H. J., and Darnell, R. B. (1995) β-NAP, a cerebellar degeneration antigen, is a neuron-specific vesicle coat protein. *Cell*, **82**, 773.
142. Dell'Angelica, E. C., Klumperman, J., Stoorvogel, W., and Bonifacino, J. S. (1998) Association of the AP-3 adaptor complex with clathrin. *Science*, **280**, 431.
143. Piper, R. C., Bryant, N. J., and Stevens, T. H. (1997) The membrane protein alkaline phosphatase is delivered to the vacuole by a route that is distinct from the VPS-dependent pathway. *J. Cell Biol.*, **138**, 531.
144. Cowles, C. R., Snyder, W. B., Burd, C. G., and Emr, S. D. (1997) Novel Golgi to vacuole delivery pathway in yeast: identification of a sorting determinant and required transport component. *EMBO J.*, **16**, 2769.
145. Cowles, C. R., Odorizzi, G., Payne, G. S., and Emr, S. D. (1997) The AP-3 adaptor complex is essential for cargo-selective transport to the yeast vacuole. *Cell*, **91**, 109.
146. Stepp, J. D., Huang, K., and Lemmon, S. K. (1997) The yeast adaptor protein complex, AP-3, is essential for the efficient delivery of alkaline phosphatase by the alternate pathway to the vacuole. *J. Cell Biol.*, **139**, 1761.
147. Rehling, P., Darsow, T., Katzmann, D. J., and Emr, S. D. (1999) Formation of AP-3 transport intermediates requires Vps41p function. *Nat. Cell Biol.*, **1**, 346.
148. Price, A., Wickner, W., and Ungermann, C. (2000) Proteins needed for vesicle budding from the Golgi complex are also required for the docking step of homotypic vacuole fusion. *J. Cell Biol.*, **148**, 1223.
149. Price, A., Seals, D., Wickner, W., and Ungermann, C. (2000) The docking stage of yeast vacuole fusion requires the transfer of proteins from a cis-SNARE complex to a Rab/Ypt protein. *J. Cell Biol.*, **148**, 1231.
150. Ooi, C. E., Moreira, J. E., Dell'Angelica, E. C., Poy, G., Wassarman, D. A., and Bonifacino, J. S. (1997) Altered expression of a novel adaptin leads to defective pigment granule biogenesis in the Drosophila eye color mutant garnet. *EMBO J.*, **16**, 4508.
151. Lloyd, V., Ramaswami, M., and Kramer, H. (1998) Not just pretty eyes: Drosophila eye-colour mutations and lysosomal delivery. *Trends Cell Biol.*, **8**, 257.
152. Swank, R. T., Novak, E. K., McGarry, M. P., Rusiniak, M. E., and Feng, L. (1998) Mouse models of Hermansky Pudlak syndrome: a review. *Pigment Cell Res.*, **11**, 60.
153. Dell'Angelica, E. C., Shotelersuk, V., Aguilar, R. C., Gahl, W. A., and Bonifacino, J. S. (1999) Altered trafficking of lysosomal proteins in Hermansky-Pudlak syndrome due to mutations in the β3A subunit of the AP-3 adaptor. *Mol. Cell*, **3**, 11.

154. Le Borgne, R., Alconada, A., Bauer, U., and Hoflack, B. (1998) The mammalian AP-3 adaptor-like complex mediates the intracellular transport of lysosomal membrane glycoproteins. *J. Biol. Chem.*, **273**, 29451.
155. Hirst, J., Bright, N. A., Rous, B., and Robinson, M. S. (1999) Characterization of a fourth adaptor-related protein complex. *Mol. Biol. Cell*, **10**, 2787.
156. Dell'Angelica, E. C., Mullins, C., and Bonifacino, J. S. (1999) AP-4, a novel protein complex related to clathrin adaptors. *J. Biol. Chem.*, **274**, 7278.
157. Noda, T. and Farquhar, M. G. (1992) A non-autophagic pathway for diversion of ER secretory proteins to lysosomes. *J. Cell Biol.*, **119**, 85.
158. Cuervo, A. M. and Dice, J. F. (1996) A receptor for the selective uptake and degradation of proteins by lysosomes. *Science*, **273**, 501.
159. Ahlberg, J. and Glaumann, H. (1985) Uptake, microautophagy and degradation of exogenous proteins by isolated rat liver lysosomes. Effects of pH, ATP and inhibitors of proteolysis. *Exp. Mol. Pathol.*, **42**, 78.
160. Mortimore, G. E., Poso, A. R., and Lardeux, B. R. (1989) Mechanism and regulation of protein degradation in liver [published erratum appears in *Diabetes Metab. Rev.* (1989) **5**, 320]. *Diabetes Metab. Rev.*, **5**, 49.
161. Dunn, W. A., Jr. (1990) Studies on the mechanisms of autophagy: Formation of the autophagic vacuole. *J. Cell Biol.*, **110**, 1923.
162. Furuno, K., Ishikawa, T., Akasaki, K., Lee, S., Nishimura, Y., Tsuji, H., *et al.* (1990) Immunocytochemical study of the surrounding envelope of autophagic vacuoles in cultured rat hepatocytes. *Exp. Cell Res.*, **189**, 261.
163. Seglen, P. O. (1987) Regulation of autophagic protein degradation in isolated liver cells. In *Lysosomes: their role in protein breakdown* (ed. H. Glaumann and F. J. Ballard), pp. 371–414. Orlando, London.
164. Fengsrud, M., Roos, N., Berg, T., Liou, W., Slot, J. W., and Seglen, P. O. (1995) Ultrastructural and immunocytochemical characterization of autophagic vacuoles in isolated hepatocytes: effects of vinblastine and asparagine on vacuole distributions. *Exp. Cell Res.*, **221**, 504.
165. Willemer, S., Bialek, R., Kohler, H., and Adler, G. (1990) Caerulein-induced acute pancreatitis in rats: changes in glycoprotein composition of subcellular membrane system in acinar cells. *Histochemistry*, **95**, 87.
166. Yamamoto, A., Masaki, R., and Tashiro, Y. (1990) Characterization of the limiting membranes of autophagosomes in rat hepatocytes by lectin cytochemistry. *J. Histochem. Cytochem.*, **38**, 573.
167. Yamamoto, A., Masaki, R., Fukui, Y., and Tashiro, Y. (1990) Absence of cytochrome P-450 and presence of autolysosomal membrane antigens on the isolation membranes and autophagosomal membranes in rat hepatocytes. *J. Histochem. Cytochem.*, **38**, 1571.
168. Dunn, W. A., Jr. (1990) Studies on the mechanisms of autophagy: Maturation of the autophagic vacuole. *J. Cell Biol.*, **110**, 1935.
169. Lawrence, B. P. and Brown, W. J. (1992) Autophagic vacuoles rapidly fuse with pre-existing lysosomes in cultured hepatocytes. *J. Cell Sci.*, **102**, 515.
170. Aplin, A., Jasionowski, T., Tuttle, D. L., Lenk, S. E., and Dunn, W. A. (1992) Cytoskeletal elements are required for the formation and maturation of autophagic vacuoles. *J. Cell. Physiol.*, **152**, 458.
171. Dunn, W. A., Jr. (1994) Autophagy and related mechanisms of lysosome-mediated protein degradation. *Trends Cell Biol.*, **4**, 139.

172. Yamamoto, A., Tagawa, Y., Yoshimori, T., Moriyama, Y., Masaki, R., and Tashiro, Y. (1998) Bafilomycin A1 prevents maturation of autophagic vacuoles by inhibiting fusion between autophagosomes and lysosomes in rat hepatoma cell line, H-4-II-E cells. *Cell Struct. Funct.*, **23**, 33.
173. Scott, S. V. and Klionsky, D. J. (1998) Delivery of proteins and organelles to the vacuole from the cytoplasm. *Curr. Opin. Cell Biol.*, **10**, 523.
174. Klionsky, D. J. and Ohsumi, Y. (1999) Vacuolar import of proteins and organelles from the cytoplasm. *Annu. Rev. Cell Dev. Biol.*, **15**, 1.
175. Klionsky, D. J. (1998) Nonclassical protein sorting to the yeast vacuole. *J. Biol. Chem.*, **273**, 10807.
176. Klionsky, D. J., Cueva, R., and Yaver, D. S. (1992) Aminopeptidase I of *Saccharomyces cerevisiae* is localized to the vacuole independent of the secretory pathway. *J. Cell Biol.*, **119**, 287.
177. Scott, S. V., Hefner-Gravink, A., Morano, K. A., Noda, T., Ohsumi, Y., and Klionsky, D. J. (1996) Cytoplasm-to-vacuole targeting and autophagy employ the same machinery to deliver proteins to the yeast vacuole. *Proc. Natl. Acad. Sci. USA*, **93**, 12304.
178. Baba, M., Osumi, M., Scott, S. V., Klionsky, D. J., and Ohsumi, Y. (1997) Two distinct pathways for targeting proteins from the cytoplasm to the vacuole/lysosome. *J. Cell Biol.*, **139**, 1687.
179. Yoshihisa, T and Anraku, Y. (1990) A novel pathway of import of α mannosidase, a marker enzyme of vacuolar membrane, in *Saccharomyces cerevisiae. J. Biol. Chem.*, **265**, 22418.
180. Scott, S. V., Baba, M., Ohsumi, Y., and Klionsky, D. J. (1997) Aminopeptidase I is targeted to the vacuole by a nonclassical vesicular mechanism. *J. Cell Biol.*, **138**, 37.
181. Kim, J., Scott, S. V., Oda, M. N., and Klionsky, D. J. (1997) Transport of a large oligomeric protein by the cytoplasm to vacuole protein targeting pathway. *J. Cell Biol.*, **137**, 609.
182. Tsukada, M. and Ohsumi, Y. (1993) Isolation and characterisation of autophagy-defective mutants of *Saccharomyces cerevisiae. FEBS Lett.*, **333**, 169.
183. Thumm, M., Egner, R., Koch, M., Schlumpberger, M., Straub, M., Veenhuis, M., *et al.* (1994) Isolation of autophagocytosis mutants of *Saccharomyces cerevisiae. FEBS Lett.*, **349**, 275.
184. Harding, T. M., Morano, K. A., Scott, S. V., and Klionsky, D. J. (1995) Isolation and characterization of yeast mutants in the cytoplasm to vacuole protein targeting pathway. *J. Cell Biol.*, **131**, 591.
185. Harding, T. M., Hefner-Gravink, A., Thumm, M., and Klionsky, D. J. (1996) Genetic and phenotypic overlap between autophagy and the cytoplasm to vacuole protein targeting pathway. *J. Biol. Chem.*, **271**, 17621.
186. Matsuura, A., Tsukada, M., Wada, Y., and Ohsumi, Y. (1997) Apg1p, a novel protein kinase required for the autophagic process in *Saccharomyces cerevisiae. Gene*, **192**, 245.
187. Tanida, I., Mizushima, N., Kiyooka, M., Ohsumi, M., Ueno, T., Ohsumi, Y., *et al.* (1999) Apg7p/Cvt2p: A novel protein-activating enzyme essential for autophagy. *Mol. Biol. Cell*, **10**, 1367.
188. Mizushima, N., Noda, T., Yoshimori, T., Tanaka, Y., Ishii, T., George, M. D., *et al.* (1998) A protein conjugation system essential for autophagy. *Nature*, **395**, 395.
189. Shintani, T., Mizushima, N., Ogawa, Y., Matsuura, A., Noda, T., and Ohsumi, Y. (1999) Apg10p, a novel protein-conjugating enzyme essential for autophagy in yeast. *EMBO J.*, **18**, 5234.
190. Mizushima, N., Noda, T., and Ohsumi, Y. (1999) Apg16p is required for the function of the Apg12p-Apg5p conjugate in the yeast autophagy pathway. *EMBO J.*, **18**, 3888.

191. George, M. D., Baba, M., Scott, S. V., Mizushima, N., Garrison, B. S., Ohsumi, Y., *et al.* (2000) Apg5p functions in the sequestration step in the cytoplasm-to-vacuole targeting and macroautophagy pathways. *Mol. Biol. Cell*, **11**, 969.
192. Kirisako, T., Baba, M., Ishihara, N., Miyazawa, K., Ohsumi, M., Yoshimori, T., *et al.* (1999) Formation process of autophagosome is traced with Apg8/Aut7p in yeast. *J. Cell Biol.*, **147**, 435.
193. Lang, T., Schaeffeler, E., Bernreuther, D., Bredschneider, M., Wolf, D. H., and Thumm, M. (1998) Aut2p and Aut7p, two novel microtubule-associated proteins are essential for delivery of autophagic vesicles to the vacuole. *EMBO J.*, **17**, 3597.
194. Kametaka, S., Okano, T., Ohsumi, M., and Ohsumi, Y. (1998) Apg14p and Apg6/Vps30p form a protein complex essential for autophagy in the yeast, *Saccharomyces cerevisiae*. *J. Biol. Chem.*, **273**, 22284.
195. Liang, X. H., Kleeman, L. K., Jiang, H. H., Gordon, G., Goldman, J. E., Berry, G., *et al.* (1998) Protection against fatal Sindbis virus encephalitis by beclin, a novel Bcl-2-interacting protein. *J .Virol.*, **72**, 8586.
196. Liang, X. H., Jackson, S., Seaman, M., Brown, K., Kempkes, B., Hibshoosh, H., *et al.* (1999) Induction of autophagy and inhibition of tumorigenesis by beclin 1. *Nature*, **402**, 672.
197. Abeliovich, H., Darsow, T., and Emr, S. D. (1999) Cytoplasm to vacuole trafficking of aminopeptidase I requires a t-SNARE-Sec1p complex composed of Tlg2p and Vps45p. *EMBO J.*, **18**, 6005.
198. Holthuis, J. C., Nichols, B. J., Dhruvakumar, S., and Pelham, H. R. (1998) Two syntaxin homologues in the TGN/endosomal system of yeast. *EMBO J.*, **17**, 113.
199. Abeliovich, H., Grote, E., Novick, P., and Ferro-Novick, S. (1998) Tlg2p, a yeast syntaxin homolog that resides on the Golgi and endocytic structures. *J. Biol. Chem.*, **273**, 11719.
200. Noda, T., Kim, J., Huang, W. P., Baba, M., Tokunaga, C., Ohsumi, Y., *et al.* (2000) Apg9p/Cvt7p is an integral membrane protein required for transport vesicle formation in the Cvt and autophagy pathways. *J. Cell Biol.*, **148**, 465.
201. Sevrioukov, E. A., He, J-P., Moghrabi, N., Sunio, A., and Kramer, H. (1999) A role for deep orange and carnation eye-color genes in lysosomal delivery in Drosophila. *Mol. Cell*, **4**, 479.
202. Kramer, H. and Phistry, M. (1999) Genetic analysis of *hook*, a gene required for endocytic trafficking in Drosophila. *Genetics*, **151**, 675.
203. Sunio, A., Metcalf, A. B., and Kramer, H. (1999) Genetic dissection of endocytic trafficking in *Drosophila* using a horseradish peroxidase-bride of sevenless chimera: *hook* is required for normal maturation of multivesicular endosomes. *Mol. Biol. Cell*, **10**, 847.
204. Martinez, I., Chakrabarti, S., Hellevik, T., Morehead, J., Fowler, K., and Andrews, N. W. (2000) Synaptotagmin VII regulates Ca^{2+} -dependent exocytosis of lysosomes in fibroblasts. *J. Cell Biol.*, **148**, 1141.
205. Perou, C. M., Leslie, J. D., Green, W., Li, L., Ward, D. M., and Kaplan, J. (1997) The Beige/Chediak-Higashi syndrome gene encodes a widely expressed cytosolic protein. *J. Biol. Chem.*, **272**, 29790.
206. Shotelersuk, V. and Gahl, W. A. (1998) Hermansky-Pudlak syndrome: models for intracellular vesicle formation. *Mol. Genet. Metab.*, **65**, 85.
207. Richards-Smith, B., Novak, E. K., Jang, E. K., He, P., Haslam, R. J., Castle, D., *et al.* (1999) Analyses of proteins involved in vesicular trafficking in platelets of mouse models of Hermansky Pudlak syndrome. *Mol. Genet. Metab.*, **68**, 14.
208. Detter, J. C., Zhang, Q., Mules, E. H., Novak, E. K., Mishra, V. S., Li, W., *et al.* (2000) Rab

geranylgeranyl transferase α mutation in the *gunmetal* mouse reduces rab prenylation and platelet synthesis. *Proc. Natl. Acad. Sci. USA*, **97**, 4144.

209. Huang, K. M., D'Hondt, K., Riezman, H., and Lemmon, S. K. (1999) Clathrin functions in the absence of heterotetrameric adaptors and AP180-related proteins in yeast. *EMBO J.*, **18**, 3897.
210. Kobayashi, T., Beuchat, M-H., Lindsay, M., Frias, S., Palmiter, R. D., Sakuraba, H., *et al.* (1999) Late endosomal membranes rich in lysobisphosphatidic acid regulate cholesterol transport. *Nat. Cell Biol.*, **1**, 113.
211. Neufeld, E. B., Wastney, M., Patel, S., Suresh, S., Cooney, A. M., Dwyer, N. K., *et al.* (1999) The Niemann-Pick C1 protein resides in a vesicular compartment linked to retrograde transport of multiple lysosomal cargo. *J. Biol. Chem.*, **274**, 9627.
212. Mukherjee, S. and Maxfield, F. R. (1999) Cholesterol: stuck in the traffic. *Nat. Cell Biol.*, **1**, E37.
213. Liscum, L. (2000) Niemann-Pick type C mutations cause lipid traffic jam. *Traffic*, **1**, 218.
214. Schmid, J. A., Mach, L., Paschke, E., and Glossl, J. (1999) Accumulation of sialic acid in endocytic compartments interferes with the formation of mature lysosomes. Impaired proteolytic processing of cathepsin B in fibroblasts of patients with lysosomal sialic acid storage disease. *J. Biol. Chem.*, **274**, 19063.
215. Chen, C-S., Bach, G., and Pagano, R. E. (1998) Abnormal transport along the lysosomal pathway in Mucolipidosis type IV disease. *Proc. Natl. Acad. Sci. USA*, **95**, 6373.
216. Zimmer, K-P., Le Coutre, P., Aerts, H. M. F. G., Harzer, K., Fukuda, M., O'Brien, J. S., *et al.* (1999) Intracellular transport of acid β-glucosidase and lysosome-associated membrane proteins is affected in Gaucher's disease (G202R mutation). *J. Pathol.*, **188**, 407.
217. Karageorgos, L. E., Isaac, E. L., Brooks, D. A., Ravenscroft, E. M., Davey, R., Hopwood, J. J., *et al.* (1997) Lysosomal biogenesis in lysosomal storage disorders. *Exp. Cell Res.*, **234**, 85.
218. Schiavo, G., Benfenati, F., Poulain, B., Rossetto, O., de Laureto, P., DasGupta, B. R., *et al.* (1992) Tetanus and botulinum-B neurotoxins block neurotransmitter release by proteolytic cleavage of synaptobrevin. *Nature*, **359**, 832.
219. Papini, E., de Bernard, M., Milia, E., Bugnoli, M., Zerial, M., Rappuoli, R., *et al.* (1994) Cellular vacuoles induced by *Helicobacter pylori* originate from late endosomal compartments. *Proc. Natl. Acad. Sci. USA*, **91**, 9720.
220. Molinari, M., Galli, C., Norais, N., Telford, J. L., Rappuoli, R., Luzio, J. P., *et al.* (1997) Vacuoles induced by *Helicobacter pylori* toxin contain both late endosomal and lysosomal markers. *J. Biol. Chem.*, **272**, 25339.
221. Reaves, B. J., Bright, N. A., Mullock, B. M., and Luzio, J. P. (1996) The effect of wortmannin on the localisation of lysosomal type I integral membrane glycoproteins suggests a role for phosphoinositide 3-kinase activity in regulating membrane traffic late in the endocytic pathway. *J. Cell Sci.*, **109**, 749.

7 Sorting and sorting signals in the endocytic pathway

STEFAN HÖNING

1. Introduction

Sorting signals located within the cytoplasmic tail of membrane proteins are required to ensure their correct targeting within the endomembrane system of a mammalian cell. Pathways which require cytoplasmic sorting signals include receptor-mediated endocytosis from the plasma membrane, the delivery from endosomes to the TGN, the recycling from endosomes to the plasma membrane, and the movement within the complex endosomal system. Since most sorting signals were first described as being required for endocytosis, the chapter will first focus on sorting at the plasma membrane, followed by the discussion of sorting within the endosomal system.

2. Sorting of membrane proteins during receptor-mediated endocytosis

It is generally accepted that the selective and efficient internalization of macromolecules occurs via receptor-mediated endocytosis (1, 2). During this process, a molecule which has to be taken up by the cell (the ligand) binds to its specific receptor at the plasma membrane forming a ligand–receptor complex. This complex is then concentrated in a specialized plasma membrane area (the coated pit), which invaginates into the cytoplasm, finally pinching off as a plasma membrane-derived clathrin-coated vesicle (CCV; see Chapter 1 and ref. 3). After the removal of the clathrin coat, the internalized vesicles are competent to fuse with early endosomes. In this context it is important to note also that membrane proteins which do not function as receptors and are normally only found on intracellular compartments, such as lysosomes or the *trans*-Golgi-Network (TGN), can under certain conditions appear at the plasma membrane. In this case, these proteins often undergo internalization followed by redirection to their respective site of function (see below for details). The most critical information for internalization is encoded by a short stretch of amino acids located within the cytoplasmic domain of membrane proteins. This was initially based on the finding that the internalization of the LDL receptor is abolished when a critical tyrosine residue in the tail of the receptor was mutated to alanine (4, 6). Up to now, the number of membrane proteins that harbour a tyrosine residue which is critical for internalization from the plasma membrane is steadily increasing. After the characterization of the tyrosine residue as an important determinant for sorting at the plasma membrane, many laboratories have analysed the importance of other amino acid residues present in a cytoplasmic tail. These analysis included the mutation of each individual tail residue (alanine scan mutagenesis), expression of the derived mutant membrane protein in a reporter cell line, and measurement of the endocytosis capacity of the protein, and have led to the concept that the tyrosine residue which mediates internalization functions together with neighbouring residues forming a sorting motif, in this case a tyrosine-based motif (7–9). A few of such motifs are listed in Fig. 1. The analysis of the importance of single amino acid residues within a tyrosine motif has revealed that frequently the tyrosine is followed by two residues (+1 and +2) that can be mutated without affecting internalization, while the residue in position +3 has to be hydrophobic (V, L, I) to mediate efficient endocytosis, thus forming a motif of the -YXXØ type (X, for any residue, Ø for a hydrophobic residue).

Apart from membrane proteins harbouring tyrosine-based sorting motifs, some proteins like the lysosomal membrane protein Limp-II do not contain a tyrosine residue within their cytoplasmic tail, or they contain tyrosine residues within the tail that were shown not to be part of a sorting signal (e.g. in the Fc receptor FcRγII). However, these proteins are also efficiently internalized. For such proteins a second

Protein	Tail length	Sequence	+1	+2	+3
Lamp-1	11	Y	Q	T	I
LAP	19	Y	R	H	V
TGN38	32	Y	Q	R	L
CD68 (macrosialin)	27	Y	Q	P	L
MPR46	67	Y	R	G	V
MPR300	163	Y	S	K	V
HIV1-env	~150	Y	S	P	L
HIV2-nef	256*	Y	S	R	F
TfR	57	Y	T	R	F
LDL-R	50	N	P	V	Y

Fig. 1 Tyrosine-based internalization motifs. The sequences of tyrosine-based sorting motifs from ten membrane proteins shown using the one letter amino acid code. Note that in all these examples a hydrophobic residue is found in the position Y +3. The tyrosine-based motif can be localized at the terminal end of a protein (such as at the COOH-terminus of Lamp-1) or somewhere in the middle of a cytoplasmic tail (e.g. residues 45–48 in MPR46). NB: The human immunodeficiency virus HIV type 2 Nef is a myristoylated cytosolic protein and not a transmembrane protein. The Y-based signal in HIV-2 is conserved in related simians viruses but not in HIV-1 Nefs. The cytoplasmic domain of the HIV-1 envelope protein is variable between different strains of virus. The Env signal is conserved in all strains of HIV-1, HIV-2, and SIV.

type of sorting motif was discovered—the di-leucine motif (10–12). Since this motif does not strictly involve two leucine residues (see Fig. 2) I prefer to use the term leucine-based sorting motif. Several of the proteins harbouring a leucine-based motif carry acid residues or serine residues in close proximity to the critical leucine-type residue which are required for efficient sorting of the respective protein, pointing to the notion that leucine-based sorting motifs are constituted not only by the two hydrophobic residues, but consist of a stretch of five to seven tail residues, such as DEXXXLI in Limp-II (13) (see Fig. 2).

The rate at which proteins are internalized via CCV-mediated endocytosis is quite variable, ranging from less than 1%/min up to more than 15%/min as detectable for the transferrin receptor (14–16). Differences in the affinity of the cytoplasmic sorting signals for the adaptor complexes that recognize the cytoplasmic sorting determinants (see Chapter 1) may contribute to these different rates. In addition, some signals such as the tyrosine-based internalization motif of CTLA can be masked by phosphorylation of the critical tyrosine residue (17, 18). Efficient internalization of CTLA only occurs after tyrosine dephosphorylation, which is thought to allow the interaction of the tyrosine motif with the plasma membrane associated AP-2 complex. On the other hand some signals such as the leucine-based motif in CD4 require phosphorylation to become active (19). In CD4, the di-leucine motif only efficiently binds AP-2 after phosphorylation of serine residues localized five residues N-terminal or one position C-terminal to the LL pair (see Fig. 2) thereby initiating rapid CD4 internalization.

Interestingly, some proteins have been found to contain more than one signal for a single sorting event. Endocytosis of MPR46 for example is only slightly effected by a

Protein	Tail length	Sequence
Limp II	19	D E R A P L I
CD3-γ	19	S D K Q T L L
Tyrosinase	32	E E R Q P L L
Tyrosinase	32	D D Y H S L L
In. chain	30	D D Q R D L L
In. chain	30	N E Q L P M L
MPR46	67	E E S E E R D D H L L
CD4	45	S Q I K R L L
FcRII-B2	47	Y S L L

Fig. 2 Leucine-based sorting motifs. Membrane proteins that contain leucine-based sorting motifs are listed. Note that some, e.g. tyrosinase and the MHC II invariant chain (Inv. chain) contain two functional signals. It some cases the critical leucine residues are replaced by other hydrophobic residues such as methionine or isoleucine. In some of the proteins, acid residues localized in close proximity to the critical leucine residues are known to be required for efficient sorting. In addition, the phosphorylation of neighbouring serine residues such as in MPR46 or CD4 have been shown to modulate the function of the motif (see text for details).

mutation of the only tail tyrosine residue. By using mutants of MPR46 in which several regions within the 67 tail amino acid were deleted revealed that MPR46 harbours three independent internalization motifs (20, 21). However it is still not clear which of the three motifs is utilized in the living cell. Other proteins that contain more than one internalization motif include MPR300, MHC II invariant chain, and furin (22–24).

3. Entry into the endocytic pathway via the TGN

Internalization from the plasma membrane is one way of entering the endocytic pathway. A second important route is by delivery of proteins directly from the TGN to endosomes. This route is taken by most lysosomal membrane proteins which are thought to be delivered from the TGN to lysosomes via the endosomal system without appearance at the plasma membrane (25). Since also other transport routes such as the secretory pathway diverge at the level of the TGN, delivery of proteins from the TGN to endosomes is thought to be signal mediated (26). Interestingly, it has been shown for several proteins that those signals originally described as internalization signals are also utilized for sorting at the TGN (Fig. 3). In case of the lysosomal membrane protein Lamp-1, substitution of the critical tyrosine tail residue to alanine resulted in cell surface delivery of the newly synthesized protein instead of delivery to lysosomes via the endosomal system (27, 28). Thus, the tyrosine motif of Lamp-1 is required for sorting at the TGN and the same signal is mediating endocytosis of the protein at the plasma membrane (see above). Two other well studied proteins that are known to be sorted at the TGN for delivery to endosomes are the mannose-6-phosphate receptors (MPRs). The two known MPRs, MPR46 and MPR300 mediate the delivery of soluble lysosomal enzymes harbouring a mannose-6-phosphate tag from the TGN to lysosomes. The receptors together with their bound enzymes are thought to be concentrated in AP-1 containing clathrin-coated pits at the TGN that bud off to transfer their cargo to endosomes (29). In this context it is important to note that for most cell surface receptors it is not known whether their biosynthetic transport from the TGN to the plasma membrane involves transient appearance in endosomes. Two proteins, the asialoglycoprotein and transferrin receptors, have been shown to be delivered from the TGN to the cell surface via endosomes (5, 30). Whether this is the general pathway for surface delivery of receptors remains to be established. Another important aspect of sorting from the TGN to endosomes which is still not well understood is the question to which endosomal subcompartment the different membrane proteins are delivered (see arrows in Fig. 3). It is for example not clear whether Lamp-1 is delivered from the TGN to early endosomes or to late endosomes. At least for MPR46 and MPR300, transport from the TGN to early endosomes is proposed (31, 32).

Here it is important to remember that the clathrin-associated adaptor complex AP-1 is thought to mediate sorting of proteins into CCVs at the TGN (33–35). Two other adaptor complexes (AP-3 and AP-4) have recently been discovered (36, 37). These also seem to be localized to some extend to the TGN. The localization of several

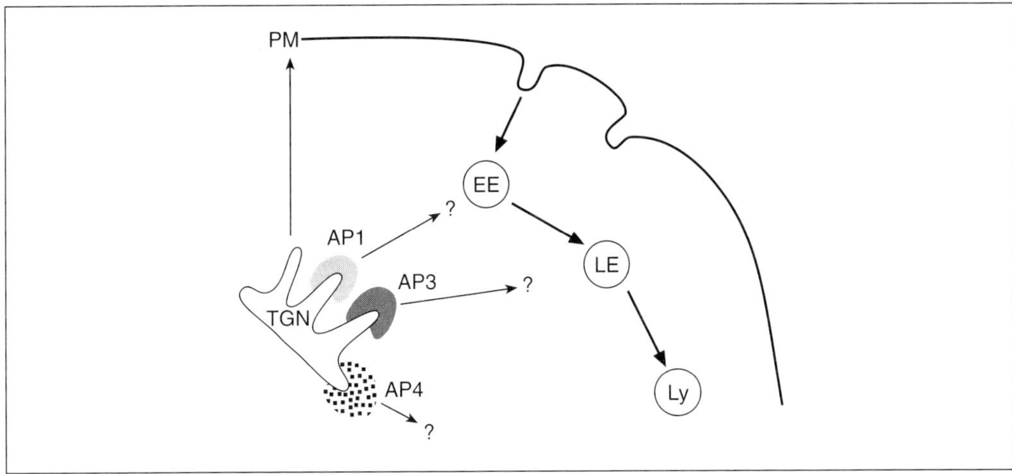

Fig. 3 Possible TGN exit sites for transport to endosomes. Several intracellular pathways diverge at the TGN and require sorting of the membrane proteins into distinct transport vesicles, such as AP-1 positive CCVs that contain the mannose-6-phosphate receptors. The recently discovered adaptor complexes AP-3 and AP-4 have also been localized to the TGN and may also participate in forming transport vesicles for a specialized subset of cargo proteins. This model however lacks the experimental proof so far. In addition, a characterization of the destination to which the TGN-derived transport vesicles are delivered is still missing. PM, plasma membrane; EE, early endosome; LE, late endosome; Ly, lysosome.

adaptors to the TGN has led to the hypothesis that each may play a role in sorting at different exit sites from the TGN; one adaptor for example mediating sorting to an early endosomal intermediate, while another may be involved in sorting to a late endosomal compartment. Although this concept is attractive, supporting experimental evidence is currently lacking.

4. Sorting from endosomes to the TGN

Trafficking from endosomes to the TGN is not regarded as a major pathway in the cell. Most receptors internalized from the plasma membrane either recycle from endosomes back to the cell surface or they are delivered along the endocytic pathway for degradation in lysosomes. However, some proteins like MPR46, MPR300, TGN38, or the endoprotease furin are selectively delivered from endosomes back to the TGN (23, 29, 38). When the intracellular distribution of these proteins is analysed by immunofluorescence at steady state, a predominant perinuclear staining is visible (as shown for MPR46 in Fig. 4A). When antibodies against the lumenal domain are added to the medium and the cells incubated at 37 °C, they can be detected in the TGN area within 30–60 minutes, indicating that the proteins can shuttle between the TGN, endosomes, and the plasma membrane (see Fig. 4B).

A kinetic analysis of the trafficking of a reporter protein composed of the lumenal domain of the interleukin-2 α chain (Tac) and the transmembrane and cyctoplasmic domains of TGN38 (TacTGN38) has shown that the bulk of TacTGN38 (82%) is in the TGN, 7% is localized to endosomes, and the remaining 11% is present at the plasma

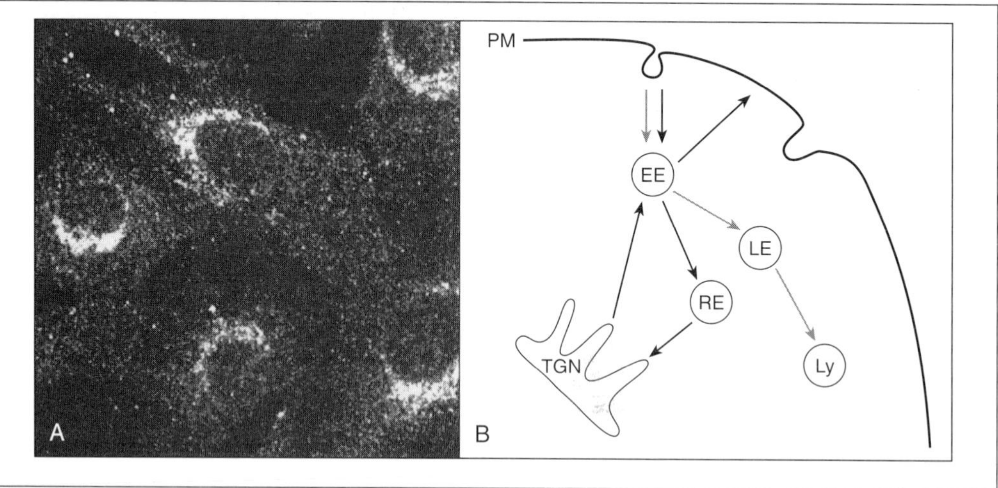

Fig. 4 Sorting from endosomes to the TGN. (A) Fibroblasts were fixed and processed for immunofluorescence to detect MPR46. The receptor is localized to the TGN and to small vesicular structures probably representing endosomes. (B) Simplified view of the trafficking of TGN38. The general endocytic pathway is indicated by grey arrows, trafficking of TGN38 is indicated by black arrows.

membrane. After internalization from the plasma membrane, 80% of the internalized protein is found to recycle back to the cell surface, while the remainder is delivered to the TGN. Further analysis revealed that TacTGN38 leaves the TGN quite slowly, resulting in its predominant TGN localization (39). The sequence motif within the cytoplasmic tail of TGN38 mediating its sorting from endosomes to the TGN is a tyrosine-based signal (38) (see Fig. 5). Interestingly, the sorting motifs in MPR46 and furin which mediate the same sorting step are totally different. For MPR46, a pair of aromatic residues (Phe18 and Trp19) is essential for the return of MPR46 from endosomes to the TGN (40–42). In case of furin, a stretch of acidic residues in close proximity to a serine residue mediate the delivery of furin from endosomes to the TGN (43).

Recognition of the latter two sorting signals by adaptors or adaptor-like proteins requires post-translational modification of the respective cytoplasmic tail. The function

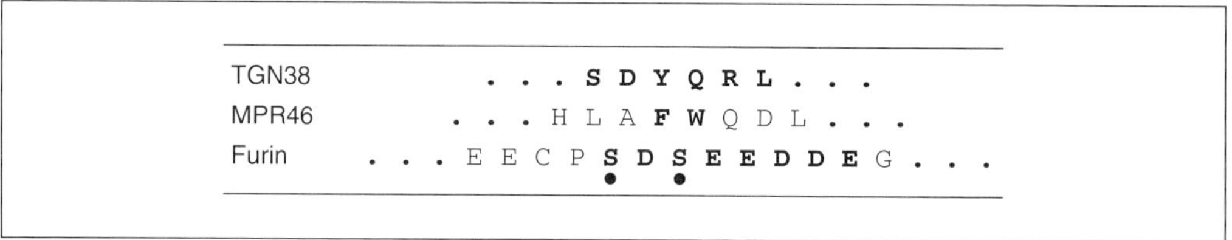

Fig. 5 Signals mediating sorting from endosomes to the TGN. Sorting motifs in TGN38, MPR46, and furin that have been shown to be required for delivery of the proteins from endosomes to the TGN. The black dots indicate serine residues in the furin cytoplasmic domain that, when dephosphorylated, allow recognition of the acidic sorting motif. The palmitoylation of the MPR46 tail which is thought to be necessary to expose the diaromatic motif is not shown (see Fig. 6).

of the diaromatic motif of MPR46 is thought to be dependent on the palmitoylation of critical cysteine residues upstream of the sorting motif (41), while the acidic cluster of furin only becomes active after dephosphorylation of neighbouring serine residues. This dephosphorylation is thought to occur in endosomes by a phosphatase 2A isoform (44, 45). The identity of the endosomal subcompartments from which delivery to the TGN can occur is still unclear. Several studies on the proteins discussed above suggest that transport to the TGN can occur from more than just one specialized endosomal subcompartment. TGN38 for example has been shown to be delivered to the TGN from recycling endosomes (39), which are often localized in close proximity to the TGN (46). In case of MPRs, late endosomes are thought to be the exit site for the delivery of the receptors to the TGN (47, 48), while the endosomal subcompartment from which the return of furin to the TGN occurs is not yet identified.

It is unclear whether the differences in the nature of the sorting signals used for TGN transport by MPR46, TGN38, and furin reflects the existence of three distinct pathways from endosomes. If so, this might imply that each of the three proteins is sorted and packaged into a different type of cargo vesicle. Formation of these vesicles

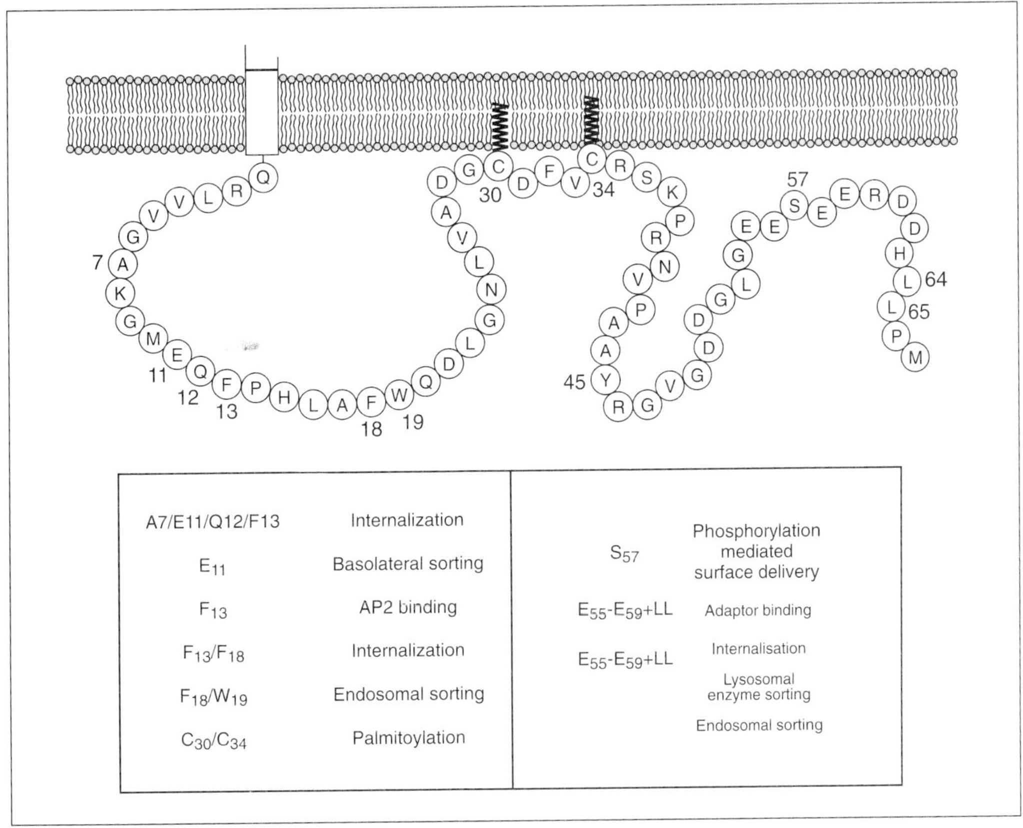

Fig. 6 MPR46 cytoplasmic sorting signals. Sorting signals in the MPR46 cytoplasmic domain are shown. The lumenal part was omitted.

would require a distinct type of sorting machinery comprising coat components or other factors facilitating the selective sorting of cargo for delivery from an endosomal subcompartment back to the TGN. One candidate sorting factor for the delivery of MPR300 (and MPR46) from endosomes to the TGN is TIP47 (49). Interestingly, MPR recycling from endosomes to the TGN is disrupted in mouse fibroblasts lacking the AP-1 medium subunit, suggesting that the AP-1 complex is also involved in the formation of endosome-derived vesicles destined for the return to the TGN (50). This notion is further supported by the observation that the retrograde delivery of shiga toxin from the cell surface to the TGN requires AP-1 (51). In addition, the retrieval of furin from endosomes which is thought to be mediated by the protein PACS-1 may also function in concert with AP-1 (23, 52). Further studies are required for a better understanding of retrograde transport from endosomes to the TGN.

5. Sorting from endosomes to the plasma membrane

Many receptors that mediate the uptake of macromolecules release their ligands in endosomes followed by recycling of the receptor to the cell surface. One typical example of a rapidly recycling receptor is the transferrin receptor. Recycling of other receptors like FcRγII depends on the type of ligand bound. Internalization of monovalent ligands bound to FcRγII is followed by rapid recycling of the receptor and ligand back to the cell surface. However, if immunocomplexes are bound to FcRγII that lead to cross-linking of the receptors, both the ligand and the receptor are delivered to lysosomes and undergo degradation (53, 54). Do either of these sorting events, or both, require an active sorting signal? When the recycling of the transferrin receptor was analysed kinetically, it turned out that the receptor recycled at a rate indistinguishable from that of fluorescent sphingolipid analogues (55). In addition, a transferrin receptor mutant lacking the entire cytoplasmic tail was found to recycle rapidly (56), suggesting that the receptor follows a default pathway also taken by bulk membrane lipids that return to the cell surface as fast as possible. In this context, we should consider that lipids can form specialized microdomains which themselves can mediate membrane protein sorting (57). It is also important to note that shortly after endocytosis, recycling receptors can be detected in tubular extensions of early endosomes, while soluble material accumulates in the vacuolar part (58, 59). Based on these early observations, one model for endosomal sorting assumes that from the tubular extensions vesicles can pinch off to return to the cell surface. These extensions may additionally also mature into other specialized endosomal subcompartments such as recycling endosomes. This latter model has recently gained some experimental proof since it was shown that Rab4 and Rab5 which are characteristic for early endosomes, gradually disappear from recycling endosomes, while Rab11 becomes more prominent, suggesting that the recycling endosome has biochemical properties distinguishable from those of early endosomes (46, 60, 61). It should also be mentioned that apart from the transferrin receptor other proteins like the glucose transporter Glut-4 in adipocytes (62) or the MHC II in antigen presenting cells (63) recycle to the plasma membrane from specialized endosomal subcompartents, strongly favouring the idea that recycling to the cell surface requires active sorting. Still the question remains, which factors of the cytoplasmic sorting machinery like adaptors or adaptor-like proteins participate in the recycling of membrane proteins from endosomes to the plasma membrane.

6. Sorting en route to lysosomes

The classical endocytic pathway includes the passage of material from early to late endosomes and the final fusion with lysosomes. The pathway to lysosomes is highly dynamic, making it difficult to determine to what extend trafficking of carrier vesicles account for sorting from endosomes to lysosomes. Alternatively, late endosomes and lysosomes may form by maturation (see Chapter 6), during which membrane components that should escape degradation such as recycling receptors are removed by vesicular carriers. This debate and the current view in the field is discussed elsewhere (see Chapters 5 and 6). It is however important to keep the model (and the uncertainties) of lysosomal trafficking/maturation in mind, since this has some important implications for the sorting of membrane proteins and receptors along this pathway. Within this pathway, the sorting of lysosomal membrane proteins and of the MPRs has been extensively studied.

The major lysosomal membrane protein constituents are members of the family of protein thought to protect the limiting membrane from degradation by lysosomal enzymes (25) and include Lamp-1, Lamp-2, Limp-I, and Limp-II. Lysosomal acid phosphatase (LAP) is a soluble lysosomal enzyme, it is transported to lysosomes as an integral membrane protein and is also discussed here (Fig. 7).

Common to all the lysosomal membrane proteins is a short cytoplasmic tail of 11–20

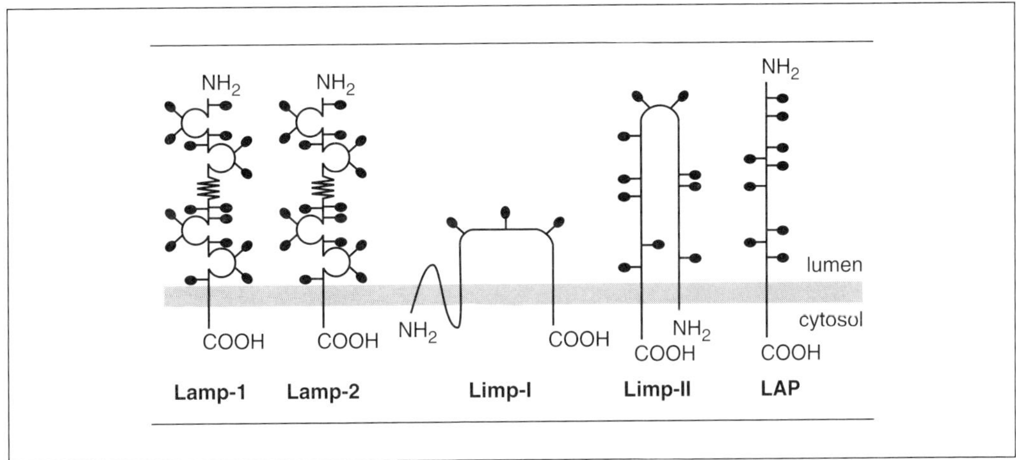

Fig. 7 Topology of lysosomal membrane proteins. The major lysosomal membrane proteins all share a high degree of glycosylation and a short carboxy terminal cytoplasmic tail which carries the information for intracellular sorting. Additional information can be found elsewhere (88). Lamp (lysosomal-associated membrane protein); Limp (lysosomal integral membrane protein); LAP (lysosomal acid phosphatase).

Protein	Sequence
Lamp-1	R K R S H A G Y Q T I
Lamp-2a	L K R H H T G Y E Q F
Lamp-2b	R R K S R T G Y Q S V
Lamp-2c	R R K T Y A G Y Q T L
Limp-I	K K S I R S G Y E V M
LAP	R M Q A Q P P G Y R H V A D G Q D H A
Limp-II	R G Q G S T D E G T A D E R A P L I R T

Fig. 8 Cytosolic tail sequences of lysosomal membrane proteins. The tail sequences of the respective mouse proteins and of human LAP are shown, starting with the first putative tail residue. Residues that were shown to be critical for correct targeting are shaded.

amino acid residues, which has been shown to contain the important intracellular sorting signals (Fig. 8). While Lamp-1, Lamp-2, Limp-I, and LAP all harbour tyrosine-based sorting motifs (28, 65, 66), the tail of Limp-II is carrying a leucine-based signal (13, 64). Although lysosomal membrane proteins are detectable at the cell surface during special developmental stages as well as in certain specialized cell types such as lymphocytes, sorting of the proteins is thought to occur mainly intracellular. In many cell types, the itinerary of these proteins after biosynthesis involves delivery from the TGN to endosomes, followed by further transport within the endosomal system to lysosomes. If however some molecules escape to the plasma membrane, they indirectly reach lysosomes after rapid internalization (25). A remarkable exception is LAP which is first delivered to the plasma membrane after biosynthesis, entering a recycling loop characterized by numerous rounds of endocytosis from and, recycling to, the cell surface, before transport to lysosomes occurs (67–69). Detailed analysis of the transport especially of Lamp proteins and LAP have revealed that the respective tyrosine motifs mediate sorting at the TGN, endocytosis at the plasma membrane, as well as basolateral sorting in polarized cells.

Although much has been learned about the targeting and signals carried by lysosomal membrane proteins, their trafficking in the endocytic system is not well characterized. Although it is clear that sorting at the TGN is required to mediate delivery to endosomes (28, 70), we do not know whether their further transport to lysosomes involves active sorting, or whether they are following the lysosomal pathway together with material destined to undergo degradation. We therefore analysed the signals in LAP and Lamp-1 with respect to endosomal sorting (S. Obermüller, K. von Figura, and S. Höning, submitted). Interestingly, the COOH terminal seven residues of LAP could be deleted without interfering with its correct targeting. The tail of this mutant, called LAPΔ7 (Fig. 9A) is very similar to the tail of Lamp-1: the

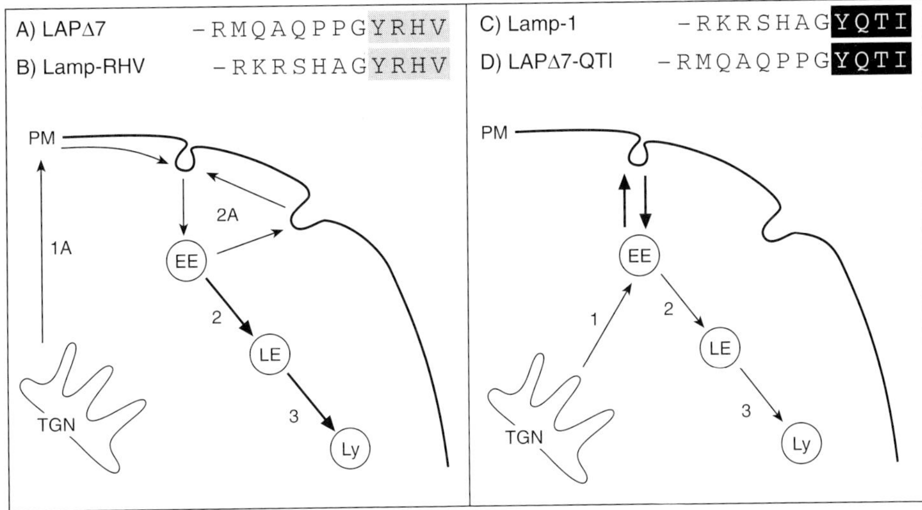

Fig. 9 Endosomal sorting of LAP depends on the composition of the tyrosine motif. A truncated form of LAP (LAPΔ7 (A), and LAP chimeras harbouring the wild-type Lamp-1 tail (C), or harbouring the first seven Lamp-1 residues and the LAP tyrosine motif (B) as well as a LAPΔ7 variant containing the tyrosine motif of Lamp-1 were generated and expressed in BHK-21 cells. The tyrosine motif of LAP is indicated by the grey box, the tyrosine motif of Lamp-1 is indicated by the black box. The pathway of wild-type LAP is shown on the left comprising delivery to the plasma membrane (A), followed by numerous rounds of recycling between endosomes and the plasma membrane (2A), and final delivery to lysosomes (2, 3). On the right, the trafficking of Lamp-1 is depicted. This involves passage from the Golgi to endosomes (1) followed by further delivery to lysosomes (2, 3). When cell surface delivery and recycling of all constructs was analysed, those chimeras containing the Lamp-1 tyrosine motif (C and D) were detectable at the cell surface of the overexpressing cells from which they were rapidly internalized (grey arrows on the right), recycling however was not detectable. On the other hand the constructs harbouring the LAP tyrosine motif (A and B) were also delivered to the cell surface, but in addition efficient recycling was detectable, showing that the tyrosine motifs of LAP and Lamp-1 mediate different endosomal sorting.

tyrosine motif is localized at the COOH terminus, they both harbour a tyrosine motif of the –YXXØ type which is separated from the membrane by only seven or eight residues. Additionally, we also constructed chimeric LAP mutants, one harbouring the wild-type Lamp-1 tail (Fig. 9C), one with the first nine residues of Lamp-1 followed by the residues +1 to +3 of the LAP tyrosine motif (Fig. 9B) and a LAPΔ7 variant in which the residues +1 to +3 of the tyrosine motif were substituted for those of Lamp-1 (Fig. 9D). All were expressed in BHK-21 cells and Lamp-1 and LAP-derived constructs analysed for their ability to recycle between endosomes and the plasma membrane.

In overexpressing cells, a fraction of each of the proteins was delivered to the cell surface, however only those constructs carrying the LAP tyrosine motif exhibited the ability to recycle between endosomes and the plasma membrane (see Fig. 9, left). In contrast, the LAP mutants carrying the Lamp-1 tyrosine motif were found to be internalized with equal efficiency, but recycling was not detectable. In summary, these results demonstrate that sorting on the level of endosomes determines the recycling of LAP and the retention of Lamp-1 in the endosomal system. Endosomal sorting of the two proteins was not dependent on the membrane-proximal tail residues, but strictly dependent on the type of tyrosine motif that was exposed. Consequently, the question arises whether the recycling of LAP to the cell surface requires an active sorting event, or whether endosomal retention of Lamp-1 is signal mediated. Considering the fact that a tail-less transferrin receptor is able to recycle (see above), one is likely to favour the idea that LAP recycling does not require active sorting. This would imply that the retention of Lamp-1 in the endosomal system is dependent on its tyrosine motif, which is recognized by cytoplasmic sorting factors. More general, the model would favour the notion that sorting (at least of some lysosomal membrane proteins) along the endocytic pathway to lysosomes is signal mediated. The tyrosine motif of LAP on the other hand would bind to the relevant sorting factor with only low affinity, thus causing low fidelity endosomal sorting and prominent recycling to the cell surface.

Sorting factors that have so far shown to bind to lysosomal membrane proteins like Lamp-1 and LAP are the adaptor complexes AP-1, AP-2, and AP-3 (71). While it is clear that AP-2 mediates the internalization of both LAP and Lamp-1, AP-1 and AP-3 binding to LAP has not been reported. The results on AP-1 and AP-3 binding to Lamp-1 are conflicting at the moment, since we reported initially *in vitro* binding of AP-1 to Lamp-1, as well as the co-localization of the two proteins in CCVs in Lamp-1 overexpressing cells (72). However, recent reports have demonstrated a mis-sorting of Lamp-1 to the cell surface in cells lacking the AP-3 β subunit and in cells with decreased expression levels of the AP-3 medium chain, pointing to the importance of the AP-3 complex for sorting of Lamp-1 (73, 74). Although the AP-3 complex has been shown to mediate the formation of synaptic-like vesicles from endosomes in PC12 cells (75), it is still unclear whether AP-3 is only functional on endosomes or whether it is also involved in sorting at the TGN. Endosomal sorting of lysosomal membrane proteins and the identification of the participating sorting factors will therefore still be a subject of intense investigation.

Another class of cytoplasmic sorting factors that has been localized to endosomes are the COP proteins. Especially COPI was thought to play a role in trafficking from early to late endosomes (76, 77). More recently, COPI was shown to be required for endosomal sorting of CD4 to lysosomes (78). First binding of the primate lentivirus protein NEF to CD4 initiate the rapid internalization of the two proteins by AP-2 containing clathrin-coated vesicles. Subsequently, the interaction of NEF with β-COP initiate the recruitment of COPI to endosomes which was shown to be required for sorting of CD4 into the degradative pathway. Recruitment of COPI was mediated by a novel endosomal sorting motif in the NEF cytoplasmic tail, which is composed of acid residues, showing that sorting of certain proteins from early endosomes to late endosomes indeed require a specific sorting signal which in the case of NEF is recognized by COPI.

In contrast to the lysosomal membrane proteins which utilize the endocytic pathway for delivery to lysosomes, the MPRs that are also detected in endosomal subcompartments escape delivery to lysosomes. In fact, one common criterium to distinguish late endosomes from lysosomes is the presence or the absence of MPRs, respectively. Because mutation of the above mentioned diaromatic motif in MPR46 leads to increased lysosomal receptor degradation, the diaromatic motif was described as a lysososmal avoidance signal (40–42). However, since the transition of late endosomes to lysosomes is very dynamic and difficult to define, the diaromatic motif in the MPR46 tail may simply be a signal for efficient sorting of the receptor into vesicles or membrane areas that give rise to vesicles recycling back to the TGN.

7. Sorting within the endosomal system of polarized cells

Sorting of membrane proteins in polarized epithelial cells has special features. In this chapter, only details of sorting mechanisms that operate within the endocytic pathway will be described. Further reading can be found elsewhere (79, 80). One special feature of epithelial cells is their ability for polarized sorting of internalized proteins. The transferrin receptor and the LDL receptor, both of which are usually restricted to the basolateral plasma membrane domain, undergo continuous cycles of internalization and recycling back to the same domain. Recycling to the respective membrane domain is of course an endosomal sorting event, that is likely to work very efficiently, since mis-sorting to the apical surface is minimal. Transcytosis—the sorting of proteins across the cell to the apical surface is also a typical feature of epithelial cells, however it seems to be only relevant for specialized transcytotic receptors such as the polymeric Ig receptor (81). Some of the molecular requirements for basolateral sorting, basolateral recycling, and transcytosis have been recently elucidated. Interestingly some proteins like the Fc receptor which harbours a di-leucine internalization motif, use exactly the same motif for basolateral targeting. Also tyrosine-based motifs mediating internalization can be used as functional basolateral sorting determinants, as for example in LAP (69). Although the internalization signal in this case is not identical with the basolateral signal, they both overlap. Other proteins like MPR46 comprise a basolateral sorting signal which is distinct from the internalization signal (82, 83 and Fig. 6). In addition, the LDL receptor is an example for a protein that has two basolateral sorting signals, one proximal signal which overlaps with the tyrosine motif required for internalization, the second distal basolateral signal is also tyrosine-based, but does not mediate internalization (84, 85). Both basolateral signals in the LDL receptor depend on adjacent acidic residues. Since basolateral sorting occurs not only at the TGN but also in endosomes after internalization, it is clear that basolateral recycling in polarized cells is signal mediated.

For a long time, the cytosolic factors recognizing epithelial specific sorting signals were unknown. However, recent work is pointing to the notion that adaptors again play a role in basolateral targeting. It has been shown that epithelial cells express a cell type-specific AP-1 μ subunit which, together with the other subunits, forms the AP-1B complex. LLC-PK1 cells lacking μ1B exhibited mis-sorting of the LDL receptor and the transferrin receptor (86). Whether the AP-1B complex is functional at the

TGN or in endosomes is not yet clear, but interestingly clathrin-coated buds on endosomes containing the recycling transferrin receptor were shown to contain AP-1. Thus basolateral sorting from endosomes may require an epithelial-specific adaptor complex. However, sorting of the Fc receptor was not affected in LLC-PK1 cells, suggesting that other adaptors or factors that participate in endosomal sorting in epithelial cells remain to be characterized.

8. Conclusions and future directions

Much has been learned about sorting and sorting signals since the discovery of clathrin-coated structures more than thirty years ago. Still it seems that not all kinds of sorting signals have been identified yet. It is also fascinating to see that the cell in many cases seems to have more than one pathway to deliver cargo to a given destination. For a long time, the endocytic pathway was regarded as a kind of one-way-street to the lysosome, but we have already begun to understand that this is far from what is real. The possibility of retrograde transport in the endocytic system, e.g. out of the lysosome or out of organelles with lysosomal characteristics still requires more attention. We are also just starting to explore the dynamics of sorting within the endocytic pathway, especially due to the availability of novel techniques such as fast high resolution video imaging and tools like the GFP technology. The number of proteins and factors that directly or indirectly participate in the sorting of cargo proteins is steadily increasing, but trying to understand the interplay between them requires more work. Moreover, many long-standing questions, such as how adaptors are targeted to membranes, remain unanswered. Nearly every compartment in the cell should contain cargo molecules that can be recognized by adaptors, thus the adaptor recruitment to their specific site of function is not trivial. To solve problems like the regulation of adaptor binding to sorting determinants will keep us busy for quite some time.

Acknowledgements

I wish to thank K. von Figura for his continuous support of my research, and the members of my group for their experimental contributions, for helpful comments and discussions on the manuscript. S. H. is supported by grants from the German Science Foundation (SFB523, A5) and the EU (XCT 960058).

References

1. Mellman, I. (1996) Endocytosis and molecular sorting. *Annu. Rev. Cell Dev. Biol.*, **12**, 575.
2. Marsh, M. and McMahon, H. T. (1999) The structural era of endocytosis. *Science*, **285**, 215.
3. Kirchhausen, T. (1999) Adaptors for clathrin-mediated traffic. *Annu. Rev. Cell Dev. Biol.*, **15**, 705.
4. Davis, C. G., Lehrman, M. A., Russell, D. W., Anderson, R. G., Brown, M. S., and Goldstein, J. L. (1986) The J. D. mutation in familial hypercholesterolemia: amino acid substitution in cytoplasmic domain impedes internalization of LDL receptors. *Cell*, **45**, 15.

5. Futter, C. E., Connolly, C. N., Cutler, D. F., and Hopkins, C. R. (1995) Newly synthesized transferrin receptors can be detected in the endosome before they appear on the cell-surface. *J. Biol. Chem.*, **270**, 10999.
6. Brown, M. S., Anderson, R. G. W., and Goldstein, J. L. (1983) Recycling receptors: The round-trip itinerary of migrant membrane proteins. *Cell*, **32**, 663.
7. Boll, W., Ohno, H., Zhou, S. Y., Rapoport, I., Cantley, L. C., Bonifacino, J. S., *et al.* (1996) Sequence requirements for the recognition of tyrosine-based endocytic signals by clathrin AP-2 complexes. *EMBO J.*, **15**, 5789.
8. Marks, M. S., Ohno, H., Kirchhausen, T., and Bonifacino, S. J. (1997) Protein sorting by tyrosine-based signals: adapting to the ys and wherefores. *Trends Cell Biol.*, **7**, 124.
9. Owen, D., Vallis, Y., Noble, M., Hunter, J., Dafforn, T., Evans, P., *et al.* (1999) A structural explanation for the binding of multiple ligands by the α-adaptin appendage domain. *Cell*, **97**, 805.
10. Haft, C., Klausner, R., and Taylor, S. (1994) Involvement of dileucine motifs in the internalization and degradation of the insulin receptor. *J. Biol. Chem.*, **269**, 26286.
11. Letourneur, F. and Klausner, R. D. (1992) A novel di-leucine motif and a tyrosine-based motif independently mediate lysosomal targeting and endoscytosis of CD3 chains. *Cell*, **69**, 1143.
12. Sandoval, I. V. and Bakke, O. (1994) Targeting of membrane proteins to endosomes and lysosomes. *Trends Cell Biol.*, **4**, 292.
13. Höning, S., Sandoval, I., and von Figura, K. (1998) A di-leucine based motif in the cytoplasmic tail of limp-II and tyrosinase mediates selective binding of AP3. *EMBO J.*, **17**, 1304.
14. Jing, S., Spencer, T., Miller, K., Hopkins, C., and Trowbridge, I. S. (1990) Role of the human transferrin receptor cytoplasmic domain in endocytosis: Localization of a specific signal sequence for internalization. *J. Cell Biol.*, **110**, 283.
15. Iacopetta, B. J., Rothenberger, S., and Kühn, L. C. (1988) A role for the cytoplasmic domain in transferrin receptor sorting and coated pit formation during endocytosis. *Cell*, **54**, 485.
16. Collawn, J. F., Stangel, M., Kuhn, L. A., Esekogwu, V., Jing, S., Trowbridge, I. S., *et al.* (1990) Transferrin receptor internalization sequence YXRF implicates a tight turn as the structural recognition motif for endocytosis. *Cell*, **63**, 1061.
17. Shiratori, T., Miyatake, S., Ohno, H., Nakaseko, C., Isono, K., Bonifacino, J. S., *et al.* (1997) Tyrosine phosphorylation controls internalization of CTLA-4 by regulating its interaction with clathrin-associated adaptor complex AP-2. *Immunity*, **6**, 583.
18. Zhang, Y. and Allison, J. P. (1997) Interaction of CTLA-4 with AP50, a clathrin-coated pit adaptor protein. *Proc. Natl. Acad. Sci. USA*, **94**, 9273.
19. Pitcher, C., Höning, S., Fingerhut, A., Bowers, K., and Marsh, M. (1999) CD4 endocytosis and adaptor complex binding require activation of the CD4 endocytosis signal by serine phosphorylation. *Mol. Biol. Cell*, **10**, 677.
20. Johnson, K. F., Chan, W., and Kornfeld, S. (1990) Cation-dependent mannose-6-phosphate receptor contains two internalization signals in its cytoplasmic domain. *Proc. Natl. Acad. Sci. USA*, **87**, 10010.
21. Denzer, K., Weber, B., Hillerehfeld, A., Vonfigura, K., and Pohlmann, R. (1997) Identification of three internalization sequences in the cytoplasmic tail of the 46 kDa mannose 6-phosphate receptor. *Biochem. J.*, **326**, 497.
22. Bremnes, B., Madsen, T., Geddedahl, M., and Bakke, O. (1994) An LI and ML motif in the cytoplasmic tail of the MHC-associated invariant chain mediate rapid internalization. *J. Cell Sci.*, **107**, 2021.

23. Molloy, S., Anderson, E., Jean, F., and Thomas, G. (1999) Bi-cycling the furin pathway from the TGN localization to pathogen activation and embryogenesis. *Trends Cell Biol.*, **9**, 28.
24. Chen, H. J., Remmler, J., Delaney, J. C., Messner, D. J., and Lobel, P. (1993) Mutational analysis of the cation-independent mannose-6-phosphate/insulin-like growth factor II receptor. *J. Biol. Chem.*, **268**, 22338.
25. Hunziker, W. and Geuze, H. (1996) Intracellular trafficking of lysosomal membrane proteins. *BioEssays*, **18**, 379.
26. Mellman, I. and Simons, K. (1992) The Golgi complex: In vitro veritas? *Cell*, **68**, 829.
27. Harter, C. and Mellman, I. (1992) Transport of the lysosomal membrane glycoprotein lgp120 (lgp-A) to lysosomes does not require appearance on the plasma membrane. *J. Cell Biol.*, **117**, 311.
28. Höning, S. and Hunziker, W. (1995) Cytoplasmic determinants involved in direct lysosomal sorting, endocytosis, and basolateral targeting of rat lgp120 (lamp-I) in MDCK cells. *J. Cell Biol.*, **128**, 321.
29. Hille-Rehfeld, A. (1995) Mannose 6-phosphate receptors in sorting and transport of lysosomal enzymes. *Biochem. Biophys. Acta.*, **1241**, 177.
30. Leitinger, B., Hillerehfeld, A., and Spiess, M. (1995) Biosynthetic transport of the asialoglycoprotein receptor h1 to the cell surface occurs via endosomes. *Proc. Natl. Acad. Sci. USA*, **92**, 10109.
31. Press, B., Feng, Y., Hoflack, B., and Wandinger-Ness, A. (1998) Mutant rab7 causes the accumulation of cathepsin D and cation-independent M6P receptor in an early endocytic compartment. *J. Cell Biol.*, **140**, 1075.
32. Tikkanen, R., Denzer, K., Pungitore, R., Geuze, H., von Figura, K., and Höning, S. (2000) The dileucine motif within the tail of MPR46 is required for sorting of the receptor in endosomes. *Traffic*, **1**, 631.
33. Traub, L. M., Ostrom, J. A., and Kornfeld, S. (1993) Biochemical dissection of AP-1 recruitment onto Golgi membranes. *J. Cell Biol.*, **123**, 561.
34. Traub, L. M., Kornfeld, S., and Ungewickell, E. (1995) Different domains of the AP-1 adaptor complex are required for golgi membrane binding and clathrin recruitment. *J. Biol. Chem.*, **270**, 4933.
35. Zhu, Y., Traub, L., and Kornfeld, S. (1999) High-affinity binding of the AP1 adaptor complex to trans-Golgi network membranes devoid of mannose-6-phosphate receptors. *Mol. Biol. Cell*, **10**, 537.
36. Dell'Angelica, E., Mullins, C., and Bonifacino, E. (1999) AP-4, a novel protein complex related to clathrin adaptors. *J. Biol. Chem.*, **274**, 7278.
37. Hirst, J., Bright, N., Rous, B., and Robinson, M. S. (1999) Characterization of a fourth adaptor-related protein complex. *Mol. Biol. Cell*, **10**, 2787.
38. Ponnambalam, S., Rabouille, C., Luzio, J. P., Nilsson, T., and Warren, G. (1994) The TGN38 glycoprotein contains two non-overlapping signals that mediate localization to the trans-Golgi network. *J. Cell Biol.*, **125**, 253.
39. Ghosh, R., Mallet, W., Soe, T., McGraw, T., and Maxfield, F. (1998) An endocytosed TGN38 chimeric protein is delivered to the TGN after trafficking through the endocytic recycling compartment in CHO cells. *J. Cell Biol.*, **142**, 923.
40. Rohrer, J., Schweizer, A., Johnson, K. F., and Kornfeld, S. (1995) A determinant in the cytoplasmic tail of the cation-dependent mannose 6-phosphate receptor prevents trafficking to lysosomes. *J. Cell Biol.*, **130**, 1297.
41. Schweizer, A., Kornfeld, S., and Rohrer, J. (1996) Cysteine(34) of the cytoplasmic tail of the

cation-dependent mannose 6-phosphate receptor is reversibly palmitoylated and required for normal trafficking and lysosomal enzyme sorting. *J. Cell Biol.*, **132**, 577.

42. Schweizer, A., Kornfeld, S., and Rohrer, J. (1997) Proper sorting of the cation-dependent mannose-6-phosphate receptor in endosomes depends on a pair of aromatic amino acids in its cytoplasmic tail. *Proc. Natl. Acad. Sci. USA*, **94**, 14471.
43. Voorhees, P., Deignan, E., Vandonselaar, E., Humphrey, J., Marks, M. S., Peters, P. J., *et al.* (1995) An acidic sequence within the cytoplasmic domain of furin functions as a determinant of trans-Golgi network localization and internalization from the cell surface. *EMBO J.*, **14**, 4961.
44. Molloy, S., Thomas, L., Kamibayashi, C., Mumby, M., and Thomas, G. (1998) Regulation of endosome sorting by a specific PP2A isoform. *J. Cell Biol.*, **142**, 1399.
45. Jones, B. G., Thomas, L., Molloy, S. S., Thulin, C. D., Fry, M. D., Walsh, K. A., *et al.* (1995) Intracellular trafficking of furin is modulated by the phosphorylation state of a casein kinase II site in its cytoplasmic tail. *EMBO J.*, **14**, 5869.
46. Ullrich, O., Reinsch, S., Urbe, S., Zerial, M., and Parton, R. G. (1996) Rab11 regulates recycling through the pericentriolar recycling endosome. *J. Cell Biol.*, **135**, 913.
47. Klumperman, J., Hille, A., Veenendaal, T., Oorschot, V., Stoorvogel, W., Vonfigura, K., *et al.* (1993) Differences in the endosomal distributions of the 2 mannose 6-phosphate receptors. *J. Cell Biol.*, **121**, 997.
48. Nicoziani, P., Vilhardt, F., Llorente, A., Hilout, L., Courtoy, P. J., Sandvig, K., *et al.* (2000) Role for dynamin in late endosome dynamics and trafficking of the cation-independent mannose 6-phosphate receptor. *Mol. Biol. Cell*, **11**, 481.
49. Diaz, E. and Pfeffer, S. (1998) TIP47: A cargo selection device for mannose 6-phosphate receptor trafficking. *Cell*, **93**, 433.
50. Meyer, C., Zizioli, D., Lausmann, S., Eskelinen, E. L., Hamann, J., Saftig, P., *et al.* (2000) μ1A-adaptin-deficient mice: lethality, loss of AP-1 binding and rerouting of mannose 6-phosphate receptors. *EMBO J.*, **19**, 2193.
51. Mallard, F., Antony, C., Tenza, C., Salamero, J., Goud, B., and Johannes, L. (1998) Direct pathway from early/recycling endosomes to the Golgi apparatus revealed through the study of shiga toxin B-fragment transport. *J. Cell Biol.*, **143**, 973.
52. Wan, L., Molloy, S., Thomas, L., Liu, G., Xiang, Y., Rybak, L., *et al.* (1998) PACS-1 defines a novel gene family of cytosolic sorting proteins required for trans-Golgi network localization. *Cell*, **94**, 205.
53. Mellman, I. and Plutner, H. (1984) Internalization and degradation of macrophage Fc receptor bound to polyvalent immune complexes. *J. Cell Biol.*, **98**, 1170.
54. Mellman, I., Plutner, H., and Ukkonen, P. (1984) Internalization and rapid recycling of macrophage Fc receptors tagged with monovalent anti-receptor antibody: possible role of a prelysosomal compartment. *J. Cell Biol.*, **98**, 1163.
55. Mayor, S., Presley, J. F., and Maxfield, F. R. (1993) Sorting of membrane components from endosomes and subsequent recycling to the cell surface occurs by a bulk flow process. *J. Cell Biol.*, **121**, 1257.
56. Marsh, E. W., Leopold, P. L., Jones, N. L., and Maxfield, F. R. (1995) Oligomerized transferrin receptors are selectively retained by a lumenal sorting signal in a long-lived endocytic recycling compartment. *J. Cell Biol.*, **129**, 1509.
57. Fiedler, K., Kobayashi, T., Kurzchalia, T. V., and Simons, K. (1993) Glycosphingolipid-enriched, detergent-insoluble complexes in protein sorting in epithelial cells. *Biochemistry*, **32**, 6365.
58. Geuze, H. J., Slot, J. W., Strous, G. J., and Schwartz, A. L. (1983) The pathway of the asialo-

glycoprotein-ligand during receptor-mediated endocytosis: a morphological study with colloidal gold/ligand in the human hepatoma cell line, Hep G2. *Eur. J. Cell Biol.*, **32**, 38.

59. Geuze, H. J., Slot, J. W., Strous, G. J., Peppard, J., von Figura, K., Hasilik, A., *et al.* (1984) Intracellular receptor sorting during endocytosis: comparative immunoelectron microscopy of multiple receptors in rat liver. *Cell*, **37**, 195.
60. Daro, E., Vandersluijs, P., Galli, T., and Mellman, I. (1996) Rab4 and cellubrevin define different early endosome populations on the pathway of transferrin receptor recycling. *Proc. Natl. Acad. Sci. USA*, **93**, 9559.
61. Sönnichsen, B., De Renzis, S., Nielsen, E., Rietdorf, J., and Zerial, M. (2000) Distinct membrane domains on endosomes in the recycling pathway visualized by multicolor imaging of rab4, rab5, and rab11 [In Process Citation]. *J. Cell Biol.*, **149**, 901.
62. Martin, S., Tellam, J., Livingstone, C., Slot, J. W., Gould, G. W., and James, D. E. (1996) The glucose transporter (GLUT-4) and vesicle-associated membrane protein-2 (VAMP-2) are segregated from recycling endosomes in insulin-sensitive cells. *J. Cell Biol.*, **134**, 625.
63. Amigorena, S., Drake, J. R., Webster, P., and Mellman, I. (1994) Transient accumulation of new class II MHC molecules in a novel endocytic compartment in B lymphocytes. *Nature*, **369**, 113.
64. Sandoval, I. V., Arredondo, J. J., Alcalde, J., Noriega, A. G., Vandekerckhove, J., Jimenez, M. A., *et al.* (1994) The residues Leu(Ile) (475)-Ile(Leu,Val,ALA) (476), contained in the extended carboxyl cytoplasmic tail, are critical for targeting of the resident lysosomal membrane protein limp II to lysosomes. *J. Biol. Chem.*, **269**, 6622.
65. Gough, N. and Fambrough, D. (1997) Different steady state subcellular distributions of the three splice variants of lysosome-associated membrane protein lamp-2 are determined largely by the COOH-terminal amino acid residue. *J. Cell Biol.*, **137**, 1161.
66. Escola, J. M., Kleijmeer, M. J., Stoorvogel, W., Griffith, J. M., Yoshie, O., and Geuze, H. J. (1998) Selective enrichment of tetraspan proteins on the internal vesicles of multivesicular endosomes and on exosomes secreted by human B- lymphocytes. *J. Biol. Chem.*, **273**, 20121.
67. Braun, M. A., Waheed, A., and von Figura, K. (1989) Lysosomal acid phosphatase is transported to lysosomes via the cell surface. *EMBO J.*, **8**, 3633.
68. Peters, C., Braun, M., Weber, B., Wendland, M., Schmidt, B., Pohlmann, R., *et al.* (1990) Targeting of a lysosomal membrane protein: a tyrosine-containing endocytosis signal in the cytoplasmic tail of lysosomal acid phosphatase is necessary and sufficient for targeting to lysosomes. *EMBO J.*, **9**, 3497.
69. Prill, V., Lehmann, L., von Figura, K., and Peters, C. (1993) The cytoplasmic tail of lysosomal acid phosphatase contains overlapping but distinct signals for basolateral sorting and rapid internalization in polarized MDCK cells. *EMBO J.*, **12**, 2181.
70. Karlsson, K. and Carlsson, S. (1998) Sorting of lysosomal membrane glycoproteins lamp-1 and lamp-2 into vesicles distinct from MPR/γ-adaptin vesicles at the trans-Golgi network. *J. Biol. Chem.*, **273**, 18966.
71. Heilker, R., Spiess, M., and Crottet, P. (1999) Recognition of sorting signals by clathrin adaptors. *BioAssays*, **21**, 558.
72. Höning, S., Griffith, J., Geuze, H. J., and Hunziker, W. (1996) The tyrosine-based lysosomal targeting signal in lamp-1 mediates sorting into golgi-derived clathrin-coated vesicles. *EMBO J.*, **15**, 5230.
73. Le Borgne, R., Alconada, A., Bauer, U., and Hoflack, B. (1998) The mammalian AP3 adaptor like complex mediates the intracellular transport of lysosomal membrane glycoproteins. *J. Biol. Chem.*, **273**, 29451.
74. Dell'Angelica, E., Shotelersuk, V., Aguilar, R., Gahl, W., and Bonifacino, J. (1999) Altered

trafficking of lysosomal proteins in Hermansky-Pudlak Syndrome due to mutations in the β3A subunit of the AP3 complex. *Mol. Cell*, **3**, 11.
75. Faundez, V., Horng, J., and Kelly, R. (1998) A function for the AP3 coat complex in synaptic vesicle formation from endosomes. *Cell*, **93**, 423.
76. Guo, Q., Penman, M., Trigatti, B. L., and Krieger, M. (1996) A single point mutation in epsilon-COP results in temperature-sensitive, lethal defects in membrane transport in a chinese hamster ovary cell mutant. *J. Biol. Chem.*, **271**, 11191.
77. Daro, E., Sheff, D., Gomez, M., Kreis, T., and Mellman, I. (1997) Inhibition of endosome function in CHO cells bearing a temperature-sensitive defect in the coatomer (COPI) component ε-COP. *J. Cell Biol.*, **139**, 1747.
78. Piguet, V., Gu, F., Foti, M., Demaurex, N., Gruenberg, J., Carpentier, J. L., *et al.* (1999) Nef-induced CD4 degradation: a diacidic-based motif in Nef functions as a lysosomal targeting signal through the binding of beta-COP in endosomes. *Cell*, **97**, 63.
79. Aroeti, B., Okhrimenko, H., Reich, V., and Orzech, E. (1998) Polarized trafficking of plasma membrane proteins: emerging roles for coats, SNAREs, GTPases and their link to the cytoskeleton. *Biochim. Biophys. Acta*, **1376**, 57.
80. Mellman, I. and Warren, G. (2000) The road taken: past and future foundations of membrane traffic. *Cell*, **100**, 99.
81. Mostov, K. E. (1994) Transepithelial transport of immunoglobulins. *Annu. Rev. Immunol.*, **12**, 63.
82. Bresciani, R., Denzer, K., Pohlmann, R., and Vonfigura, K. (1997) The 46 kDa mannose-6-phosphate receptor contains a signal for basolateral sorting within the 19 juxtamembrane cytosolic residues. *Biochem. J.*, **327**, 811.
83. Distel, B., Bauer, U., Le Borgne, R., and Hoflack, B. (1998) Basolateral sorting of the cation-dependent mannose-6-phosphate receptor in MDCK cells. Identification of a basolateral determinant unrelated to clathrin-coated pit localization signals. *J. Biol. Chem.*, **273**, 186.
84. Matter, K., Hunziker, W., and Mellman, I. (1992) Basolateral sorting of LDL receptor in MDCK cells—the cytoplasmic domain contains 2 tyrosine-dependent targeting determinants. *Cell*, **71**, 741.
85. Matter, K., Whitney, J. A., Yamamoto, E. M., and Mellman, I. (1993) Common signals control low density lipoprotein receptor sorting in endosomes and the golgi complex of MDCK cells. *Cell*, **74**, 1053.
86. Fölsch, H., Ohno, H., Bonifacino, J., and Mellman, I. (1999) A novel clathrin adaptor complex mediates basolateral targeting in polarized epithelial cells. *Cell*, **99**, 189.
87. Futter, C., Gibson, A., Allchin, E., Maxwell, S., Ruddock, L., Odorizzi, G., *et al.* (1998) In polarized MDCK cells basolateral vesicles arise from clathrin γ-adaptin coated domains on endosomal tubules. *J. Cell Biol.*, **141**, 611.
88. Fukuda, M. (1991) Lysosomal membrane glycoproteins. Structure, biosynthesis, and intracellular trafficking. *J. Biol. Chem.*, **266**, 21327.

8 Endocytosis and antigen presentation

MONIQUE J. KLEIJMEER and GRAÇA RAPOSO

1. Introduction

The presentation of antigen required to initiate most immune responses results from the intracellular generation of protein fragments, and their binding and transport to the cell surface in stable association with the polymorphic products of the class I and class II loci of the major histocompatibility complex (MHC) (1, 2). The ability of MHC molecules to associate with and present antigenic peptides generated in distinct intracellular sites is related to their pattern of cellular expression (3). Ubiquitously expressed MHC I molecules bind peptides derived primarily from endogenous antigens in the endoplasmic reticulum (ER), resulting in their presentation to $CD8^+$ cytotoxic T cells. MHC II molecules are expressed predominantly in so-called antigen presenting cells (APCs: macrophages, B lymphocytes, dendritic cells (DCs), and thymic epithelial cells), and cell types such as mast cells, melanocytes, and epithelial cells exposed to specific cytokines. Newly synthesized MHC II molecules escape the ER and sample peptides located in endocytic compartments for presentation to $CD4^+$ helper T cells. These peptides arise from the proteolysis of self-proteins, present within the endocytic pathway, or exogenous antigens internalized by endocytosis (4, 5). For cell biologists the use of MHC products as reporter molecules has been a fascinating paradigm for analysing protein transport and the intracellular organization of the endocytic and exocytic pathway of eukaryotic cells. The study of these molecules was greatly facilitated by their high expression at the cellular level and the availability of established cell lines and a large panel of antibodies specific for distinct conformations. A major breakthrough towards understanding how antigens are handled and where MHC molecules meet antigenic peptides has been accomplished in the past years by the combination of high-resolution immunoelectron microscopy (IEM), subcellular fractionation, and biochemical methods.

In this chapter we outline how APCs apply the endocytic system to achieve antigen presentation. First we focus on the different entry routes of antigens in APCs and their processing into antigenic peptides. Then the different steps in biosynthesis of MHC II molecules are discussed, including the transfer to the endocytic pathway with its multiple compartments involved in peptide loading. Next, we will summarize the current knowledge on the subcellular mechanisms allowing the escape of MHC II molecules to the cell surface or to the extracellular environment in association with exosomes. Finally, we briefly point out that endocytosis may also play a role in MHC I-restricted antigen presentation.

2. Entry of antigens into antigen presenting cells

Presentation of exogenous antigens by MHC II molecules requires degradation of these antigens into small peptides within endosomal and lysosomal compartments (6). Only a small percentage of the whole set of antigenic peptides that is generated, the so-called immunodominant T cell epitopes, will induce an immune response. The MHC does not discriminate between peptides derived from foreign or self-proteins, and the selection is mainly based on the amino acid sequence of the peptide with specific residues favoured in certain positions (7, 8). The peptide-binding groove of MHC II is open at both ends, which enables MHC II to bind a broad spectrum of peptides, with an optimal length that varies from 13–17 residues. The generation of peptides is influenced by several factors, such as the mode of antigen uptake, proteolytic conditions, and physical form of the antigen. These issues will be discussed in Sections 2 and 3 below.

2.1 Uptake via the B cell antigen receptor

APCs make use of different internalization mechanisms, i.e. receptor-mediated endocytosis (R-ME), phagocytosis, and macropinocytosis (5) (Figs 1 and 2) (see Chapters 1–4). R-ME uptake, i.e. via the complement receptor, B cell antigen receptor (BCR), Fc receptor (FcR), and mannose receptor (MR) allows presentation of antigens at relatively low antigen concentrations. The MR belongs to the family of multilectin protein receptors, and internalization of this receptor is mediated both via clathrin-coated vesicles (CCVs) and phagocytosis (9) (see below). Complement receptors, like the receptor for C3d (CD21), provide an important link between the innate (non-specific) and acquired (specific) immune response (10). Both complement receptors and BCR enhance the efficiency of antigen presentation by thousands of fold. In the case of the BCR, uptake and presentation is 10^3 to 10^4 fold more efficient compared to fluid phase endocytosis (11). The BCR complex consists of a surface immunoglobulin (sIg) non-covalently associated with a disulfide-linked heterodimer of Igα and Igβ subunits (12–14). Naïve B cells use IgM and IgD as sIg, whereas memory B cells use IgG, IgA, or IgE. Via the immune receptor tyrosine activation motifs (ITAMs) in the Igα/Igβ complex, which become phosphorylated upon BCR cross-linking (15), the sIgs mediate signal transduction and hence B cell activation.

In addition to B cell activation, the BCR mediates antigen uptake and transport to processing compartments (5, 16–18). Regulation of antigen presentation via the BCR can be achieved in two ways. First, it was shown that epitope formation from antigens is influenced by the determinant recognized by the Ig on the B cell (19, 20). Moreover, cross-linking of the BCR accelerates the rate of internalization (16), and the strength of

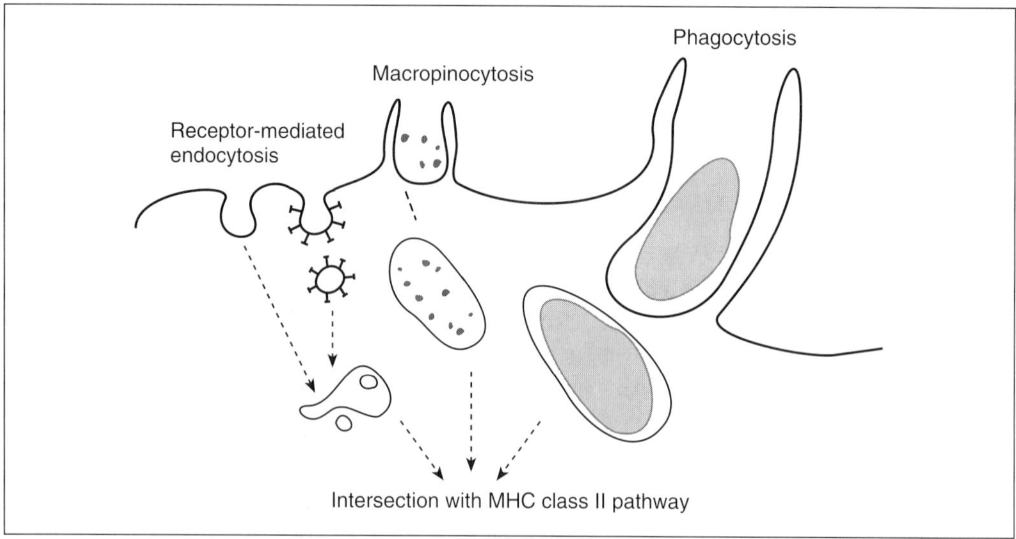

Fig. 1 Modes of antigen entry. Schematic drawing of the different modes of antigen uptake. Receptor-mediated endocytosis, i.e. via the BCR, MR, or FcR, involves clathrin-coated pits and vesicles, which mediate transport to early endosomes. Macropinocyotosis enables cells to take up large fluid volumes that are concentrated intracellularly. Entry of large particles, like bacteria, is mediated by phagocytosis. Phagosomes mature into phagolysosomes by transient interaction with components of the early endocytic pathway and eventually fuse with lysosomes. Although all entry pathways eventually converge with the MHC II pathway, the exact meeting point may vary and result in the generation of different MHC II–peptide complexes.

the BCR–antigen interaction can determine the outcome of the immune response by altering the speed of antigen processing (21). Secondly, a major role is found for targeting signals that are either present in the tails of Igs or in the Igα and Igβ subunits of the BCR. For IgG1 it has been shown that endocytosis and presentation can occur in the absence of the Igα / Igβ heterodimer (22), indicating that targeting information is present within the cytoplasmic tail of IgGs. The influence of the BCR on antigen sorting and loading is further demonstrated by the fact that a point mutation in the transmembrane domain of IgM was shown to abolish antigen presentation, although association with the Igα / Igβ complex was normal and antigen uptake and degradation was still occurring (23). A study on the function of the separate Igα and Igβ subunits showed that each subunit has a different effect on intracellular targeting and the meeting point with MHC II molecules (24). Several studies indicate that the BCR delivers its ligand primarily to a pool of newly synthesized MHC II molecules (25, 26), and this trafficking is dependent on activation of a tyrosine kinase, Syk, by the Igα subunit (27). In addition to a direct effect of the BCR complex on targeting, BCR cross-linking induces signal transduction cascades, which ultimately have an effect on the processing machinery (28, 29). Thus post-endocytic sorting events, controlled by the composition of the BCR and the nature of the antigen–antibody interaction, may target the BCR to distinct compartments with different degradation kinetics that influence the formation of T cell epitopes.

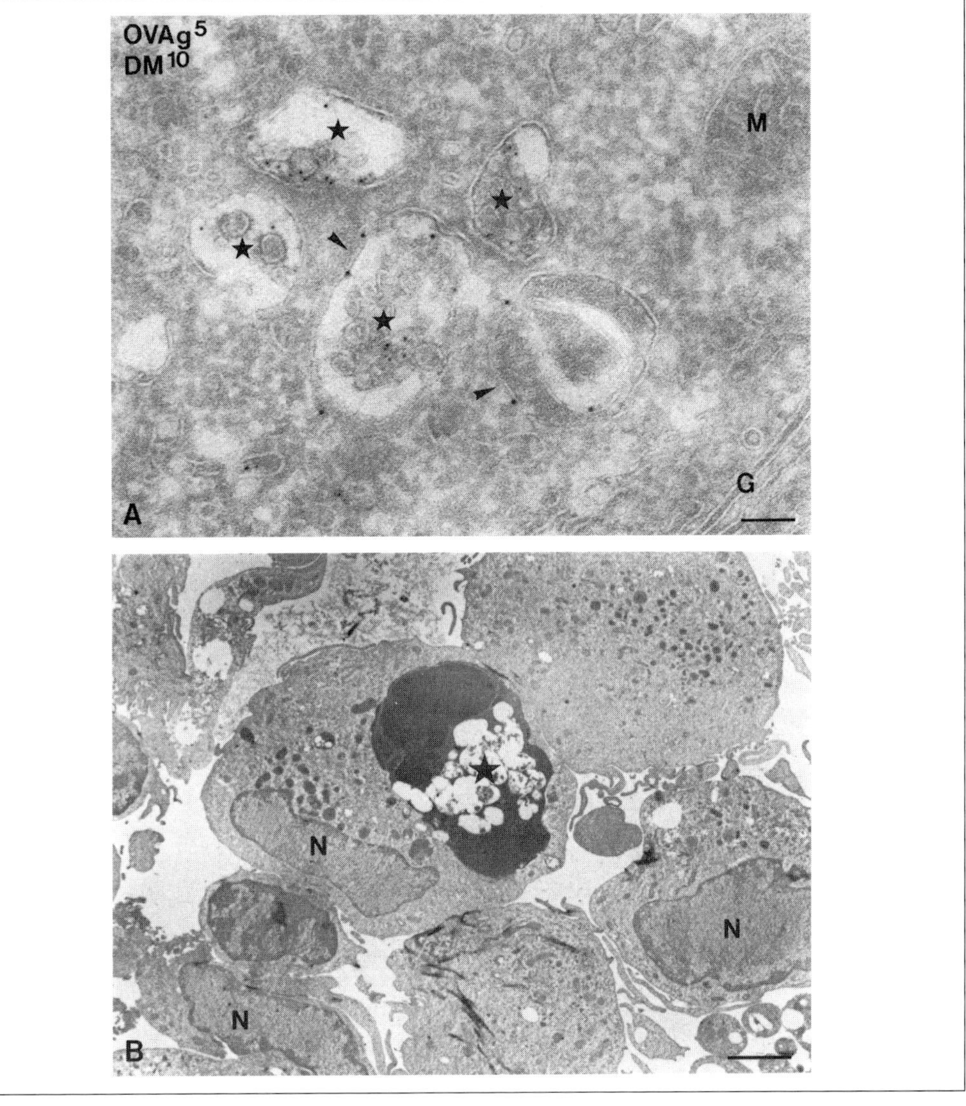

Fig. 2 Visualization of FcγRI mediated endocytosis and phagocytosis. (A) To visualize the transport route of FcγRI, murine B cells, expressing a γ chain mutant without a functional targeting motif, were incubated with rabbit IgG anti-ovalbumin (OVA) for 15 min at 4°C, followed by 5 nm colloidal gold particles coupled to OVA (OVAg). The cells were then washed and chased for 60 min at 37 °C. Subsequently, the cells were fixed, embedded in gelatin, infiltrated with sucrose, and ultrathin cryosections were immunolabelled for H2-DM. OVAg is detected in several multivesicular compartments (stars). The arrowheads point to compartments that are labelled for DM (10 nm gold particles) on the limiting membrane of these compartments, indicating that FcγRI plus OVAg/anti-OVA Ig is directed to antigen loading compartments, without the requirement for a functional targeting signal in the γ chain. G, Golgi complex; M, mitochondrion; bar = 100 nm. (B) Human immature blood-derived DCs were incubated with apoptotic bodies for 1 h. The electron micrograph shows an ultrathin cryosection of several DCs, of which one shows a large intracellular compartment with phagocytosed apoptotic material (star). N, nucleus; bar = 2.5 μm.

2.2 Properties of FcRs

FcRs comprise an elaborate family of receptors, of which primarily the FcγRs and FcεRs are involved in antigen presentation (30). These receptors recognize the Fc portion of IgG and IgE, respectively, and provide a link between specific humoral responses, which are mediated by antibodies, and the cellular part of the immune system. All types of APCs constitutively express FcγRs, and three subclasses have been distinguished: FcγRI (CD64), FcγRII (CD32), and FcγRIII (CD16). FcγRI is expressed on monocytes and macrophages and is a high affinity receptor for IgG. In contrast, FcγRIII, a low affinity receptor for IgG, is expressed on mast cells, DCs, neutrophils, and natural killer cells. Like the BCR, the FcR is a multichain complex, containing a ligand binding part and a transducing part with a γ chain. As described for the BCR, the γ chain in FcγRI and III contains an ITAM to couple ligand binding to intracellular stimuli for cell activation and receptor down-modulation (31). In contrast, FcγRIIB is a single chain receptor that contains an inhibitory motif ITIM in its cytoplasmic tail. Upon co-ligation of FcγRIIB with ITAM containing receptors, like BCR, cell activation is blocked (32, 33).

For antigen presentation to occur FcRs need to be internalized to allow degradation of its ligand (34). In the case of FcγRI it was found that the receptor itself contains a targeting signal for efficient transport to endosomes, and does not require the γ chain to mediate antigen processing (35) (Fig. 2A). When there is a requirement for the γ chain in proper targeting, ligand binding results in phosphorylation of the ITAM and recruitment of Syk family protein tyrosine kinases to activate different signalling pathways. Analyses of mutations in the γ chain-tail indicate that internalization and signalling can occur independently. Notably, transport to late endocytic compartments is influenced when the signalling function is lost (36). In the case of FcγRIII presentation of certain epitopes was impaired, whereas formation of other epitopes was not effected. These findings may reflect the use of different endocytic compartments in the formation of epitopes (37), but further studies are required to elucidate the molecular basis for the link between signalling and trafficking.

2.3 Internalization mechanisms of macrophages and DCs

Macrophages and DCs make use of antigen receptors, such as FcRs and MRs, often in combination with phagocytosis and macropinocytosis to accumulate large particles and large fluid volumes, respectively (5, 9, 38). In macrophages, after formation, phagosomes undergo progressive maturation, whereby they acquire endosomal markers such as Rab5 and mannose-6-phosphate receptor (MPR) (39, 41). Ultimately, these markers are lost and mature phagosomes fuse with lysosomes. Peptide loading onto MHC II molecules can occur in phagolysosomes, which appear to be fully competent peptide-loading compartments even after isolation (42, 43). Alternatively, degraded antigen may traffic from the phagolysosomes to MHC II-positive compartments for loading onto MHC II. Some microorganisms can escape processing and presentation by preventing maturation of the phagosome that harbors them (41, 44–47) (see Chapter 11).

The MR is a 180 kDa C-type lectin belonging to the family of multilectin mannose receptors (9, 48). MR has broad specificity for sugars but recognition of mannose and fucose is primarily restricted to a series of eight tandem lectin-like domains. As proposed by Janeway (49), the MR can be considered as a pattern recognition receptor, which recognizes molecular structures that are shared by a large number of pathogens. Different factors can modulate MR expression, like FcRs, cytokines, and pathogens. For example, interferon γ-treated macrophages show a decreased uptake via MR-dependent endocytosis, whereas MR-dependent phagocytosis is enhanced in these cells (50).

The two main mechanisms for antigen uptake by DCs are via macropinocytosis and the MR. (38, 51). In contrast to other cell types, DCs display constitutive macropinocytosis, which allows them to continuously internalize large volumes of fluid (51). As will be discussed below, the high endocytic capacities only apply to immature DCs. Macropinocytosis is non-clathrin dependent and associated with membrane ruffling (52, 53). By shrinking of the macropinosome the pinocytosed macromolecules are concentrated and ultimately delivered to MHC II-containing compartments (51). Whereas in fibroblasts membrane ruffling is dependent on the actin filament regulatory protein gelsolin, in DCs both ruffling and antigen processing are not affected in DCs from gelsolin knock-out mice (54). To increase the efficiency of uptake DCs make use of the MR, which adds selectivity to the internalization process. After antigen capture, MR delivers its ligand to compartments with low pH and recycles back to the cell surface, allowing consecutive rounds of uptake (51, 55). By its presence on DCs, the MR provides an important link between innate and adaptive immunity. Another aspect of the MR is that it can be cleaved into a soluble form and as such deliver bound antigen to lymphoid organs (56). More recently, is has been demonstrated that MR is involved in the uptake and delivery of glycolipids to endosomes and lysosomes that contain CD1b and MHC II molecules (57). CD1b is a non-polymorphic MHC I-like molecule expressed on most professional APCs and is unique in its ability to activate T cells against lipids and lipoglycans, which are present on pathogens like mycobacteria (58, 59). A DC-specific member of the multilectin family is DEC-205 (60). Although its precise role in antigen presentation has yet to be determined, it is clear that this family of receptors plays an important role in host defence and antigen presentation. Another recent example of efficient antigen capture by DCs is the combination of phagocytosis with the scavenger receptor CD36 (61, 62) (Fig. 2B).

As already mentioned, the strong endocytic capacity of DCs is restricted to the immature state of these cells. DCs that are derived from the myeloid lineage migrate as immature cells to the peripheral organs, such as the skin (where they are known as Langerhans cells) and mucosa, where they search for the presence of incoming pathogens (63, 64). An encounter with antigen triggers maturation and the DC moves from the periphery to lymphoid organs to activate primary T lymphocytes. Previously, DCs studied *ex vivo* were thought to display only low antigen uptake capacity, which turned out to be a result of their activated state. Currently, DCs can be obtained from $CD34^+$ precursor cells or monocytes (63, 65). By culturing these cells in a combination of cytokines, like GM-CSF and IL-4, DCs with excellent endocytic capacities are obtained. After activation by cytokines like TNFα, bacterial lipopolysaccharide (LPS), CD40-ligand, or FcR cross-linking (66, 67), the ability of these DCs to endocytose and process antigens is mostly lost. Mature DCs display high surface levels of peptide-loaded MHC II, together with co-stimulatory molecules, on their surface, which leads to their unique capacity to activate primary T cells.

In summary, each APC type has developed distinct modes of antigen uptake to protect the body against a broad spectrum of invading pathogens. B cells primarily use R-ME through the BCR, which greatly enhances the efficiency of uptake. Macrophages use R-ME, through the FcR and MR, and phagocytosis, which allows them to internalize large particles with broad specificity. DCs have multiple uptake mechanisms, including R-ME through the FcR and MR, macropinocytosis, and also phagocytosis. Each entry pathway affects the trafficking of antigens through endocytic compartments. In addition, the activation of specific signalling pathways can alter antigen processing and presentation.

3. Processing of antigens in the endocytic pathway

As described above, the composition of antigen receptors influences antigen degradation and presentation. Although all the entry pathways that operate within one cell type eventually converge at lysosomes, prior targeting to early endocytic compartments soon after uptake may generate different antigenic determinants. In this section we will describe how the available proteases and the form of the antigen may alter antigen processing and epitope formation. Importantly, despite the fact that the MHC does not distinguish between self and foreign antigens, the pool of antigenic peptides displayed to the T cells must ensure that pathogens are adequately detected. Already the first step in uncovering antigenic epitopes is achieved by the acidic environment of endocytic compartments, which helps to unfold proteins. The enzymes that are involved in protein degradation within the endocytic pathway also require acidic pH for their activity. These enzymes include aspartate proteases, such as cathepsin (Cat) D and E, and cysteine proteases, such as Cat B, L, S, and H (5, 68). Knock-out mice have revealed that Cat B and D, which are both ubiquitously expressed, are not essential to generate the peptide repertoire for MHC II presentation (69). Recently, in particular Cat S (70), mostly expressed in APCs, and Cat L (71) were found to be important for the degradation of the MHC II-associated invariant chain (Ii) (see Section 4). In addition to cathepsins, an asparagine-specific cysteine endopeptidase (AEP) was shown to perform a key step in the generation of presentable peptides from the microbial tetanus toxin antigen (72). This enzyme is similar to the mammalian homologue of the legumain/haemoglobinase asparaginyl endopeptidase found in plants and parasites and its action is blocked by the *N*-glycosylation of asparagine residues. AEP could discriminate between differentially *N*-glycosylated polypeptides, favouring the processing of microbial agents versus mammalian proteins.

For most antigens a balance is needed between the destructive capacities of proteolytic enzymes and their ability to expose antigenic determinants. In this respect pH is an important factor to influence degradation efficiency and epitope formation. The release of determinants from core regions of proteins requires very acidic pH for optimal activity of proteases. However, some proteins require only mild digestion to liberate fragments with T cell activating capacity. These epitopes may be present in more superficial regions of the protein and some proteases may destroy such epitopes and therefore impair efficient antigen presentation (5). To make certain regions of a protein better accessible to protease activity, denaturation may be necessary before protease action (73). An enzyme constitutively expressed by APCs is the gamma-interferon-inducible lysosomal thiol reductase (GILT), which is able to reduce disulfide bonds, suggesting a potentially important role in antigen processing (74) (Fig. 3). Thus, different antigens require distinct proteolytic environments avail-

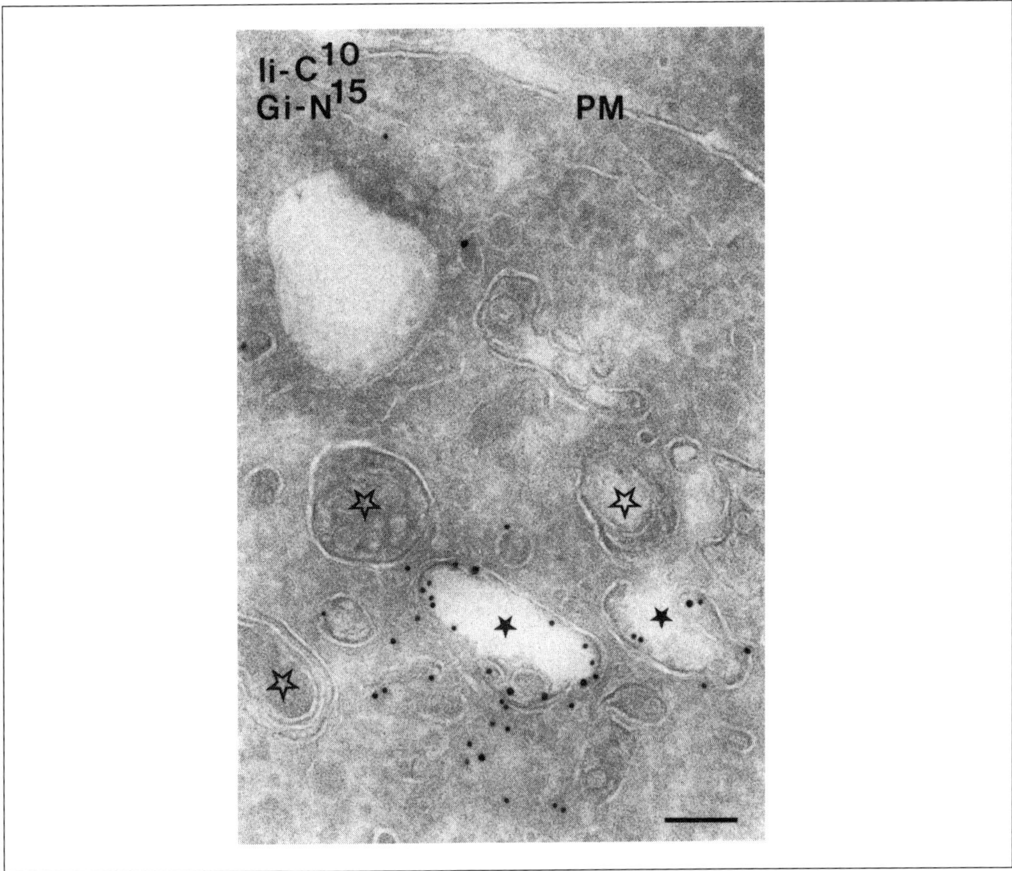

Fig. 3 Intracellular localization of GILT. Ultrathin cryosections of the human B cell line Raji were double-immunolabelled for the C-terminus of Ii (10 nm) and the proform of GILT (Gi) (15 nm). ProGILT is located in early endosomes that also contain Ii (filled stars). Late endocytic compartments (open stars) are negative for proGILT. They only contain the active, mature form of GILT (not shown). Bar = 200 nm. Reproduced from ref. 74 by permission of the National Academy of Sciences, USA.

able in different types of endocytic compartments, ranging from early endosomes to lysosomes. This phenomenon has been extensively studied for the antigen hen egg lysozyme (HEL) (75, 76), which contains multiple T cell epitopes that are formed under distinct circumstances. It has been demonstrated that different HEL-derived peptides can be generated and bound to MHC II in early endosomes and in lysosomal compartments, respectively (77). Similarly, antigenic peptides from ribonuclease A are generated and bound to MHC II molecules in relative early endocytic compartments (78). These studies indicate that there is not one specific type of peptide-loading compartment. Most likely, MHC II molecules that are involved in peptide binding in early endosomes are recycling molecules derived from the cell surface (79).

Other factors that influence epitope formation are complexion of antigens to other proteins, flanking sequences, and Ii. When tetanus toxin is associated with antibody some regions are protected from degradation (19, 80). However, formation of such complexes allows the exposure of other antigenic domains promoting the presentation of new T cell epitopes (81). Thus, certain T cell epitopes are only revealed when present in a complex. These so-called cryptic determinants revealed by protein/protein interaction are processed in an unpredictable fashion by APCs. In addition to complex formation with other proteins, sequences distant from T cell epitopes may have an effect on presentation, possibly by influencing the folding of a protein or regulating an interaction with other molecules (82). It has also been shown that MHC II molecules can associate with long peptide sequences, which undergo additional trimming after binding to the groove (76, 83, 84). This can be accomplished because, as mentioned earlier, the peptide-binding groove of MHC II is open at both ends and still accessible to proteases. A regulatory role for Ii is also implied, whereby the p41 isoform of Ii (see Section 5) contains a domain that is structurally related to a novel family of protease inhibitors, called thyropins. It was shown that this domain could specifically inhibit Cat L, but not Cat B or Cat S, implicating a regulatory role for Iip41 in antigen processing (85).

Together these data indicate that it is very difficult to predict dominant T cell epitopes by examination of the antigen alone. It may well be that processing events that occur *in vivo* strongly influence the formation of these epitopes.

4. An overview of the biosynthetic pathway of MHC II molecules: association with invariant chain, HLA-DM, and HLA-DO

As described in the preceding sections foreign antigens are broken down within endocytic organelles generating a set of peptides that can be presented to T cells. The subsequent recognition of these peptides by a T cell receptor is strictly dependent on their association with MHC II molecules.

MHC II molecules are a family of related polymorphic proteins encoded within the MHC: HLA-DR, DP, DQ in human and I-A and I-E in mouse (86, 87). In human and in mouse, due to their high cellular expression, mostly HLA-DR and I-A, respectively, have been studied in detail. Much less is known about the cell biology of DP, DQ, and I-E, in particular the early steps of their biosynthesis, association with other molecules, and intracellular localization. Thus it is important to keep in mind that, in general, when referring to human MHC II molecules it corresponds to HLA-DR and in mouse mostly to I-A. However, since I-E and I-A are structurally similar to HLA-DR and HLA-DQ, respectively, and since I-A and DR have similar characteristics, the presented studies are likely to reflect general principles.

4.1 Biosynthesis of MHC II molecules

MHC II molecules consist of two transmembrane type I polypeptides, the α chain (32–35 kDa) and the β chain (27–29 kDa). Detailed biochemical analysis of newly synthesized polypeptides has unravelled the events occurring during synthesis of MHC II molecules in the ER (for reviews see refs 4, 5, 88, 89). Soon after their synthesis, three αβ dimers associate with non-covalently associated trimers of the polypeptide Ii (90–93). The Ii or invariant chain is a non-polymorphic type II protein (NH_2-terminus in the cytoplasm, COOH-terminus in the ER lumen) and corresponds to a family of proteins generated by alternative splicing and in human, but not in mouse, by alternative translational initiation. An important role of Ii is to occupy the peptide-binding groove of MHC II molecules to prevent them from sampling peptides in the ER, which is the site where MHC I molecules usually bind peptides (94–96).

Approximately 30–60 minutes after synthesis, $(\alpha\beta Ii)^3$ complexes move out of the ER and are transferred to the Golgi complex where they undergo complex *N*-glycosylation (*O*-glycosylation for Ii) and terminal sialylation. Reaching the *trans*-Golgi network (TGN) MHC II/Ii complexes deviate from the 'classical' constitutive secretory pathway used by other transmembrane proteins (e.g. peptide-loaded MHC I molecules or the transferrin receptor). A lag time of 2–6 hours is observed between the sorting of MHC II molecules at the TGN and their arrival with associated peptides at the cell surface. During this time most newly synthesized MHC II molecules are delivered to an intracellular compartment, representing a meeting point between the biosynthetic route and the endocytic pathway (97–99) (Fig. 4A). However, MHC II/Ii complexes can be delivered to the cell surface, undergo endocytosis, and subsequently play a role in antigen presentation (100–105).

4.2 Delivery of MHC II molecules to the endocytic pathway by Ii

Numerous studies, including the development of Ii knock-out mice, revealed the critical role of the Ii as a chaperone helping the egress of αβ dimers from the ER (90, 93, 106)and their delivery to the endocytic pathway and the cell surface (107–110), reviewed in ref. 88. The cytoplasmic domain of Ii has been analysed in detail and

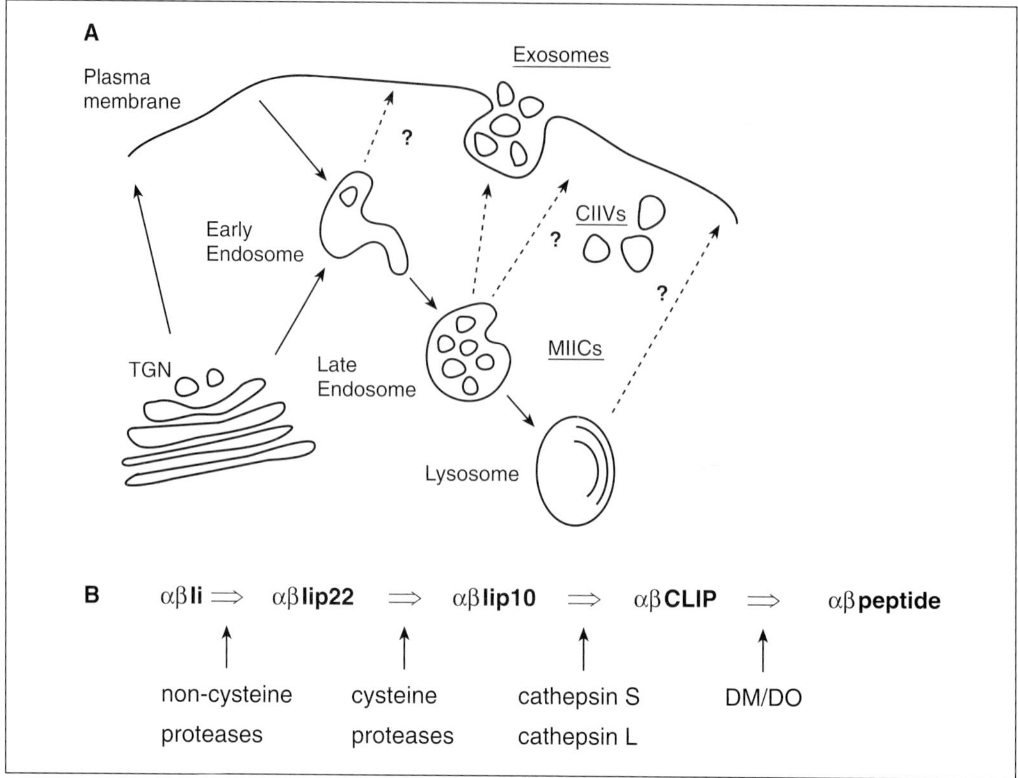

Fig. 4 Intracellular trafficking of MHC II molecules. (A) Schematic drawing of the intracellular transport routes and compartments that contain MHC II molecules. From the TGN the majority of MHC II molecules is directly transported to (early) endocytic compartments. Only a minority escapes to the cell surface and is subsequently internalized. Depending on the cell type, MHC II molecules accumulate in late endosomes and lysosomes (MIICs) where peptide loading is occurring. Transport to the cell surface is mediated by direct fusion of multivesicular MIICs with the plasma membrane, resulting in the secretion of exosomes. An alternative pathway to reach the cell surface is via a transport intermediate, possibly the CIIV. Peptide loading and transport to the cell surface are tightly regulated processes, influenced by the activation state of the cells, as shown for DCs. (B) Different steps in the processing of Ii and peptide loading. In the ER, trimers of αβ-dimers associate with Ii-trimers. From the TGN these complexes are sorted to endosomes where aspartate proteases start degrading Ii. Subsequent fragments of p22 and p10 are formed by action of non-cysteine and cysteine proteases. Generation of CLIP is accomplished by cleavage through Cat S (Cat L in thymic epithelial cells). Finally, CLIP is removed from the peptide-binding groove, a process catalysed by DM and in some cell types by DO.

dual di-leucine-like motifs were identified as the residues involved in this targeting (111–114) (see Chapter 7). Despite the fact that an enormous amount of data points out the chaperoning role of Ii, it appears that in some cell models, e.g. transfected fibroblastic cells, the association of αβ dimers with Ii is not a prerequisite for targeting to the endocytic pathway (115, 116).

Once multimeric αβIi heterotrimers are delivered to an intracellular compartment(s), a complex series of proteolytic events sequentially degrade the Ii starting from its C-terminus (Fig. 4B). The use of different protease inhibitors and mice genetically deficient in the lysosomal proteases (cathepsins) facilitated the identification of the enzymes involved in the controlled degradation of Ii and the different Ii intermediates (reviewed in ref. 68). Differences in protease use may be a rule among distinct APCs. Cat L was found to be necessary for Ii degradation in cortical thymic epithelial cells (71). In macrophages, Cat F is highly expressed and can substitute for other enzymes (117). Cat S, however has appeared as the most crucial enzyme in Ii degradation in B cells and DCs and consequently in the regulation of MHC II trafficking (5, 69, 70, 118–120). In these cells, and for most MHC II haplotypes, Cat S collaborates with other leupeptin-insensitive proteases to convert Ii into a small peptide called CLIP (class II-associated invariant chain peptide) corresponding to residues 81–105 (121–124) (see Fig. 4B). In DCs the degradation of the Ii is dependent on an extra step of control by the protease inhibitor cystatin C, which has been identified in immature DCs as an inhibitor of Cat S (125). The signalling cascade induced during DCs maturation (see above) down-regulates cystatin C, and allows further degradation of Ii. Such regulation of the degradation of the Ii and the consequent accumulation of its intermediates may play a crucial role in the localization of MHC II molecules along the endocytic pathway in different APCs (see Section 6).

4.3 Role of DM and DO in peptide loading

The next critical event in the process of antigen presentation is the removal of CLIP from the peptide-binding pocket of MHC II molecules and the association with antigenic peptides. Two molecules encoded into the MHC, named DM and DO are directly involved in this step (126–128). As a key molecule, HLA-DM (in humans) and H2-M or H2-DM (in mouse) has been shown to transiently associate with the MHC II–CLIP complex (129–132). During this interaction, DM is thought to catalyse not only the release of CLIP, but also of other peptides of low affinity at low pH, favouring the capture and binding of high affinity antigenic peptides (133, 134). Thus, DM is thought to act as a 'peptide editor', in which high affinity peptides are favoured over low affinity peptides. The localization of DM in late endocytic compartments (135–137) is mediated by a targeting signal in its cytoplasmic tail (138, 139), which allows rapid internalization from the cell surface (140). However, the presence of HLA-DM on the cell surface has been demonstrated, where it may help in the association of exogenously processed peptides to MHC II molecules (141). The role of DO, which is only expressed in B cells, is less clear. It is not yet fully understood whether it may function as an inhibitor of HLA-DM function, transiently stabilizing the $\alpha\beta$-CLIP intermediate (142), or as an enhancer of DM activity (143, 144). In addition, DO may directly influence the peptide loading of MHC II molecules (145) and inhibit presentation of peptides present in early endosomal compartments, thereby restricting their binding in the late endocytic pathway (146).

In summary, the concerted action of different proteases, together with DM and DO, results in the degradation of Ii and formation of a stable MHC II–peptide complex, the so-called mature form, ready to be transferred to the cell surface for recognition by the T cell receptor.

5. Targeting MHC II molecules to the endocytic pathway

The studies reviewed above revealed the involvement of Ii in targeting MHC II molecules to the endocytic pathway. An important aspect has been the identification of the endocytic compartment that is the first to receive newly synthesized MHC II molecules (αβIi complexes). Early experiments indicated that newly synthesized MHC II/Ii complexes in B cells intersect early endosomal compartments that are accessed by transferrin–neuraminidase conjugates (97). Although subsequent work favoured the idea that the majority of MHC II traffics directly from the TGN to late endosomes and lysosomes (98, 147, 148), accumulating evidence strongly supports the transit of MHC II through early endosomes, before their accumulation in late endocytic compartments. Cross-linking experiments utilizing transferrin conjugated to horse-radish peroxidase (Tf-HRP), similar to those developed to analyse transport from late endosomes to lysosomes (149), have been used to specifically inactivate the early endosomes of B cells (150, 151). These studies strongly suggest that, in these cells, early endosomes are an intermediate in the transfer of MHC II molecules and the associated Ii from the TGN to the endocytic pathway. Emphasizing the transient passage through the early endocytic system, it was recently reported that MHC II molecules might mature in early endosomal compartments independently of Cat S and DM (152). These biochemical approaches are in agreement with previous morphological studies, which identified an early endosome-like compartment as the first endocytic compartment reached by the MHC II-Ii heterodimer after sorting in TGN (153–155). In human B cells this compartment shows a high expression of Ii (both luminal and cytoplasmic epitopes) (154–156), and displays an irregular morphology with few internal vesicles (Fig. 5A and 5B). The presence of Tf, after 10 minutes of uptake, in this compartment suggests that it is placed early in the endocytic pathway and could play a crucial role in the delivery of newly synthesized MHC II molecules to the endocytic pathway (155).

If in B cells early endocytic compartments are the recipients for newly synthesized MHC II molecules, what are the vesicular intermediates responsible for their transfer from the TGN to endosomes? One well-studied protein known to traffic directly from the TGN to the endocytic pathway is the MPR. Both the CI (cation independent)- and the CD (cation dependent)-MPRs are known to be enriched in AP-1/CCVs in the TGN and may use these vesicles as transport intermediates (157, 158). Interestingly, this does not seem to be the case for MHC II/Ii complexes. Despite their abundant expression, these molecules are not detected in CCVs in the *trans*-Golgi region in either I-cell disease B cells (154) or other murine and human B cell lines (150). In agreement, transport of MHC II molecules to the endocytic pathway is unaffected by dominant negative clathrin mutants (140). Though this does not exclude the possibility that MHC II/Ii complexes, or even Ii-free αβ dimers, may use AP-1/CCVs during their intracellular transport (159–162). It seems that APCs have developed a mechanism, possibly involving clathrin-independent adaptors or lipid-driven sorting,

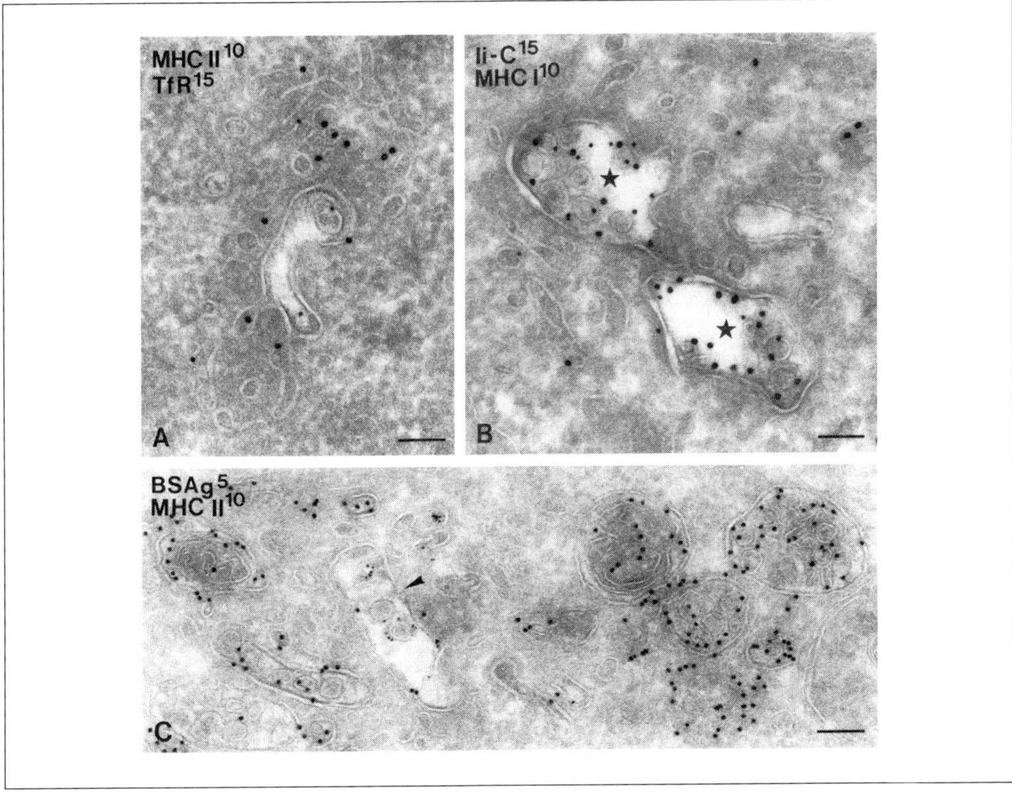

Fig. 5 MHC II molecules in early endosomes. (A) Double-immunolabelling of TfR (15 nm) and MHC II (10 nm) on early endosomes in a human B cell. Bar = 100 nm. (B) The cytoplasmic domain (luminal) of Ii (15 nm) co-localizes with the heavy chain of MHC I in multivesicular endosomes (stars) in human B cells. Bar = 100 nm. (C) Human B cells were incubated for 3 min at 37 °C with 5 nm bovine serum albumin (BSA) coupled to 5 nm colloidal gold (BSAg). Ultrathin cryosections were immunolabelled for MHC II. Multiple MIICs with internal membrane vesicles and sheets are shown, strongly labelled for MHC II (10 nm). The arrowhead points at an early endosome that contains both internalized BSAgold and MHC II. Bar = 200 nm. Reproduced from ref. 155 by permission of The Rockefeller University Press.

to transfer MHC II/Ii complexes to the endocytic pathway. Future studies, stimulated by findings on new adaptors and sorting machineries, are needed to elucidate these issues.

Less is known about the targeting of MHC II molecules to the endocytic pathway in DCs. Compared to B lymphocytes it seems that a higher proportion of newly synthesized MHC II molecules transits through the cell surface before being re-internalized (163). It is not clear whether this fraction of newly synthesized molecules are directly transferred to the plasma membrane by the so-called constitutive route or if they are rapidly sorted from a recipient endocytic compartment in the absence of a retention mechanism. The differences observed between APCs may be partially explained by the existence of different isoforms of Ii. In humans, there are four primary forms of Ii (p33, p35, p41, and p43), whereas in mouse only two forms (p33 and p41) have been described (4, 88, 164). These isoforms of Ii, differentially expressed in cells, may independently contribute to the egress of MHC II molecules to distinct intracellular compartments (105). The p41 isoform is expressed at higher levels in DCs compared to B cells, and may play a role in regulating the trafficking of MHC II in these cells (165) (see Section 3).

6. The intracellular compartments involved in antigen processing and presentation: an adaptation of the endocytic pathway?

One cannot address the question of where in the endocytic pathway MHC II molecules meet antigenic peptides without taking a look at the work of the 1990s. The identification of organelle 'marker' proteins as well as endocytic tracers greatly facilitated the identification of distinct MHC II positive compartments and their position in the endocytic pathway. In general, the TfR can be used to identify early sorting and recycling endosomes, the MPR to characterize late endosomes, and lysosomal-associated membrane proteins (Lamps/CD63) and lysosomal hydrolases (β-hexosaminidase, Cat D) to distinguish lysosomal compartments. In addition, the well-determined kinetics of fluid phase tracer accessibility to different endocytic compartments is very useful in distinguishing between early and late compartments. Finally, the accumulation of weak bases (for example 3-(2,4-dinitroanilino)-3′-amino-*N*-methyldipropylamine (DAMP) can demonstrate the relative acidity of organelles (156, 166, 167). Since peptides are available in endosomes and lysosomes, it is plausible that MHC II molecules use the pre-existing endocytic system to fulfil their antigen presenting function by adapting it to their advantage. First they protect themselves against proteases and low pH ranges to which they can be exposed during their passage in the endocytic pathway. Secondly, they travel together with the above-described partners (Ii, DM, DO) or with other molecules (co-stimulatory and adhesion molecules, tetraspanins) that will help them to fulfil their task.

6.1 Identification of MHC II-enriched compartments (MIICs)

The first attempts to identify a 'MHC II compartment' using immunocytochemistry and the EM reported the presence of MHC II molecules in the early endosomes of B cells (153) and macrophages (168). These compartments seemed related to the CURL (compartment of uncoupling of receptor and ligand) (169), and they could mediate the generation of peptides, loading onto MHC II molecules, and the transport of peptide-loaded complexes to the cell surface. In the following years, however, additional morphological studies on the human Burkitt lymphoma cell line JY revealed that MHC II molecules accumulated in pre-lysosomal and lysosomal compartments (98, 148). These compartments were designated MHC II-enriched compartments (MIICs) (148), and were characterized by internal, typical concentrically arranged, multilaminar membranes (Fig. 5C and 6), mild acidic content, and the presence of lysosomal

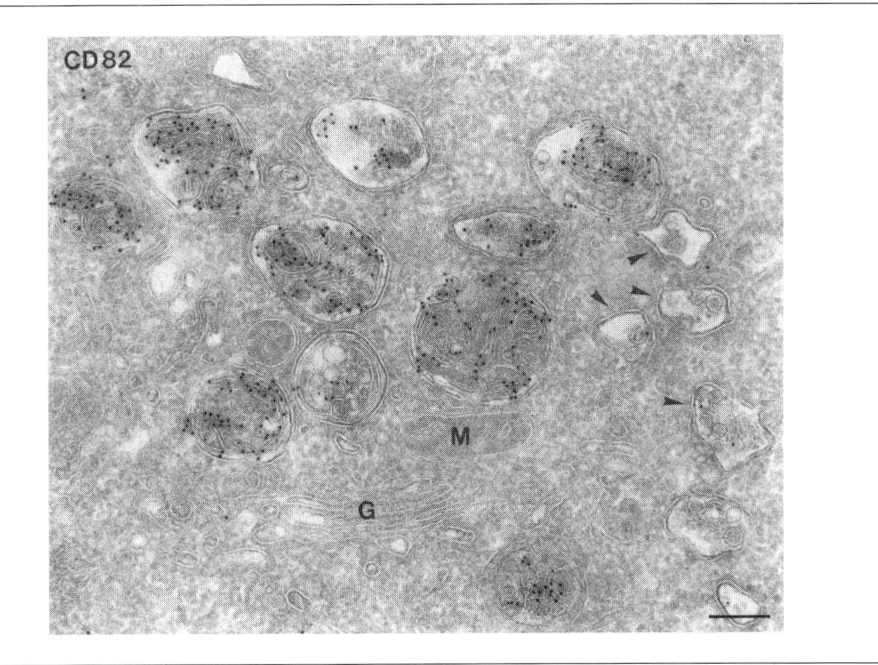

Fig. 6 MIICs. Ultrathin cryosection of a human B cell line RN were immunolabelled for the tetraspan protein CD82. Abundant labelling for CD82 is observed on the internal membranes of multivesicular and multilaminar MIICs. The arrowheads point at early endosomes. G, Golgi complex; M, mitochondrion; bar = 200 nm.

enzymes (β-hexosaminidase, cathepsin D) and lysosomal membrane proteins (Lamps, CD63).

Further investigations of MHC II localization using IEM revealed the heterogeneity of lysosomal compartments and indicated that late endocytic compartments, distinct from terminal lysosomes, might also be involved in the intracellular retention of MHC II molecules. In melanoma cells (Mel Juso) late endosomal compartments accumulated Ii and MHC II molecules (170), whereas in macrophages, early tubular lysosomes (pre-lysosomes), morphologically different from the MIICs in B cells, were reported as the possible meeting points between MHC II molecules and antigenic peptides (171). Detailed IEM studies, on human and mouse DCs, including the Langerhans cells of the skin, and a number of EBV transformed B cell lines (HOM-2, PALA, RN), revealed that MIICs are a more heterogeneous population of structures than originally thought. In these studies, which were much improved by modifications in the immunogold labelling and cryosectioning technique (172, 173), at least three types of MIICs could be discerned, multivesicular, multilaminar, and intermediate types of MIICs possessing both membrane sheets and vesicles (136, 174–176). All subtypes of MIICs contain lysosomal components, although they show different accessibility to endocytic tracers and accumulate different amounts of Ii and HLA-DM (156). Endocytic tracers (BSA coupled to gold) access MIICs containing internal vesicles, reminiscent of multivesicular bodies (MVBs), after 20 minutes. Whereas 1–2 hours of internalization are necessary to reach MIICs containing membrane sheets (167, 177). Apart from their morphological appearance, the kinetics of accessibility to fluid phase tracers point out the similarity between these MIICs and late endosomes and lysosomes, respectively. The luminal epitope (C-terminus) of Ii is present in multivesicular MIICs but not in multilaminar compartments, whereas a cytoplasmic epitope can be detected in the majority of both compartments (175). These studies, together with the observation that DM and MHC II–peptide complexes can be visualized in MIICs (135, 136, 178), suggested that the acidic and protease-rich MIICs may indeed represent the perfect environment for the degradation of Ii, the degradation of antigens, and the meeting between MHC II molecules and antigenic peptides.

6.2 Specialized peptide loading compartments

Ongoing investigations were strengthened, but also challenged, by the use of subcellular fractionation combined with functional assays to analyse the intracellular compartments of APCs (79, 171, 179–183). A number of these studies pointed out that (specialized) compartments co-fractionating with late endosomes and lysosomes play a key role in antigen processing and presentation (180, 182). Using free flow electrophoresis (FFE), a unique compartment containing MHC II molecules was identified in murine B lymphoma cells (A20, and called class II vesicle; CIIV) (179). This compartment appeared different from classical endosomes and lysosomes and was proposed to be a special peptide-loading requisite. CIIVs contained low levels of TfR, were devoid of lysosomal markers, only poorly accessible by fluid phase endocytic tracers and, in contrast to the other intracellular compartments, they could be easily isolated because of a shift towards the anode in FFE. These observations introduced in the field the idea of the existence of a 'special' and 'unique' compartment involved in antigen processing and presentation.

The discrepancy between the results regarding the existence of unique peptide-loading compartments has been subject of debate. Recently, a number of observations have been made that help to understand the differences reported and suggest that MIICs and CIIVs may coexist in APCs depending on their state of activation and antigen presenting function (155, 184). It is evident that the overall distribution of MHC II molecules varies between cells. In human EBV-transformed B cell lines the majority of MHC II is localized intracellularly (80%), in murine B lymphoma cells the majority is present on the cell surface (90% versus 10% intracellularly). In addition, the amount of Ii detectable at steady state varies between cells. At any time in human B cells substantial amounts of Ii can be detected in the MIICs. In contrast, the degradation of Ii seems to be very efficient in A20 cells, which results in a low detection of Ii at steady state. These observations lead to the assumption that Ii could play a critical role in the retention of MHC II molecules in late endocytic compartments. As reported by Brachet and colleagues (185), when B4-14 (A20-derived) cells are incubated with the cysteine protease inhibitor leupeptin, which inhibits degradation of Ii, MHC II molecules accumulate in late endosomes and lysosomes giving raise to a distribution very similar to that observed in human B cells. Thus Ii seems to have, apart from its role in protecting the peptide-binding groove and targeting of MHC II to the endocytic pathway, an important function in the intracellular retention of these molecules in lysosomal compartments. This may have important consequences in the immune response allowing MHC II to bind peptides that are generated in late endocytic compartments.

These findings were consolidated by the analysis of the intracellular distribution of MHC II molecules versus endosomal and lysosomal markers in immature DCs and the changes accompanying their activation and migration (184). In immature DCs the bulk of MHC II molecules is localized in late endosomal and lysosomal MIICs, corresponding mostly to multivesicular and multilaminar compartments, respectively (136). This lysosomal localization is certainly maintained by the low rate of Ii degradation due to the expression of the Cat S inhibitor cystatin C (125). DCs from Cat S deficient mice MHC II molecules remain accumulated in lysosomal compartments in the absence of a complete degradation of the Ii (120). Once the activation process of DCs has started, cystatin C is down-regulated, which allows the final degradation of Ii and the binding of peptides. MHC II–peptide complexes are only produced during DC differentiation induced by inflammatory mediators (TNFα, CD40 ligand, or LPS) (186), conditions known to induce a rapid redistribution of mature MHC II molecules to the cell surface. Thus, MHC II–peptide complexes may arise in late endosomal and lysosomal MIICs, concomitantly with their transport to CIIVs together with co-stimulatory molecules (187). Subsequently, CIIVs may act as intermediates in the transport of mature MHC II molecules to the cell surface in activated DCs (see Fig. 4A), and in B cells in the absence of an Ii retention signal.

Altogether these studies indicate that the majority of efficient peptide loading takes place in late endosomal/lysosomal compartments (B cells and DCs). Still, it should be kept in mind that peptide loading also occurs elsewhere in the cells, namely in early endocytic structures (77, 78, 188). This may depend on the APC, the MHC II allele, and, as discussed in Sections 2 and 3, on the antigen and on the environment required to generate a certain set of antigenic peptides. Consistent with this, it has been shown recently that even if most MHC II molecules in the early endocytic pathway are tightly associated with the Ii, some free αβ-dimers are able to bind longer polypeptide precursors independently of Cat S and DM (152). These long polypeptide precursors could be either presented to T cells directly or, subsequently routed to the late endocytic pathway for further processing.

7. Transport of MHC II molecules to the cell surface

Any step in MHC II trafficking is crucial for the presentation process. Nevertheless, the ability of any APC to stimulate an immune response once MHC II–peptides are formed, is tightly dependent on their ability to correctly expose them at their surface. How are MHC II molecules transferred to the cell surface? Whereas MHC II–peptide complexes formed in early endosomal compartments can follow the well-defined recycling pathway to the plasma membrane, little information is available about transport routes allowing egress from late endosomes and lysosomes, thought to be the dead-end stages of the endocytic route, to the cell surface. Lysosomal membrane proteins, like LEP 100 or Lamp-1, have been detected on the plasma membrane suggesting that a transport route to the cell surface exists (189, 190). However, it is not known whether this is due to biosynthetic transit to lysosomes or to fusion of lysosomal membranes with the cell surface. MHC II molecules may use a similar transport as well. Alternatively, lysosomal components may be recycled first to the TGN and then to the cell surface by the so-called constitutive pathway or they may be transferred to early endosomes from which recycling to the plasma membrane can take place.

7.1 Direct fusion of multivesicular MIICs with the cell surface

Our ultrastructural studies (177), together with *in vivo* time lapse video of green fluorescent protein (GFP)-tagged MHC II constructs (191), have recently revealed that a direct fusion of MIICs with the plasma membrane occurs. In particular, the multivesicular type of MIIC in B cells and melanoma cells can fuse with the plasma membrane in an exocytic manner analogous to that described for secretory lysosomes (192, 193). As proposed by Wubbolts and collaborators (191), fusion of MIICs with the plasma membrane may allow MHC II present in the limiting membrane to be transferred directly to the plasma membrane. However, our studies indicate that direct fusion is probably not a major pathway for transfer to the cell surface (156, 177). The major consequence of exocytic fusion event of multivesicular MIICs is the release of exosomes into the extracellular environment.

7.2 Characteristics of exosomes

As observed for MVBs in differentiating reticulocytes (194–196), the small 60–80 nm intralumenal vesicles are released into the extracellular environment as exosomes during the exocytic fusion of MIICs (Fig. 7). These exosomes carry MHC II molecules in their membrane with the peptide-binding site on the outside and are able to stimulate T cell proliferation (177). In addition to reticulocytes, cytotoxic T cells, platelets, B cells, DCs, and MHC II-expressing mast cells have been shown to secrete exosomes (177, 197–199). Thus, APCs and other haematopoietic cells seem to take advantage of the duality of their endocytic and exocytic pathway to achieve their function (193, 197).

Interestingly, for cell biologists studying the biogenesis of intracellular compartments, exosomes represent released internal vesicles of late endosomal MVBs and can be obtained from APC culture media in relatively high quantities. This reveals the unique opportunity to analyse the molecular composition of these vesicles, and thus to investigate their formation and the sorting events occurring in late endosomes. The distribution of proteins within the MVB, either on the internal vesicles or the limiting membrane, already demonstrates that active sorting must be occurring. Proteins like Lamp and DM appear to be more abundant on the limiting membrane (Fig. 2A), whereas tetraspan proteins, like CD63 and CD82 (Fig. 6), are selectively enriched on internal membranes. The characteristics of exosomes fit well with those

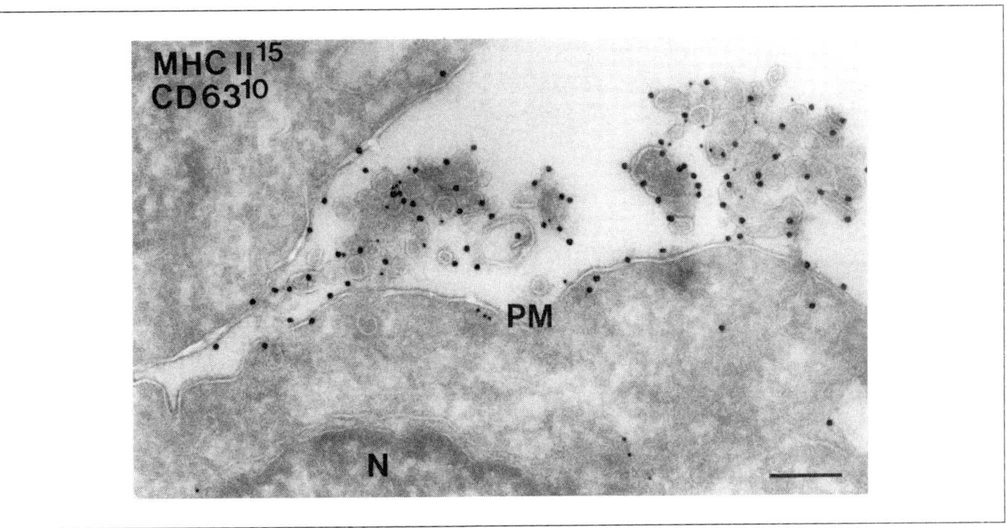

Fig. 7 Exosomes secreted by DCs. Immuno-double labelling on ultrathin cryosection of blood-derived DCs shows many small membrane vesicles (exosomes) in close vicinity of the plasma membrane (PM). MHC II labelling is present on the plasma membrane and on exosomes, whereas CD63 is primarily found on exosomes. N, nucleus; bar = 200 nm.

of the internal MVB vesicles, being devoid of plasma membrane proteins and early endosome markers (177, 200). When analysed by IEM, membranes pelleted at 70 000 *g* represented a homogeneous population of vesicles that were highly enriched in MHC II molecules. Western blot analysis further showed that MHC II molecules in exosome preparations are mature SDS-stable molecules associated with peptides. When analysed by Western blotting, the major components of the exosomal membranes of B cells are MHC II and molecules that could play a critical role in adhesion and co-stimulation (tetraspanins such as CD81, CD82, and CD63 and the co-stimulatory molecule B7.2) (201) (Table 1). In agreement, Théry and colleagues used matrix assisted laser desorption ionization-time of flight mass spectrometry (MALDI-TOF) to analyse the protein composition of exosomes released from the murine DC line D1 (201). They reported that some of the major bands appearing after Coomassie blue staining corresponded to proteins involved in adhesion: MFG-E8 or lactadherine, a protein known to be present in the milk fat globule interacting with the integrin $\alpha V\beta 5$; Mac 1, a known ligand of ICAM-1; and the tetraspanin CD9. In addition to the adhesion molecules, proteins were revealed that could play a crucial role in exosome formation and function. For example, the heat shock protein HSC-73, also know as the uncoating-ATPase, could be directly implicated in exosome biogenesis. By promoting the uncoating of endosomes it may therefore facilitate budding towards the intralumenal domain of endosomes. Alternatively, HSC-73 could, by its ability to bind peptides, be involved in the efficient transfer of antigenic peptides onto MHC molecules thereby contributing to their antigen presenting function and their potent anti-tumour effects (199, 201).

Table 1 Proteins enriched in exosomes and their putative function[a]

Exosomal proteins	Mr (kDa)	Function
MHC class II (α/β)	35/29	Antigen presentation
MHC class I (HC/β_2m)	43/12	
Tetraspanins		Signal transduction / adhesion / complex formation with MHC/DM/DO
CD9	25	
CD37	40–52	
CD53	32–40	
CD63	53	
CD81	26	
CD82	35–65	
CD86 (B7.2)	80	Co-stimulation
CD54 (ICAM-1)	90	Adhesion to target cells
Mac 1	127	
MFG-E8 (lactadherin)	45	
HSC 73	71	Uncoating ATPase / peptide transfer to MHC
Annexin II	39	MVB formation
Gi2 α subunit	40	Fusion

[a] Some of these proteins have been found in exosomes from B cells by Western blotting and immunogold labelling (MHC class I and II; CD37, 53, 63, 81, 82; CD86; CD54). Analysis by MALDI-TOF of exosomes from the DC line D1 revealed MHC class I and II, CD 9, Mac 1, MFG-E8, HSC 73, annexin II, and Gi2 α as the major exosomal proteins.

Together, these findings reinforce the idea that exosomes may have extracellular functions that involves their interaction with target membranes. How exosomes interact with these targets and whether they have a function in the immune response *in vivo* is not known yet. Recent observations open the possibility that exosome release is regulated and that exosomes released from B cells and DCs may not have the same targets and function. It appears that only activated B cells release exosomes (177) (our unpublished observations), whereas DCs lose their ability to secrete exosomes during their activation and maturation process (201). B cell exosomes can be detected in germinal centres in close apposition to the cell surface of follicular dendritic cells (FDCs), which have no endogenous expression of MHC II (202). By binding exosomes, FDCs could acquire preformed MHC II–peptide complexes and present them to surrounding T cells. Further experiments designed to follow the fate of exosomes *in vivo* and *in vitro* are needed to test these hypothesis and to determine the mode of interaction of exosomes with target cells.

7.3 Additional pathways for the transfer of MHC II to the cell surface

The observation that DCs lose their ability to secrete exosomes during maturation (201) is consistent with the idea that other transport routes operate for the transfer of MHC II–peptide complexes to the cell surface (151). In B cells, cross-linking experiments (203) have shown that MHC II–peptide complexes seem to be transferred directly from late endocytic compartments to the cell surface in a brefeldin A-sensitive manner, suggesting the involvement of a coated vesicular intermediate. Whether this intermediate is derived directly from the MIIC is unclear. Although clathrin coats are observed on MIICs, there seems to be no enrichment of peptide-loaded MHC II molecules in these coated areas (204).

Recent observations in DCs indicate that the originally described CIIVs may indeed correspond to vesicles bearing MHC II–peptide complexes and co-stimulatory molecules, positioned between lysosomal compartments and the cell surface (187). These vesicles may directly originate from lysosomal compartments and it is tempting to propose that a similar transport mechanism also operates in B cells as well as in other APCs. This raises the important issue of how differential sorting of MHC II molecules from lysosomal membrane proteins is achieved. A goal for the future is to clarify the subcellular mechanisms underlying the massive redistribution of MHC II molecules to the cell surface, and thereby unravel transport routes occurring from lysosomal compartments to the plasma membrane.

8. Conclusions

In the past years, dissecting the trafficking of antigens and MHC II molecules in APCs contributed to the generation of new concepts in the definition and biogenesis of compartments of the endocytic pathway and in the transport occurring into and out of these organelles. The very sophisticated way APCs exploit the endocytic system is also relevant for other molecules in different specialized cell systems.

In this chapter we have focused on the trafficking of MHC II molecules through the endocytic pathway where antigenic peptides from exogenous pathogens are generated. In the past years, however, accumulating data indicates that the presentation of exogenous versus endogenous antigens by MHC II and MHC I molecules respectively, may not be as strict as it originally appeared. MHC II molecules may also present endogenous peptides and MHC I molecules may present exogenous antigens raising a number of interesting issues. We do not understand how endogenous antigens present in the cytosol access the endocytic pathway to associate with MHC II molecules independently of macroautophagy (205). Mechanisms similar to those operating in yeast to deliver cytosolic constituents to the vacuole may also exist in APCs (206, 207). However, it is tempting to propose that microautophagy at the level of the multivesicular late endosome may allow the meeting between MHC II molecules and cytosolic antigens (156). Much more is known about presentation of exogenous antigens in the context of MHC I molecules (208–211). This presentation process can either be dependent or independent on the transporters associated with antigen processing (TAP) responsible for peptide translocation into the ER lumen (212). The presentation of exogenous antigens in the context of MHC I molecules has major implications for the peptide repertoire that can be presented to $CD8^+$ T cells and play a role in so-called cross-presentation. Exogenous antigens have different options to access MHC I. First, they may be delivered to the cytosol to reach the ER for loading onto MHC I, by a TAP-dependent manner (213–215). This could be achieved by leakage of the endosomal membrane (213, 214) or by the existence of a subcellular machinery allowing translocation of antigens from the endosome lumen to the cytosol (215). The second possibility for antigens to meet MHC I molecules appears from the recent observations that these molecules are also present in the endocytic pathway which leads to a TAP-independent loading (216). Interestingly in B cells, DCs, and melanoma cells, MHC I molecules are enriched in the multivesicular type of MIICs (200, 216, 217) and in exosomes (200, 217). MHC I molecules reach multivesicular MIICs following internalization from the plasma membrane (218). Therefore, the number of MHC I molecules in these late endosomal compartments could be directly related to the endocytic capacity and state of activation of a given cell. This may explain why MIICs in the B cell line JY are MHC I negative (98, 148).

From the numerous studies published on antigen processing and presentation it is evident that immune responses can be regulated at many levels. For MHC II restricted responses, regulation can take place at two major sites, first at the site of antigen entry, and second at the site of peptide loading. On one hand, the mode of antigen uptake and processing by regulated proteases plays a major role in the formation of antigenic determinants. On the other hand, binding of these determinants is controlled by Ii and DM/DO, which ultimately leads to their display at the cell surface for recognition by T cells. The distinct environments offered by the different compartments of the endocytic pathway can influence this regulation. It is remarkable that distinct APCs have adapted their endocytic capacities and intracellular compartments to perform their particular duties in the immune response.

Acknowledgements

We thank Dr Michael Marks (University of Pennsylvania, Philadelphia) for critical comments and many suggestions on this chapter. We are indebted to Dr Hans Geuze (Department of Cell Biology, UMC Utrecht, The Netherlands) for his continuous support in the past years and for reading the manuscript. R. Schriwaneck (Department of Cell Biology, UMC Utrecht, the Netherlands) is acknowledged for his photographical expertise. M. J. Kleijmeer is supported by the Nederlandse Organisatie voor Wetenschappelijk Onderzoek (grant 805–48–014). Finally, we apologize to all colleagues in the field whose work could not be cited.

References

1. Chesnut, R. W. and Grey, H. M. (1985) Antigen presenting cells and mechanisms of antigen presentation. *Crit. Rev. Immunol.*, **5**, 263.
2. Germain, R. N. (1994) MHC-dependent antigen processing and peptide presentation: providing ligands for T lymphocyte activation. *Cell*, **76**, 287.
3. Morrison, L. A., Lukacher, A. E., Braciale, V. L., Fan, D. P., and Braciale, T. J. (1986) Differences in antigen presentation to MHC class I-and class II- restricted influenza virus-specific cytolytic T lymphocyte clones. *J. Exp. Med.*, **163**, 903.

4. Germain, R. N. and Margulies, D. H. (1993) The biochemistry and cell biology of antigen processing and presentation. *Annu. Rev. Immunol.*, **11**, 403.
5. Watts, C. (1997) Capture and processing of exogenous antigens for presentation on MHC molecules. *Annu. Rev. Immunol.*, **15**, 821.
6. Ziegler, K. and Unanue, E. R. (1981) Identification of a macrophage antigen-processing event required for I-region-restricted antigen presentation to T lymphocytes. *J. Immunol.*, **127**, 1869.
7. Rudensky, A., Preston-Hurlburt, P., Hong, S. C., Barlow, A., and Janeway, C. A., Jr. (1991) Sequence analysis of peptides bound to MHC class II molecules [see comments]. *Nature*, **353**, 622.
8. Rammensee, H. G. (1995) Chemistry of peptides associated with MHC class I and class II molecules. *Curr. Opin. Immunol.*, **7**, 85.
9. Stahl, P. D. and Ezekowitz, R. A. (1998) The mannose receptor is a pattern recognition receptor involved in host defense. *Curr. Opin. Immunol.*, **10**, 50.
10. Dempsey, P. W., Allison, M. E., Akkaraju, S., Goodnow, C. C., and Fearon, D. T. (1996) C3d of complement as a molecular adjuvant: bridging innate and acquired immunity. *Science*, **271**, 348.
11. Lanzavecchia, A. (1990) Receptor-mediated antigen uptake and its effect on antigen presentation to class II-restricted T lymphocytes. *Annu. Rev. Immunol.*, **8**, 773.
12. Hombach, J., Lottspeich, F., and Reth, M. (1990) Identification of the genes encoding the IgM-alpha and Ig-beta components of the IgM antigen receptor complex by amino-terminal sequencing. *Eur. J. Immunol.*, **20**, 2795.
13. Venkitaraman, A. R., Williams, G. T., Dariavach, P., and Neuberger, M. S. (1991) The B-cell antigen receptor of the five immunoglobulin classes. *Nature*, **352**, 777.
14. Reth, M. (1992) Antigen receptors on B lymphocytes. *Annu. Rev. Immunol.*, **10**, 97.
15. Reth, M. (1989) Antigen receptor tail clue [letter]. *Nature*, **338**, 383.
16. Song, W., Cho, H., Cheng, P., and Pierce, S. K. (1995) Entry of B cell antigen receptor and antigen into class II peptide- loading compartment is independent of receptor cross-linking. *J. Immunol.*, **155**, 4255.
17. Drake, J. R., Lewis, T. A., Condon, K. B., Mitchell, R. N., and Webster, P. (1999) Involvement of MIIC-like late endosomes in B cell receptor-mediated antigen processing in murine B cells. *J. Immunol.*, **162**, 1150.
18. Wagle, N. M., Kim, J. H., and Pierce, S. K. (1998) Signaling through the B cell antigen receptor regulates discrete steps in the antigen processing pathway. *Cell. Immunol.*, **184**, 1.
19. Davidson, H. W. and Watts, C. (1989) Epitope-directed processing of specific antigen by B lymphocytes. *J. Cell Biol.*, **109**, 85.
20. Watts, C. and Lanzavecchia, A. (1993) Suppressive effect of antibody on processing of T cell epitopes. *J. Exp. Med.*, **178**, 1459.
21. Aluvihare, V. R., Khamlichi, A. A., Williams, G. T., Adorini, L., and Neuberger, M. S. (1997) Acceleration of intracellular targeting of antigen by the B-cell antigen receptor: importance depends on the nature of the antigen- antibody interaction. *EMBO J.*, **16**, 3553.
22. Knight, A. M., Lucocq, J. M., Prescott, A. R., Ponnambalam, S., and Watts, C. (1997) Antigen endocytosis and presentation mediated by human membrane IgG1 in the absence of the Ig(alpha)/Ig(beta) dimer. *EMBO J.*, **16**, 3842.
23. Mitchell, R. N., Barnes, K. A., Grupp, S. A., Sanchez, M., Misulovin, Z., Nussenzweig, M. C., *et al.* (1995) Intracellular targeting of antigens internalized by membrane immunoglobulin in B lymphocytes. *J. Exp. Med.*, **181**, 1705.

24. Bonnerot, C., Lankar, D., Hanau, D., Spehner, D., Davoust, J., Salamero, J., *et al.* (1995) Role of B cell receptor Ig alpha and Ig beta subunits in MHC class II- restricted antigen presentation. *Immunity*, **3**, 335.
25. Zimmermann, V. S., Rovere, P., Trucy, J., Serre, K., Machy, P., Forquet, F., *et al.* (1999) Engagement of B cell receptor regulates the invariant chain-dependent MHC class II presentation pathway. *J. Immunol.*, **162**, 2495.
26. Forquet, F., Barois, N., Machy, P., Trucy, J., Zimmermann, V. S., Leserman, L., *et al.* (1999) Presentation of antigens internalized through the B cell receptor requires newly synthesized MHC class II molecules. *J. Immunol.*, **162**, 3408.
27. Lankar, D., Briken, V., Adler, K., Weiser, P., Cassard, S., Blank, U., *et al.* (1998) Syk tyrosine kinase and B cell antigen receptor (BCR) immunoglobulin- alpha subunit determine BCR-mediated major histocompatibility complex class II-restricted antigen presentation. *J. Exp. Med.*, **188**, 819.
28. Xu, X., Press, B., Wagle, N. M., Cho, H., Wandinger-Ness, A., and Pierce, S. K. (1996) B cell antigen receptor signaling links biochemical changes in the class II peptide-loading compartment to enhanced processing. *Int. Immunol.*, **8**, 1867.
29. Barois, N., Forquet, F., and Davoust, J. (1997) Selective modulation of the major histocompatibility complex class II antigen presentation pathway following B cell receptor ligation and protein kinase C activation. *J. Biol. Chem.*, **272**, 3641.
30. Daeron, M. (1997) Fc receptor biology. *Annu. Rev. Immunol.*, **15**, 203.
31. Cambier, J. C. (1995) Antigen and Fc receptor signaling. The awesome power of the immunoreceptor tyrosine-based activation motif (ITAM). *J. Immunol.*, **155**, 3281.
32. Amigorena, S., Bonnerot, C., Drake, J. R., Choquet, D., Hunziker, W., Guillet, J. G., *et al.* (1992) Cytoplasmic domain heterogeneity and functions of IgG Fc receptors in B lymphocytes. *Science*, **256**, 1808.
33. Wagle, N. M., Faassen, A. E., Kim, J. H., and Pierce, S. K. (1999) Regulation of B cell receptor-mediated MHC class II antigen processing by FcgammaRIIB1. *J. Immunol.*, **162**, 2732.
34. Amigorena, S. and Bonnerot, C. (1999) Fc receptors for IgG and antigen presentation on MHC class I and class II molecules. *Semin. Immunol.*, **11**, 385.
35. van Vugt, M. J., Kleijmeer, M. J., Keler, T., Zeelenberg, I., van Dijk, M. A., Leusen, J. H., *et al.* (1999) The FcgammaRIa (CD64) ligand binding chain triggers major histocompatibility complex class II antigen presentation independently of its associated FcR gamma-chain. *Blood*, **94**, 808.
36. Bonnerot, C., Briken, V., Brachet, V., Lankar, D., Cassard, S., Jabri, B., *et al.* (1998) syk protein tyrosine kinase regulates Fc receptor gamma-chain-mediated transport to lysosomes. *EMBO J.*, **17**, 4606.
37. Lindner, R. and Unanue, E. R. (1996) Distinct antigen MHC class II complexes generated by separate processing pathways. *EMBO J.*, **15**, 6910.
38. Steinman, R. M. and Swanson, J. (1995) The endocytic activity of dendritic cells [comment]. *J. Exp. Med.*, **182**, 283.
39. Desjardins, M., Celis, J. E., van Meer, G., Dieplinger, H., Jahraus, A., Griffiths, G., *et al.* (1994) Molecular characterization of phagosomes. *J. Biol. Chem.*, **269**, 32194.
40. Alvarez-Dominguez, C., Barbieri, A. M., Beron, W., Wandinger-Ness, A., and Stahl, P. D. (1996) Phagocytosed live Listeria monocytogenes influences Rab5-regulated in vitro phagosome-endosome fusion. *J. Biol. Chem.*, **271**, 13834.
41. Ramachandra, L., Chu, R. S., Askew, D., Noss, E. H., Canaday, D. H., Potter, N. S., *et al.* (1999) Phagocytic antigen processing and effects of microbial products on antigen processing and T-cell responses. *Immunol. Rev.*, **168**, 217.

42. Harding, C. V. and Geuze, H. J. (1993) Antigen processing and intracellular traffic of antigens and MHC molecules. *Curr. Opin. Cell Biol.*, **5**, 596.
43. Ramachandra, L., Song, R., and Harding, C. V. (1999) Phagosomes are fully competent antigen-processing organelles that mediate the formation of peptide:class II MHC complexes. *J. Immunol.*, **162**, 3263.
44. Clemens, D. L. and Horwitz, M. A. (1995) Characterization of the Mycobacterium tuberculosis phagosome and evidence that phagosomal maturation is inhibited. *J. Exp. Med.*, **181**, 257.
45. Lang, T., Hellio, R., Kaye, P. M., and Antoine, J. C. (1994) Leishmania donovani-infected macrophages: characterization of the parasitophorous vacuole and potential role of this organelle in antigen presentation. *J. Cell Sci.*, **107**, 2137.
46. Ojcius, D. M., Gachelin, G., and Dautry-Varsat, A. (1996) Presentation of antigens derived from microorganisms residing in host-cell vacuoles. *Trends Microbiol.*, **4**, 53.
47. Sturgill-Koszycki, S., Schlesinger, P. H., Chakraborty, P., Haddix, P. L., Collins, H. L., Fok, A. K., *et al.* (1994) Lack of acidification in Mycobacterium phagosomes produced by exclusion of the vesicular proton-ATPase. *Science*, **263**, 678.
48. Stahl, P. and Gordon, S. (1982) Expression of a mannosyl-fucosyl receptor for endocytosis on cultured primary macrophages and their hybrids. *J. Cell Biol.*, **93**, 49.
49. Janeway, C. A., Jr. (1992) The immune system evolved to discriminate infectious nonself from noninfectious self. *Immunol. Today*, **13**, 11.
50. Marodi, L., Schreiber, S., Anderson, D. C., MacDermott, R. P., Korchak, H. M., and Johnston, R. B., Jr. (1993) Enhancement of macrophage candidacidal activity by interferon-gamma. Increased phagocytosis, killing, and calcium signal mediated by a decreased number of mannose receptors. *J. Clin. Invest.*, **91**, 2596.
51. Sallusto, F., Cella, M., Danieli, C., and Lanzavecchia, A. (1995) Dendritic cells use macropinocytosis and the mannose receptor to concentrate macromolecules in the major histocompatibility complex class II compartment: downregulation by cytokines and bacterial products [see comments]. *J. Exp. Med.*, **182**, 389.
52. Swanson, J. A. (1989) Phorbol esters stimulate macropinocytosis and solute flow through macrophages. *J. Cell Sci.*, **94**, 135.
53. Araki, N., Johnson, M. T., and Swanson, J. A. (1996) A role for phosphoinositide 3-kinase in the completion of macropinocytosis and phagocytosis by macrophages. *J. Cell Biol.*, **135**, 1249.
54. West, M. A., Antoniou, A. N., Prescott, A. R., Azuma, T., Kwiatkowski, D. J., and Watts, C. (1999) Membrane ruffling, macropinocytosis and antigen presentation in the absence of gelsolin in murine dendritic cells. *Eur. J. Immunol.*, **29**, 3450.
55. Engering, A. J., Cella, M., Fluitsma, D. M., Hoefsmit, E. C., Lanzavecchia, A., and Pieters, J. (1997) Mannose receptor mediated antigen uptake and presentation in human dendritic cells. *Adv. Exp. Med. Biol.*, **417**, 183.
56. Martinez-Pomares, L., Kosco-Vilbois, M., Darley, E., Tree, P., Herren, S., Bonnefoy, J. Y., *et al.* (1996) Fc chimeric protein containing the cysteine-rich domain of the murine mannose receptor binds to macrophages from splenic marginal zone and lymph node subcapsular sinus and to germinal centers. *J. Exp. Med.*, **184**, 1927.
57. Prigozy, T. I., Sieling, P. A., Clemens, D., Stewart, P. L., Behar, S. M., Porcelli, S. A., *et al.* (1997) The mannose receptor delivers lipoglycan antigens to endosomes for presentation to T cells by CD1b molecules. *Immunity*, **6**, 187.
58. Sieling, P. A., Chatterjee, D., Porcelli, S. A., Prigozy, T. I., Mazzaccaro, R. J., Soriano, T., *et al.* (1995) CD1-restricted T cell recognition of microbial lipoglycan antigens [see comments]. *Science*, **269**, 227.

59. Beckman, E. M., Porcelli, S. A., Morita, C. T., Behar, S. M., Furlong, S. T., and Brenner, M. B. (1994) Recognition of a lipid antigen by CD1-restricted alpha beta+ T cells [see comments]. *Nature*, **372**, 691.
60. Jiang, W., Swiggard, W. J., Heufler, C., Peng, M., Mirza, A., Steinman, R. M., *et al.* (1995) The receptor DEC-205 expressed by dendritic cells and thymic epithelial cells is involved in antigen processing. *Nature*, **375**, 151.
61. Albert, M. L., Sauter, B., and Bhardwaj, N. (1998) Dendritic cells acquire antigen from apoptotic cells and induce class I- restricted CTLs. *Nature*, **392**, 86.
62. Albert, M. L., Pearce, S. F., Francisco, L. M., Sauter, B., Roy, P., Silverstein, R. L., *et al.* (1998) Immature dendritic cells phagocytose apoptotic cells via alphavbeta5 and CD36, and cross-present antigens to cytotoxic T lymphocytes. *J. Exp. Med.*, **188**, 1359.
63. Hart, D. N. (1997) Dendritic cells: unique leukocyte populations, which control the primary immune response. *Blood*, **90**, 3245.
64. Banchereau, J. and Steinman, R. M. (1998) Dendritic cells and the control of immunity. *Nature*, **392**, 245.
65. Romani, N., Gruner, S., Brang, D., Kampgen, E., Lenz, A., Trockenbacher, B., *et al.* (1994) Proliferating dendritic cell progenitors in human blood. *J. Exp. Med.*, **180**, 83.
66. Cella, M., Sallusto, F., and Lanzavecchia, A. (1997) Origin, maturation and antigen presenting function of dendritic cells. *Curr. Opin. Immunol.*, **9**, 10.
67. Regnault, A., Lankar, D., Lacabanne, V., Rodriguez, A., Thery, C., Rescigno, M., *et al.* (1999) Fcgamma receptor-mediated induction of dendritic cell maturation and major histocompatibility complex class I-restricted antigen presentation after immune complex internalization. *J. Exp. Med.*, **189**, 371.
68. Villadangos, J. A. and Ploegh, H. L. (2000) Proteolysis in MHC class II antigen presentation: who's in charge? *Immunity*, **12**, 233.
69. Villadangos, J. A., Riese, R. J., Peters, C., Chapman, H. A., and Ploegh, H. L. (1997) Degradation of mouse invariant chain: roles of cathepsins S and D and the influence of major histocompatibility complex polymorphism. *J. Exp. Med.*, **186**, 549.
70. Riese, R. J., Wolf, P. R., Bromme, D., Natkin, L. R., Villadangos, J. A., Ploegh, H. L., *et al.* (1996) Essential role for cathepsin S in MHC class II-associated invariant chain processing and peptide loading. *Immunity*, **4**, 357.
71. Nakagawa, T., Roth, W., Wong, P., Nelson, A., Farr, A., Deussing, J., *et al.* (1998) Cathepsin L: critical role in Ii degradation and CD4 T cell selection in the thymus. *Science*, **280**, 450.
72. Manoury, B., Hewitt, E. W., Morrice, N., Dando, P. M., Barrett, A. J., and Watts, C. (1998) An asparaginyl endopeptidase processes a microbial antigen for class II MHC presentation. *Nature*, **396**, 695.
73. Collins, D. S., Unanue, E. R., and Harding, C. V. (1991) Reduction of disulfide bonds within lysosomes is a key step in antigen processing. *J. Immunol.*, **147**, 4054.
74. Arunachalam, B., Phan, U. T., Geuze, H. J., and Cresswell, P. (2000) Enzymatic reduction of disulfide bonds in lysosomes: characterization of a gamma-interferon-inducible lysosomal thiol reductase (GILT). *Proc. Natl. Acad. Sci. USA*, **97**, 745.
75. Dadaglio, G., Nelson, C. A., Deck, M. B., Petzold, S. J., and Unanue, E. R. (1997) Characterization and quantitation of peptide-MHC complexes produced from hen egg lysozyme using a monoclonal antibody. *Immunity*, **6**, 727.
76. Latek, R. R. and Unanue, E. R. (1999) Mechanisms and consequences of peptide selection by the I-Ak class II molecule. *Immunol. Rev.*, **172**, 209.
77. Griffin, J. P., Chu, R., and Harding, C. V. (1997) Early endosomes and a late endocytic

compartment generate different peptide-class II MHC complexes via distinct processing mechanisms. *J. Immunol.*, **158**, 1523.

78. Escola, J. M., Grivel, J. C., Chavrier, P., and Gorvel, J. P. (1995) Different endocytic compartments are involved in the tight association of class II molecules with processed hen egg lysozyme and ribonuclease A in B cells. *J. Cell Sci.*, **108**, 2337.
79. Castellino, F. and Germain, R. N. (1995) Extensive trafficking of MHC class II-invariant chain complexes in the endocytic pathway and appearance of peptide-loaded class II in multiple compartments. *Immunity*, **2**, 73.
80. Simitsek, P. D., Campbell, D. G., Lanzavecchia, A., Fairweather, N., and Watts, C. (1995) Modulation of antigen processing by bound antibodies can boost or suppress class II major histocompatibility complex presentation of different T cell determinants [see comments]. *J. Exp. Med.*, **181**, 1957.
81. Watts, C., Antoniou, A., Manoury, B., Hewitt, E. W., McKay, L. M., Grayson, L., *et al.* (1998) Modulation by epitope-specific antibodies of class II MHC-restricted presentation of the tetanus toxin antigen. *Immunol. Rev.*, **164**, 11.
82. Sercarz, E. E., Lehmann, P. V., Ametani, A., Benichou, G., Miller, A., and Moudgil, K. (1993) Dominance and crypticity of T cell antigenic determinants. *Annu. Rev. Immunol.*, **11**, 729.
83. Germain, R. N., Castellino, F., Han, R., Reis e Sousa, C., Romagnoli, P., Sadegh-Nasseri, S., *et al.* (1996) Processing and presentation of endocytically acquired protein antigens by MHC class II and class I molecules. *Immunol. Rev.*, **151**, 5.
84. Castellino, F., Zappacosta, F., Coligan, J. E., and Germain, R. N. (1998) Large protein fragments as substrates for endocytic antigen capture by MHC class II molecules. *J. Immunol.*, **161**, 4048.
85. Bevec, T., Stoka, V., Pungercic, G., Dolenc, I., and Turk, V. (1996) Major histocompatibility complex class II-associated p41 invariant chain fragment is a strong inhibitor of lysosomal cathepsin L. *J. Exp. Med.*, **183**, 1331.
86. Ploegh, H. L., Orr, H. T., and Strominger, J. L. (1981) Major histocompatibility antigens: the human (HLA-A, -B, -C) and murine (H-2K, H-2D) class I molecules. *Cell*, **24**, 287.
87. Kaufman, J. F., Auffray, C., Korman, A. J., Shackelford, D. A., and Strominger, J. (1984) The class II molecules of the human and murine major histocompatibility complex. *Cell*, **36**, 1.
88. Cresswell, P. (1994) Assembly, transport, and function of MHC class II molecules. *Annu. Rev. Immunol.*, **12**, 259.
89. Wolf, P. R. and Ploegh, H. L. (1995) How MHC class II molecules acquire peptide cargo: biosynthesis and trafficking through the endocytic pathway. *Annu. Rev. Cell Dev. Biol.*, **11**, 267.
90. Marks, M. S., Blum, J. S., and Cresswell, P. (1990) Invariant chain trimers are sequestered in the rough endoplasmic reticulum in the absence of association with HLA class II antigens. *J. Cell Biol.*, **111**, 839.
91. Roche, P. A., Marks, M. S., and Cresswell, P. (1991) Formation of a nine-subunit complex by HLA class II glycoproteins and the invariant chain. *Nature*, **354**, 392.
92. Cresswell, P. (1992) Chemistry and functional role of the invariant chain. *Curr. Opin. Immunol.*, **4**, 87.
93. Lamb, C. A. and Cresswell, P. (1992) Assembly and transport properties of invariant chain trimers and HLA-DR- invariant chain complexes. *J. Immunol.*, **148**, 3478.
94. Roche, P. A. and Cresswell, P. (1990) Invariant chain association with HLA-DR molecules inhibits immunogenic peptide binding. *Nature*, **345**, 615.

95. Teyton, L., O'Sullivan, D., Dickson, P. W., Lotteau, V., Sette, A., Fink, P., *et al.* (1990) Invariant chain distinguishes between the exogenous and endogenous antigen presentation pathways. *Nature*, **348**, 39.
96. Roche, P. A. and Cresswell, P. (1991) Proteolysis of the class II-associated invariant chain generates a peptide binding site in intracellular HLA-DR molecules. *Proc. Natl. Acad. Sci. USA*, **88**, 3150.
97. Cresswell, P. (1985) Intracellular class II HLA antigens are accessible to transferrin-neuraminidase conjugates internalized by receptor-mediated endocytosis. *Proc. Natl. Acad. Sci. USA*, **82**, 8188.
98. Neefjes, J. J., Stollorz, V., Peters, P. J., Geuze, H. J., and Ploegh, H. L. (1990) The biosynthetic pathway of MHC class II but not class I molecules intersects the endocytic route. *Cell*, **61**, 171.
99. Davidson, H. W., Reid, P. A., Lanzavecchia, A., and Watts, C. (1991) Processed antigen binds to newly synthesized MHC class II molecules in antigen-specific B lymphocytes. *Cell*, **67**, 105.
100. Harding, C. V. and Unanue, E. R. (1989) Antigen processing and intracellular Ia. Possible roles of endocytosis and protein synthesis in Ia function. *J. Immunol.*, **142**, 12.
101. Reid, P. A. and Watts, C. (1992) Constitutive endocytosis and recycling of major histocompatibility complex class II glycoproteins in human B-lymphoblastoid cells. *Immunology*, **77**, 539.
102. Salamero, J., Humbert, M., Cosson, P., and Davoust, J. (1990) Mouse B lymphocyte specific endocytosis and recycling of MHC class II molecules. *EMBO J.*, **9**, 3489.
103. Pinet, V. M. and Long, E. O. (1998) Peptide loading onto recycling HLA-DR molecules occurs in early endosomes. *Eur. J. Immunol.*, **28**, 799.
104. Roche, P. A., Teletski, C. L., Stang, E., Bakke, O., and Long, E. O. (1993) Cell surface HLA-DR-invariant chain complexes are targeted to endosomes by rapid internalization. *Proc. Natl. Acad. Sci. USA*, **90**, 8581.
105. Warmerdam, P. A., Long, E. O., and Roche, P. A. (1996) Isoforms of the invariant chain regulate transport of MHC class II molecules to antigen processing compartments. *J. Cell Biol.*, **133**, 281.
106. Bonnerot, C., Marks, M. S., Cosson, P., Robertson, E. J., Bikoff, E. K., Germain, R. N., *et al.* (1994) Association with BiP and aggregation of class II MHC molecules synthesized in the absence of invariant chain. *EMBO J.*, **13**, 934.
107. Lamb, C. A. and Cresswell, P. (1992) Assembly and transport properties of invariant chain trimers and HLA-DR-invariant chain complexes. *J. Immunol.*, **148**, 3478.
108. Bikoff, E. K., Huang, L. Y., Episkopou, V., van Meerwijk, J., Germain, R. N., and Robertson, E. J. (1993) Defective major histocompatibility complex class II assembly, transport, peptide acquisition, and CD4+ T cell selection in mice lacking invariant chain expression. *J. Exp. Med.*, **177**, 1699.
109. Viville, S., Neefjes, J., Lotteau, V., Dierich, A., Lemeur, M., Ploegh, H., *et al.* (1993) Mice lacking the MHC class II-associated invariant chain. *Cell*, **72**, 635.
110. Elliott, E. A., Drake, J. R., Amigorena, S., Elsemore, J., Webster, P., Mellman, I., *et al.* (1994) The invariant chain is required for intracellular transport and function of major histocompatibility complex class II molecules. *J. Exp. Med.*, **179**, 681.
111. Lotteau, V., Teyton, L., Peleraux, A., Nilsson, T., Karlsson, L., Schmid, S. L., *et al.* (1990) Intracellular transport of class II MHC molecules directed by invariant chain. *Nature*, **348**, 600.
112. Bakke, O. and Dobberstein, B. (1990) MHC class II-associated invariant chain contains a sorting signal for endosomal compartments. *Cell*, **63**, 707.

113. Pieters, J., Bakke, O., and Dobberstein, B. (1993) The MHC class II-associated invariant chain contains two endosomal targeting signals within its cytoplasmic tail. *J. Cell Sci.*, **106**, 831.
114. Odorizzi, C. G., Trowbridge, I. S., Xue, L., Hopkins, C. R., Davis, C. D., and Collawn, J. F. (1994) Sorting signals in the MHC class II invariant chain cytoplasmic tail and transmembrane region determine trafficking to an endocytic processing compartment. *J. Cell Biol.*, **126**, 317.
115. Miller, J. (1994) Endosomal localization of MHC class II-invariant chain complexes. *Immunol. Res.*, **13**, 244.
116. Humbert, M., Raposo, G., Cosson, P., Reggio, H., Davoust, J., and Salamero, J. (1993) The invariant chain induces compact forms of class II molecules localized in late endosomal compartments. *Eur. J. Immunol.*, **23**, 3158.
117. Shi, G. P., Bryant, R. A., Riese, R., Verhelst, S., Driessen, C., Li, Z., *et al.* (2000) Role for cathepsin F in invariant chain processing and major histocompatibility complex class II peptide loading by macrophages. *J. Exp. Med.*, **191**, 1177.
118. Maric, M. A., Taylor, M. D., and Blum, J. S. (1994) Endosomal aspartic proteinases are required for invariant-chain processing. *Proc. Natl. Acad. Sci. USA*, **91**, 2171.
119. Chapman, H. A. (1998) Endosomal proteolysis and MHC class II function. *Curr. Opin. Immunol.*, **10**, 93.
120. Driessen, C., Bryant, R. A., Lennon-Dumenil, A. M., Villadangos, J. A., Bryant, P. W., Shi, G. P., *et al.* (1999) Cathepsin S controls the trafficking and maturation of MHC class II molecules in dendritic cells. *J. Cell Biol.*, **147**, 775.
121. Chicz, R. M., Urban, R. G., Gorga, J. C., Vignali, D. A., Lane, W. S., and Strominger, J. L. (1993) Specificity and promiscuity among naturally processed peptides bound to HLA-DR alleles. *J. Exp. Med.*, **178**, 27.
122. Newcomb, J. R. and Cresswell, P. (1993) Structural analysis of proteolytic products of MHC class II-invariant chain complexes generated *in vivo*. *J. Immunol.*, **151**, 4153.
123. Riberdy, J. M., Newcomb, J. R., Surman, M. J., Barbosa, J. A., and Cresswell, P. (1992) HLA-DR molecules from an antigen-processing mutant cell line are associated with invariant chain peptides. *Nature*, **360**, 474.
124. Cresswell, P. (1996) Invariant chain structure and MHC class II function. *Cell*, **84**, 505.
125. Pierre, P. and Mellman, I. (1998) Developmental regulation of invariant chain proteolysis controls MHC class II trafficking in mouse dendritic cells. *Cell*, **93**, 1135.
126. Morris, P., Shaman, J., Attaya, M., Amaya, M., Goodman, S., Bergman, C., *et al.* (1994) An essential role for HLA-DM in antigen presentation by class II major histocompatibility molecules. *Nature*, **368**, 551.
127. Roche, P. A. (1995) HLA-DM: an *in vivo* facilitator of MHC class II peptide loading. *Immunity*, **3**, 259.
128. Kropshofer, H., Hammerling, G. J., and Vogt, A. B. (1999) The impact of the non-classical MHC proteins HLA-DM and HLA-DO on loading of MHC class II molecules. *Immunol. Rev.*, **172**, 267.
129. Sanderson, S., Frauwirth, K., and Shastri, N. (1995) Expression of endogenous peptide-major histocompatibility complex class II complexes derived from invariant chain-antigen fusion proteins. *Proc. Natl. Acad. Sci. USA*, **92**, 7217.
130. Denzin, L. K. and Cresswell, P. (1995) HLA-DM induces CLIP dissociation from MHC class II alpha beta dimers and facilitates peptide loading. *Cell*, **82**, 155.
131. Sloan, V. S., Cameron, P., Porter, G., Gammon, M., Amaya, M., Mellins, E., *et al.* (1995) Mediation by HLA-DM of dissociation of peptides from HLA-DR. *Nature*, **375**, 802.

132. Martin, W. D., Hicks, G. G., Mendiratta, S. K., Leva, H. I., Ruley, H. E., and Van Kaer, L. (1996) H2-M mutant mice are defective in the peptide loading of class II molecules, antigen presentation, and T cell repertoire selection. *Cell*, **84**, 543.
133. Kropshofer, H., Hammerling, G. J., and Vogt, A. B. (1997) How HLA-DM edits the MHC class II peptide repertoire: survival of the fittest? *Immunol. Today*, **18**, 77.
134. Denzin, L. K., Hammond, C., and Cresswell, P. (1996) HLA-DM interactions with intermediates in HLA-DR maturation and a role for HLA-DM in stabilizing empty HLA-DR molecules. *J. Exp. Med.*, **184**, 2153.
135. Sanderson, F., Kleijmeer, M. J., Kelly, A., Verwoerd, D., Tulp, A., Neefjes, J. J., *et al.* (1994) Accumulation of HLA-DM, a regulator of antigen presentation, in MHC class II compartments. *Science*, **266**, 1566.
136. Nijman, H. W., Kleijmeer, M. J., Ossevoort, M. A., Oorschot, V. M., Vierboom, M. P., van de Keur, M., *et al.* (1995) Antigen capture and major histocompatibility class II compartments of freshly isolated and cultured human blood dendritic cells. *J. Exp. Med.*, **182**, 163.
137. Pierre, P., Denzin, L. K., Hammond, C., Drake, J. R., Amigorena, S., Cresswell, P., *et al.* (1996) HLA-DM is localized to conventional and unconventional MHC class II- containing endocytic compartments. *Immunity*, **4**, 229.
138. Marks, M. S., Roche, P. A., van Donselaar, E., Woodruff, L., Peters, P. J., and Bonifacino, J. S. (1995) A lysosomal targeting signal in the cytoplasmic tail of the beta chain directs HLA-DM to MHC class II compartments. *J. Cell Biol.*, **131**, 351.
139. Copier, J., Kleijmeer, M. J., Ponnambalam, S., Oorschot, V., Potter, P., Trowsdale, J., *et al.* (1996) Targeting signal and subcellular compartments involved in the intracellular trafficking of HLA-DMB. *J. Immunol.*, **157**, 1017.
140. Liu, S. H., Marks, M. S., and Brodsky, F. M. (1998) A dominant-negative clathrin mutant differentially affects trafficking of molecules with distinct sorting motifs in the class II major histocompatibility complex (MHC) pathway. *J. Cell Biol.*, **140**, 1023.
141. Arndt, S. O., Vogt, A. B., Markovic-Plese, S., Martin, R., Moldenhauer, G., Wolpl, A., *et al.* (2000) Functional HLA-DM on the surface of B cells and immature dendritic cells. *EMBO J.*, **19**, 1241.
142. Denzin, L. K., Sant'Angelo, D. B., Hammond, C., Surman, M. J., and Cresswell, P. (1997) Negative regulation by HLA-DO of MHC class II-restricted antigen processing. *Science*, **278**, 106.
143. Kropshofer, H., Vogt, A. B., Thery, C., Armandola, E. A., Li, B. C., Moldenhauer, G., *et al.* (1998) A role for HLA-DO as a co-chaperone of HLA-DM in peptide loading of MHC class II molecules. *EMBO J.*, **17**, 2971.
144. van Ham, S. M., Tjin, E. P. M., Lillemeier, B. F., Gruneberg, U., van Meijgaarden, K. E., Pastoors, L., *et al.* (1997) HLA-DO is a negative modulator of HLA-DM-mediated MHC class II peptide loading. *Curr. Biol.*, **7**, 950.
145. Liljedahl, M., Winqvist, O., Surh, C. D., Wong, P., Ngo, K., Teyton, L., *et al.* (1998) Altered antigen presentation in mice lacking H2-O. *Immunity*, **8**, 233.
146. van Ham, M., van Lith, M., Lillemeier, B., Tjin, E., Gruneberg, U., Rahman, D., *et al.* (2000) Modulation of the major histocompatibility complex class II-associated peptide repertoire by human histocompatibility leukocyte antigen (HLA)-DO [In Process Citation]. *J. Exp. Med.*, **191**, 1127.
147. Benaroch, P., Yilla, M., Raposo, G., Ito, K., Miwa, K., Geuze, H. J., *et al.* (1995) How MHC class II molecules reach the endocytic pathway. *EMBO J.*, **14**, 37.

148. Peters, P. J., Neefjes, J. J., Oorschot, V., Ploegh, H. L., and Geuze, H. J. (1991) Segregation of MHC class II molecules from MHC class I molecules in the Golgi complex for transport to lysosomal compartments. *Nature*, **349**, 669.
149. Futter, C. E., Pearse, A., Hewlett, L. J., and Hopkins, C. R. (1996) Multivesicular endosomes containing internalized EGF-EGF receptor complexes mature and then fuse directly with lysosomes. *J. Cell Biol.*, **132**, 1011.
150. Brachet, V., Pehau-Arnaudet, G., Desaymard, C., Raposo, G., and Amigorena, S. (1999) Early endosomes are required for major histocompatiblity complex class II transport to peptide-loading compartments. *Mol. Biol. Cell*, **10**, 2891.
151. Pond, L. and Watts, C. (1999) Functional early endosomes are required for maturation of major histocompatibility complex class II molecules in human B lymphoblastoid cells. *J. Biol. Chem.*, **274**, 18049.
152. Villadangos, J. A., Driessen, C., Shi, G. P., Chapman, H. A., and Ploegh, H. L. (2000) Early endosomal maturation of MHC class II molecules independently of cysteine proteases and H-2DM. *EMBO J.*, **19**, 882.
153. Guagliardi, L. E., Koppelman, B., Blum, J. S., Marks, M. S., Cresswell, P., and Brodsky, F. M. (1990) Co-localization of molecules involved in antigen processing and presentation in an early endocytic compartment. *Nature*, **343**, 133.
154. Glickman, J. N., Morton, P. A., Slot, J. W., Kornfeld, S., and Geuze, H. J. (1996) The biogenesis of the MHC class II compartment in human I-cell disease B lymphoblasts. *J. Cell Biol.*, **132**, 769.
155. Kleijmeer, M. J., Morkowski, S., Griffith, J. M., Rudensky, A. Y., and Geuze, H. J. (1997) Major histocompatibility complex class II compartments in human and mouse B lymphoblasts represent conventional endocytic compartments. *J. Cell Biol.*, **139**, 639.
156. Geuze, H. J. (1998) The role of endosomes and lysosomes in MHC class II functioning. *Immunol. Today*, **19**, 282.
157. Le Borgne, R., Griffiths, G., and Hoflack, B. (1996) Mannose 6-phosphate receptors and ADP-ribosylation factors cooperate for high affinity interaction of the AP-1 Golgi assembly proteins with membranes. *J. Biol. Chem.*, **271**, 2162.
158. Kirchhausen, T., Bonifacino, J. S., and Riezman, H. (1997) Linking cargo to vesicle formation: receptor tail interactions with coat proteins. *Curr. Opin. Cell Biol.*, **9**, 488.
159. Futter, C. E., Gibson, A., Allchin, E. H., Maxwell, S., Ruddock, L. J., Odorizzi, G., *et al.* (1998) In polarized MDCK cells basolateral vesicles arise from clathrin-gamma-adaptin-coated domains on endosomal tubules. *J. Cell Biol.*, **141**, 611.
160. Mallard, F., Antony, C., Tenza, D., Salamero, J., Goud, B., and Johannes, L. (1998) Direct pathway from early/recycling endosomes to the Golgi apparatus revealed through the study of shiga toxin B-fragment transport. *J. Cell Biol.*, **143**, 973.
161. Stoorvogel, W., Oorschot, V., and Geuze, H. J. (1996) A novel class of clathrin-coated vesicles budding from endosomes. *J. Cell Biol.*, **132**, 21.
162. Salamero, J., Le Borgne, R., Saudrais, C., Goud, B., and Hoflack, B. (1996) Expression of major histocompatibility complex class II molecules in HeLa cells promotes the recruitment of AP-1 Golgi-specific assembly proteins on Golgi membranes. *J. Biol. Chem.*, **271**, 30318.
163. Saudrais, C., Spehner, D., de la Salle, H., Bohbot, A., Cazenave, J. P., Goud, B., *et al.* (1998) Intracellular pathway for the generation of functional MHC class II peptide complexes in immature human dendritic cells. *J. Immunol.*, **160**, 2597.
164. Pieters, J. (1997) MHC class II restricted antigen presentation. *Curr. Opin. Immunol.*, **9**, 89.
165. Kampgen, E., Koch, N., Koch, F., Stoger, P., Heufler, C., Schuler, G., *et al.* (1991) Class II

major histocompatibility complex molecules of murine dendritic cells: synthesis, sialylation of invariant chain, and antigen processing capacity are down-regulated upon culture. *Proc. Natl. Acad. Sci. USA*, **88**, 3014.

166. Anderson, R. G., Falck, J. R., Goldstein, J. L., and Brown, M. S. (1984) Visualization of acidic organelles in intact cells by electron microscopy. *Proc. Natl. Acad. Sci. USA*, **81**, 4838.
167. Kleijmeer, M. J., Raposo, G., and Geuze, H. J. (1996) Characterization of MHC class II compartments by immunoelectron microscopy. *Methods*, **10**, 191.
168. Harding, C. V., Unanue, E. R., Slot, J. W., Schwartz, A. L., and Geuze, H. J. (1990) Functional and ultrastructural evidence for intracellular formation of major histocompatibility complex class II-peptide complexes during antigen processing. *Proc. Natl. Acad. Sci. USA*, **87**, 5553.
169. Geuze, H. J., Slot, J. W., Strous, G. J., Lodish, H. F., and Schwartz, A. L. (1983) Intracellular site of asialoglycoprotein receptor-ligand uncoupling: double-label immunoelectron microscopy during receptor-mediated endocytosis. *Cell*, **32**, 277.
170. Pieters, J., Horstmann, H., Bakke, O., Griffiths, G., and Lipp, J. (1991) Intracellular transport and localization of major histocompatibility complex class II molecules and associated invariant chain. *J. Cell Biol.*, **115**, 1213.
171. Harding, C. V. and Geuze, H. J. (1993) Immunogenic peptides bind to class II MHC molecules in an early lysosomal compartment. *J. Immunol.*, **151**, 3988.
172. Liou, W., Geuze, H. J., and Slot, J. W. (1996) Improving structural integrity of cryosections for immunogold labeling. *Histochem. Cell Biol.*, **106**, 41.
173. Raposo, G., Kleijmeer, M. J., Posthuma, G., Slot, J. W., and Geuze, H. J. (1997) Immunogold labeling of ultrathin cryosections: application in immunology. In *Weirs handbook of experimental immunology* (ed. L. A. Herzenberg, D. M. Weir, L. A. Herzenberg, and C. Blackwell), p. 208.1. Blackwell Science, Inc., Malden, MA.
174. Kleijmeer, M. J., Oorschot, V. M., and Geuze, H. J. (1994) Human resident langerhans cells display a lysosomal compartment enriched in MHC class II. *J. Invest. Dermatol.*, **103**, 516.
175. Peters, P. J., Raposo, G., Neefjes, J. J., Oorschot, V., Leijendekker, R. L., Geuze, H. J., *et al.* (1995) Major histocompatibility complex class II compartments in human B lymphoblastoid cells are distinct from early endosomes. *J. Exp. Med.*, **182**, 325.
176. Kleijmeer, M. J., Ossevoort, M. A., van Veen, C. J., van Hellemond, J. J., Neefjes, J. J., Kast, W. M., *et al.* (1995) MHC class II compartments and the kinetics of antigen presentation in activated mouse spleen dendritic cells. *J. Immunol.*, **154**, 5715.
177. Raposo, G., Nijman, H. W., Stoorvogel, W., Leijendekker, R., Harding, C. V., Melief, C. J., *et al.* (1996) B lymphocytes secrete antigen-presenting vesicles. *J. Exp. Med.*, **183**, 1161.
178. Morkowski, S., Raposo, G., Kleijmeer, M., Geuze, H. J., and Rudensky, A. Y. (1997) Assembly of an abundant endogenous major histocompatibility complex class II/ peptide complex in class II compartments. *Eur. J. Immunol.*, **27**, 609.
179. Amigorena, S., Drake, J. R., Webster, P., and Mellman, I. (1994) Transient accumulation of new class II MHC molecules in a novel endocytic compartment in B lymphocytes. *Nature*, **369**, 113.
180. West, M. A., Lucocq, J. M., and Watts, C. (1994) Antigen processing and class II MHC peptide-loading compartments in human B-lymphoblastoid cells. *Nature*, **369**, 147.
181. Tulp, A., Verwoerd, D., Dobberstein, B., Ploegh, H. L., and Pieters, J. (1994) Isolation and characterization of the intracellular MHC class II compartment [see comments]. *Nature*, **369**, 120.

182. Qiu, Y., Xu, X., Wandinger-Ness, A., Dalke, D. P., and Pierce, S. K. (1994) Separation of subcellular compartments containing distinct functional forms of MHC class II. *J. Cell Biol.*, **125**, 595.
183. Pierre, P. and Mellman, I. (1998) Exploring the mechanisms of antigen processing by cell fractionation. *Curr. Opin. Immunol.*, **10**, 145.
184. Pierre, P., Turley, S. J., Gatti, E., Hull, M., Meltzer, J., Mirza, A., *et al.* (1997) Developmental regulation of MHC class II transport in mouse dendritic cells [see comments]. *Nature*, **388**, 787.
185. Brachet, V., Raposo, G., Amigorena, S., and Mellman, I. (1997) Ii chain controls the transport of major histocompatibility complex class II molecules to and from lysosomes. *J. Cell Biol.*, **137**, 51.
186. Inaba, K., Turley, S., Iyoda, T., Yamaide, F., Shimoyama, S., Reis e Sousa, C., *et al.* (2000) The formation of immunogenic major histocompatibility complex class II-peptide ligands in lysosomal compartments of dendritic cells is regulated by inflammatory stimuli. *J. Exp. Med.*, **191**, 927.
187. Turley, S. J., Inaba, K., Garrett, W. S., Ebersold, M., Unternaehrer, J., Steinman, R. M., *et al.* (2000) Transport of peptide-MHC class II complexes in developing dendritic cells. *Science*, **288**, 522.
188. Pinet, V., Malnati, M. S., and Long, E. O. (1994) Two processing pathways for the MHC class II-restricted presentation of exogenous influenza virus antigen. *J. Immunol.*, **152**, 4852.
189. Lippincott-Schwartz, J. and Fambrough, D. M. (1987) Cycling of the integral membrane glycoprotein, LEP100, between plasma membrane and lysosomes: kinetic and morphological analysis. *Cell*, **49**, 669.
190. Hunziker, W. and Geuze, H. J. (1996) Intracellular trafficking of lysosomal membrane proteins. *Bioessays*, **18**, 379.
191. Wubbolts, R., Fernandez-Borja, M., Oomen, L., Verwoerd, D., Janssen, H., Calafat, J., *et al.* (1996) Direct vesicular transport of MHC class II molecules from lysosomal structures to the cell surface. *J. Cell Biol.*, **135**, 611.
192. Griffiths, G. M. (1997) Protein sorting and secretion during CTL killing. *Semin. Immunol.*, **9**, 109.
193. Stinchcombe, J. C. and Griffiths, G. M. (1999) Regulated secretion from hemopoietic cells. *J. Cell Biol.*, **147**, 1.
194. Johnstone, R. M., Adam, M., Hammond, J. R., Orr, L., and Turbide, C. (1987) Vesicle formation during reticulocyte maturation. Association of plasma membrane activities with released vesicles (exosomes). *J. Biol. Chem.*, **262**, 9412.
195. Pan, B. T., Teng, K., Wu, C., Adam, M., and Johnstone, R. M. (1985) Electron microscopic evidence for externalization of the transferrin receptor in vesicular form in sheep reticulocytes. *J. Cell Biol.*, **101**, 942.
196. Harding, C., Heuser, J., and Stahl, P. (1984) Endocytosis and intracellular processing of transferrin and colloidal gold-transferrin in rat reticulocytes: demonstration of a pathway for receptor shedding. *Eur. J. Cell Biol.*, **35**, 256.
197. Raposo, G., Vidal, M., and Geuze, H. J. (1997) Secretory lysosomes and the production of exosomes. In *Unusual secretory pathways: from batceria to man* (ed. K. Kuchler, A. Rubartelli, and B. Holland), p. 161. R.G. Landes Company, Georgetown.
198. Zitvogel, L., Regnault, A., Lozier, A., Wolfers, J., Flament, C., Tenza, D., *et al.* (1998) Eradication of established murine tumors using a novel cell-free vaccine: dendritic cell-derived exosomes. *Nat. Med.*, **4**, 594.

199. Heijnen, H. F., Schiel, A. E., Fijnheer, R., Geuze, H. J., and Sixma, J. J. (1999) Activated platelets release two types of membrane vesicles: microvesicles by surface shedding and exosomes derived from exocytosis of multivesicular bodies and alpha-granules. *Blood*, **94**, 3791.
200. Escola, J. M., Kleijmeer, M. J., Stoorvogel, W., Griffith, J. M., Yoshie, O., and Geuze, H. J. (1998) Selective enrichment of tetraspan proteins on the internal vesicles of multivesicular endosomes and on exosomes secreted by human B- lymphocytes. *J. Biol. Chem.*, **273**, 20121.
201. Thery, C., Regnault, A., Garin, J., Wolfers, J., Zitvogel, L., Ricciardi-Castagnoli, P., *et al.* (1999) Molecular characterization of dendritic cell-derived exosomes. Selective accumulation of the heat shock protein hsc73. *J. Cell Biol.*, **147**, 599.
202. Denzer, K., van Eijk, M., Kleijmeer, M. J., Jakobson, E., de Groot, C., and H. J. Geuze. (2000) Follicular dendritic cells Carry MHC class II-expressing microvesicles at their surface. *J. Immunol.*, **165**, 1259.
203. Pond, L. and Watts, C. (1997) Characterization of transport of newly assembled, T cell-stimulatory MHC class II-peptide complexes from MHC class II compartments to the cell surface. *J. Immunol.*, **159**, 543.
204. Ramm, G., Pond, L., Watts, C., and Stoorvogel, W. (2000) Clathrin-coated lattices and buds on MHC class II compartments do not selectively recruit mature MHC-II. *J. Cell Sci.*, **113**, 303.
205. Lich, J. D., Elliott, J. F., and Blum, J. S. (2000) Cytoplasmic processing is a prerequisite for presentation of an endogenous antigen by major histocompatibility complex class II proteins. *J. Exp. Med.*, **191**, 1513.
206. Klionsky, D. J. (1997) Protein transport from the cytoplasm into the vacuole. *J. Membr. Biol.*, **157**, 105.
207. Baba, M., Osumi, M., Scott, S. V., Klionsky, D. J., and Ohsumi, Y. (1997) Two distinct pathways for targeting proteins from the cytoplasm to the vacuole/lysosome. *J. Cell Biol.*, **139**, 1687.
208. Reimann, J. and Schirmbeck, R. (1999) Alternative pathways for processing exogenous and endogenous antigens that can generate peptides for MHC class I-restricted presentation. *Immunol. Rev.*, **172**, 131.
209. Jondal, M., Schirmbeck, R., and Reimann, J. (1996) MHC class I-restricted CTL responses to exogenous antigens. *Immunity*, **5**, 295.
210. Rock, K. L. (1996) A new foreign policy: MHC class I molecules monitor the outside world. *Immunol. Today*, **17**, 131.
211. Yewdell, J. W., Norbury, C. C., and Bennink, J. R. (1999) Mechanisms of exogenous antigen presentation by MHC class I molecules *in vitro* and *in vivo*: implications for generating CD8+ T cell responses to infectious agents, tumors, transplants, and vaccines. *Adv. Immunol.*, **73**, 1.
212. Cresswell, P., Bangia, N., Dick, T., and Diedrich, G. (1999) The nature of the MHC class I peptide loading complex. *Immunol. Rev.*, **172**, 21.
213. Norbury, C. C., Chambers, B. J., Prescott, A. R., Ljunggren, H. G., and Watts, C. (1997) Constitutive macropinocytosis allows TAP-dependent major histocompatibility complex class I presentation of exogenous soluble antigen by bone marrow-derived dendritic cells. *Eur. J. Immunol.*, **27**, 280.
214. Kovacsovics-Bankowski, M. and Rock, K. L. (1995) A phagosome-to-cytosol pathway for exogenous antigens presented on MHC class I molecules. *Science*, **267**, 243.
215. Rodriguez, A., Regnault, A., Kleijmeer, M., Ricciardi-Castagnoli, P., and Amigorena, S.

(1999) Selective transport of internalized antigens to the cytosol for MHC class I presentation in dendritic cells. *Nat. Cell Biol.*, **1**, 362.

216. Gromme, M., Uytdehaag, F. G., Janssen, H., Calafat, J., van Binnendijk, R. S., Kenter, M. J., *et al.* (1999) Recycling MHC class I molecules and endosomal peptide loading. *Proc. Natl. Acad. Sci. USA*, **96**, 10326.
217. Kleijmeer, M. J., Escola, J. M., Uytde-Haag, F. G. C. M., Jakobsen, J., Griffiths, J., Osterhaus, A. D. M. E., Stoorvogel, W., Melief, C. J. M., Rabouille, C., and Geurje, H. J. (2001) Antigen loading of MHC class 1 molecules in the endocytic tract. *Traffic*, **2**, In press.
218. Reid, P. A. and Watts, C. (1990) Cycling of cell-surface MHC glycoproteins through primaquine-sensitive intracellular compartments. *Nature*, **346**, 655.

9 | Synaptic vesicle recycling: multiple pathways and functions in the nervous system

PATRICK WIGGE and HARVEY T. McMAHON

1. Overview: potential functions of synaptic vesicle recycling in the nervous system

Amongst the roughly 100 billion neurons in the human brain how does synaptic transmission make possible so many different adaptable behaviours, learning, memory, thoughts, emotions? Each neuron is connected to thousands of others by chemical synapses, slowing down the speed of communication between neurons but providing a greatly increased number of ways in which neurons can 'talk to each other'—by choosing amongst a diverse array of different transmitters, different firing patterns, and different circuits. The brain is complex, but flexible too: It is now generally realized that not only can new synapses form and old ones retract, but that, even in adult humans, new neurons can be born to replace older, redundant ones. The formation of new memories might thus require the creation of whole new circuits. Studies showing experience-dependent neuronal replacement on a large scale in the songbird brain first dispelled the notion of an adult brain fully formed and static in terms of neuronal turnover (1). Even in the primate cortex, neurogenesis as well as synaptogenesis occurs in adulthood (2), although the functions of these new neurons are still unclear. Are they important, for example, in laying down new memories, or in the learning and execution of new behaviours?

At the level of individual neurons and synapses, it is still a mystery how new neurons develop from stem cell precursors, differentiate, and—in particular relevance to this chapter—form synaptic contacts with their correct partners. Once a synapse has formed, input of the right stimuli will determine whether it survives or not. This also is very dependent on an animal's environment; rats reared in more complex surroundings develop a far greater number of synaptic connections in their cortex than do animals exposed to a less challenging environment. Their cortex actually grows in mass by several percent. The 'decision' a neuron must make to grow new axonal processes and synapse with other neurons, or retract its neurites, may depend a great deal on the stimulation it receives from other neurons. This stimulation seems to act, at least in part, by directly triggering the rapid endocytic down-regulation of surface adhesion molecules, thus changing the adherence properties of the axon surface and therefore its ability to extend new neurites. Thus, endocytosis can serve to alter synaptic plasticity in a very basic way, and might help to shape the patterning of the nervous system during development, as well as serving its most obvious role in replenishing the synaptic vesicle pool at the nerve terminal following exocytosis.

At the molecular level, it is still not clear how the molecular machinery of synaptic vesicle cycling is orchestrated to allow such rapid, regulated transmission of information at the nerve terminal. Much progress has been made in identifying the individual components of the protein complexes responsible for exocytosis and endocytosis, but the problem of how they fit together is more difficult. To further complicate matters, synaptic vesicles can release their transmitter and be recycled by more than one route: a fast 'kiss-and-run' pathway, and a slower process in which vesicles fully fuse with the presynaptic plasma membrane and are then re-internalized by a clathrin- and dynamin-mediated pathway. Much of the molecular machinery for the latter pathway is now known, though little is known of the former. The two processes might well share features in common, but they are thought to be modulated in very different ways by the prevailing environment of the synapse, for example by the frequency of depolarizing stimuli. And with such different kinetics, it is likely that the two pathways fulfil different functions in the nervous system.

Synaptic vesicle recycling therefore has at least two main functions in the nervous system. First, it is absolutely essential for synaptic transmission—if synaptic vesicles were not re-internalized from the presynaptic plasma membrane, the nerve terminal would become rapidly depleted of new synaptic vesicles. This is clearly illustrated by the phenotype of *Drosophila* carrying the *shibire* mutation in dynamin: at the restrictive temperature, the dynamin 'motor' cannot function, resulting in rapid paralysis (3, 4). Secondly, endocytosis at many regions of the neuronal surface (not restricted to synaptic terminals) may have the potential to greatly modulate synaptic remodelling and plasticity (5). Recent work has shown that endocytosis occurs postsynaptically as well: AMPA receptors are continually internalized at the postsynaptic density (6), with changing levels of such ion channels being vital in synaptic strengthening events such as long-term potentiation (LTP).

2. Pathways of synaptic vesicle recycling

The role of clathrin-coated vesicles (CCVn) in synaptic vesicle recycling has been questioned often and persistently (7, 8). Electrophysiological studies have provided support for the existence of a potentially much faster mechanism, that of 'kiss-and-run' endocytosis. This hypothesis states that synaptic vesicles dock with the membrane but do not fully fuse; instead they release their neurotransmitter through a narrow (< 3 nm) fusion pore that transiently 'flickers' open and shut (9) (Plate 5). This would not necessitate the use of any clathrin coats, adaptors, or associated sorting machinery. Against these sound theoretical arguments, it is important to ask what the evidence is that clathrin-mediated mechanisms are used at all in synaptic vesicle recycling?

2.1 Evidence for clathrin-mediated endocytosis

The 'CCV hypothesis' dates back to 1973. CCVs were the first type of transport vesicle discovered, and were purified from mammalian cells in the 1970s (10, 11). Morphological experiments on the frog neuromuscular junction constitute the first and still among the most convincing evidence for clathrin-mediated vesicle recycling in the brain. Electron microscopy showed the presence of CCVs in nerve terminals, which were reported to increase in number following tetanic stimulation (12). Subsequent studies showed the existence of synaptic vesicles in different stages of vesicle fusion with some collapsing fully with the membrane (13). Since then, with the advent of molecular biology and the purification of synaptic vesicles *en masse*, nearly all the proteins making up the synaptic vesicle have been identified (14, 15). CCVs isolated from synaptosomes (a preparation of purified nerve terminals) primarily contain synaptic vesicle proteins (16). Many of these membrane proteins (e.g. synaptophysin, p38) are not present at all on the presynaptic plasma membrane, being specific vesicle markers. Further, much of the machinery for clathrin-coated endocytosis, including clathrin, adaptors, and dynamin, is greatly enriched in brain synaptosomes. Thus one good test of the CCV hypothesis would be to examine whether antibody labelling can detect p38 on the plasma membrane after stimulation of exo-endocytosis. This experiment has been done and shows that, while p38 can hardly be detected under conditions of mild stimulation, the levels of this protein increase markedly after more intense stimuli (17, 18). This suggests that synaptic vesicles fully collapse with the presynaptic plasma membrane when neurons are stimulated to fire repeatedly, but under milder, more transient conditions of stimulation, the CCV route may not be the major pathway of synaptic vesicle recycling. Although controversial, the concept of two distinct pools of synaptic vesicles, a 'readily-releasable pool' (that is internalized rapidly (in milliseconds) by flicker-fusion close to the active zone), and a reserve pool (which is triggered by more intense stimuli and recycles more slowly by a clathrin-mediated route at sites distinct from the active zone), is consistent with experiments in fruitflies (19) and membrane capacitance measurements in chromaffin cells (20) (see below).

Genetic evidence is perhaps the most persuasive data supporting an essential role of CCVs in synaptic transmission. *Drosophila* containing a temperature-sensitive mutation in the fly homologue of dynamin, *shibire*, become rapidly and reversibly paralysed at the restrictive temperature of 29 °C (4, 21). While exocytosis is normal, there is a total block in synaptic vesicle re-formation. Recently, mutants in the *Drosophila* α-adaptin gene have been created (22). Synaptic vesicle recycling in these nerve terminals was impaired, but not abolished, suggesting that AP-2 adaptors are fulfilling an essential role in CCV formation similar to their known function in receptor-mediated endocytosis.

2.2 Evidence for kiss-and-run endocytosis

One of the most dramatic distinctions between classical receptor-mediated endocytosis in non-neuronal cells and the synaptic retrieval pathway in neurons is the speed of the two events. The uptake of transferrin receptor in fibroblasts, considered a paradigm of efficiency, has a half-time of 3–5 minutes (23) and clathrin-mediated internalization of Semliki Forest virus in BHK cells occurs with a similar slow time constant (24). The recent development of fluorescent vesicle tracers, including the styryl dye FM-143, has allowed the accurate determination of just how fast synaptic vesicle recycling really is (5). FM dyes bind to lipid bilayers with fast 'on' and 'off' rates; only when in the membrane do they fluoresce. Thus, the rate and extent of dye internalized (and therefore protected) in synaptic vesicles is a good measure of endocytosis. Such experiments have shown that in the frog neuromuscular junction, retrieval occurs with a time constant of 30 seconds (26). Similar estimates have been obtained from hippocampal synapses (27), and frog hair cells (28). After endocytosis, the processes of recycling, refilling, and docking that makes a new vesicle available for the next round of neurotransmitter release are also rapid (occupying a minimum of 15–45 seconds) (29). Thus, the whole synaptic vesicle cycle can be completed in about a minute.

Does synaptic vesicle endocytosis represent a speeded-up version of the classical clathrin-mediated route? As well as an approximately 100-fold enrichment of CCV components in nerve terminals, the synapse contains a number of apparently brain-specific proteins, e.g. the clathrin adaptor AP180, and for much of the rest of the general endocytosis machinery, unique splice variants exist that are only expressed in the nervous system (23). Also, the endosomal 'intermediate' may be omitted as a sorting station obligatory in other classical RME pathways (30). Thus, following vesicle fission, CCVs are uncoated, refilled, and directly form new synaptic vesicles without fusing with another membrane. Such a simplified cycle could account for the at least tenfold greater speed of endocytosis in the brain, as measured by FM dyes.

If the already fast clathrin-mediated pathway coexists with an even faster kiss-and-run retrieval mechanism (as is thought to be the case in many cells), it would not necessarily be detected by the use of styryl dyes. However, measurement of the release of styryl dyes from vesicles that have previously been dye-loaded show that there may be a restricted exchange of the dye on stimulation of exocytosis (35). This should not be the case if the vesicles collapse into the plasma membrane and so point to the existence of a 'kiss-and-run' pathway in hippocampal neurons. Also, capacitance measurements in retinal bipolar synapses have shown that after brief stimuli, vesicles are retrieved in only 1–2 seconds (20, 31, 32). This is good support for a very fast retrieval pathway in these cells that might not be clathrin-mediated. This pathway may require dynamin because of the complete lack of endocytosis in *shibire* flies.

2.3 How might the two pathways coexist?

Although most experiments have examined only one endocytic pathway in isolation, a very detailed study by Koenig and Ikeda has suggested a model in which fast and slow recycling pathways might coexist at the nerve terminal (19) (Plate 5).

The evidence for this model is mainly based on detailed microscopic analysis of *shibire* synapses, and suggests that the 'fast' and 'slow' pathways may correspond to kiss-and-run and clathrin-mediated pathways, respectively. They differ in form, function, kinetics, and relative responsiveness to stimuli strength (19). Though speculative, their model suggests that after relatively weak depolarizing stimuli, the prevailing route for synaptic vesicle recycling is one in which the readily releasable pool of vesicles, already docked at the presynaptic plasma membrane, transiently open their fusion pores. Within only fractions of a millisecond, the transmitter will have escaped, and the vesicle reseals, detaching from the membrane to be once again refilled with transmitter. The reserve pool is thought to be mobilized only during more intense, prolonged stimuli, such as long trains of action potentials. Under these conditions, the model predicts that vesicle cycling will rely predominantly on the clathrin-mediated pathway, retrieving those vesicles that have fully collapsed into the plasma membrane at sites distinct from the active zone. This process probably occurs three orders of magnitude more slowly, in agreement with recent electrophysiological studies at the retinal bipolar synapse showing longer time constants of endocytosis after heavier stimuli (33).

Of course, this Koenig-Ikeda model is only one of many that might describe how two different pathways of vesicle cycling could coexist at the nerve terminal. Not all the data fits. One recent study, conducted in neuroendocrine cells, has also concluded that kiss-and-run and clathrin-mediated recycling routes occur in the same cell, but responding to stimulation levels in different ways. In this cell type, it was found that clathrin-mediated endocytosis predominates in the normal cellular conditions of low calcium. By contrast, kiss-and-run cycling occurs only at much higher extracellular calcium levels (10–90 mM) (concentrations thought to occur locally only near Ca^{2+} channels immediately after depolarization) (34). Much work will be needed before we can be confident of just how many recycling pathways exist in the brain, and their different forms and functions.

3. Regulatory cues influencing synaptic endocytosis

The retrieval of synaptic vesicles must occur in a very temporally and spatially controlled fashion for optimally rapid synaptic communication: some neurons (those involved in relaying sound stimuli, for example) propagate action potentials at over 500 times per second. Regulation of the endocytic machinery, in particular by regulatory kinases and phosphatases, seems to be one way vesicle traffic is 'tuned'. In cultured hippocampal synapses for example, fluorescent dye labelling studies have indicated that staurosporine, a protein kinase C inhibitor, increases rates of retrieval (35, 36). And the opposite treatment—inhibition of the calcium-dependent phosphatase calcineurin—produces a potent inhibition of synaptic vesicle uptake in synaptosomes (37). These data suggest that protein dephosphorylation drives endocytosis. Any treatment that increases the proportion of proteins in their dephosphorylated state, or decreases the phosphorylated pool, would be predicted to increase rates of endocytosis. (Alternatively, proteins might cycle continuously between these two states, according to their step in the endocytic cycle. But if this were true, then any treatment causing an imbalance between kinases and phosphatases should reduce endocytosis.)

What might be the synaptic targets of these kinases (like PKC) and phosphatases (principally calcineurin)? Many—in fact most—of the synaptic proteins cloned and characterized to date are phosphoproteins: these include dynamin, amphiphysin, and synaptojanin (38). All three are dephosphorylated upon stimulation of nerve terminals, and interestingly, all three share a common phosphatase, calcineurin (37). *In vitro* experiments suggest that only in their dephosphorylated state can these proteins associate with one another, interactions governed primarily by SH3-PRD interactions (39). Because the calcium influx that follows depolarization is known to activate calcineurin, a picture is beginning to emerge in which synaptic vesicle endocytosis is triggered, at least in part, by the calcium-stimulated dephosphorylation of these components of the endocytic machinery. This may be an oversimplification, because prevention of re-phosphorylation (e.g. by staurosporine) does not, in all systems, prevent endocytosis as one might expect; fast and slow endocytosis might have different regulatory requirements, especially if they use distinct endocytic proteins. More refined dissection of the two pathways will be important to know exactly which one is being measured in a given experiment.

In other parts of the neuron away from the nerve terminal, endocytosis may be regulated by a different set of factors. The uptake of adhesion molecules in *Aplysia* axons, so important in neurite outgrowth and synaptic plasticity (see below), only occurs after serotonergic stimulation. This internalization event also requires phosphorylation, but this time it is the cytoplasmic 'tail' of the adhesion molecules that are phosphorylated (by MAP kinase) (5). This resembles the mechanism of receptor-mediated endocytosis in non-neuronal cells, where the dimerization and phosphorylation of growth factor receptors is linked to their uptake. It is highly possible that this regulatory mechanism extends to other synaptic vesicle cargo as well—integral membrane proteins such as synaptophysin, for example, are phosphorylated on their cytoplasmic domains by tyrosine as well as serine/threonine kinases. Synaptophysin has recently been shown to play a role in synaptic endocytosis via its interaction with cholesterol in specific lipid microdomains (40).

Other regulatory cues will doubtless be involved in ensuring that vesicle uptake is precisely co-ordinated with neuronal activity. Calcium microdomains, as well as influencing activities of kinases and phosphatases, may directly modulate protein–protein interactions, as several synaptic proteins have high-affinity calcium-binding sites. Various extracellular factors such as nitric oxide and other signalling molecules may act directly on the endocytic machinery; and more speculatively, it would be interesting to know if membrane depolarization itself (i.e. voltage gradients) had any effect on synaptic vesicle recycling. As well as triggering ion channels such as the NMDA receptor, voltage differences might feasibly modulate the internalization of certain transmembrane vesicle cargo proteins.

4. Potential roles of endocytosis in learning and memory

In addition to ensuring the continuance of synaptic transmission by replenishing a vesicle pool which would otherwise be rapidly depleted, a great deal of recent work has shown how important endocytosis is in synaptic remodelling. A great mystery in neuroscience concerns just how the short-term, fleeting, synaptic firing patterns of distinct subsets of neurons are converted into the formation of long-term memories. Below, we discuss the internalization of the family of cell adhesion molecules, the CAMs, and the role this might serve in various forms of synaptic plasticity, with reference to one particularly well-studied example, that in *Aplysia*.

4.1 Endocytosis determines cell-surface expression of neuronal CAMs

It is believed that many of the processes underlying learning and memory involve making new synaptic connections, or the strengthening of existing ones. An important set of transmembrane glycoproteins, containing immunoglobulin-like domains, are expressed on the surface membrane of axons and dendrites, having the potential to form homophilic or heterophilic interactions with other adhesion molecules. Kandel and others have shown that, in the sea-slug *Aplysia*, internalization of ApCAM, an important member of this family, is likely to be a key molecular event underlying memory acquisition during the gill-withdrawal reflex (41). ApCAM exists in two isoforms, a transmembrane form and a GPI-linked alternative. Only the former is internalized in response to stimulation by serotonin (5-HT). This internalization is dependent on two MAP kinase consensus sites; microinjection of a specific MAP kinase inhibitor is sufficient to prevent endocytic down-regulation of ApCAM (5). Serotonin is known to stimulate MAP kinase activity in sensory neurons (42).

The precise mechanism of ApCAM internalization is thought to be one of clathrin-mediated endocytosis: serotonin has been shown to increase expression of clathrin light chain and numbers of clathrin-coated pits and coated vesicles are increased too (43). But the involvement of MAP kinase sites suggests that it is a variation of the mechanism seen in non-neuronal cells. Tyrosine-based sequences are apparently not required or present on the cytoplasmic tail of ApCAM but instead, there seems to be an involvement of ubiquitin. Indeed, the ultimate destiny of the internalized ApCAM is to one of proteolytic degradation by the proteasome. This is reminiscent of endocytosis in yeast (44). By contrast, the related adhesion molecule, LI, is internalized by a clathrin-mediated pathway that *does* require a tyrosine-based motif, namely YRSL, and this has been shown to bind to AP-2 adaptors (45). Whether this molecule is degraded too by a ubiquitin-mediated pathway, or has a different destiny, would be interesting to know in the future.

What is the purpose of CAM internalization in learning and memory? It is thought that during synaptogenesis, inhibitory constraints between neighbouring axons in the bundle must be relieved before new neurites and axonal processes can extend outwards to form new connections. These inhibitory constraints have been postulated to require the transmembrane form of ApCAM, while the GPI-linked form (which is not endocytosed) might instead serve to anchor axonal terminals to their postsynaptic contacts, the dendritic spines. For this hypothesis to be correct, the two isoforms of ApCAM would need to maintain distinct spatial distributions on the neuronal surface membrane, with the transmembrane form concentrated along the length of neurite processes (maintaining inhibitory fasciculation), whilst the GPI-linked form is localized at the presynaptic terminals, where it could serve to mediate heterophilic adhesion with other CAMs on the postsynaptic neuron (Fig. 1). According to this framework, a burst of serotonin would result in a selective down-regulation of those ApCAM molecules involved in homophilic interactions between adjacent neuronal processes, leading to axonal defasciculation. Following this, the neurite processes would be liberated to form, perhaps, new synaptic connections. Thus, the role of endocytosis of cell adhesion molecules might be viewed as a way of relieving 'inhibitory constraints on memory storage that normally are active within the cell' (5).

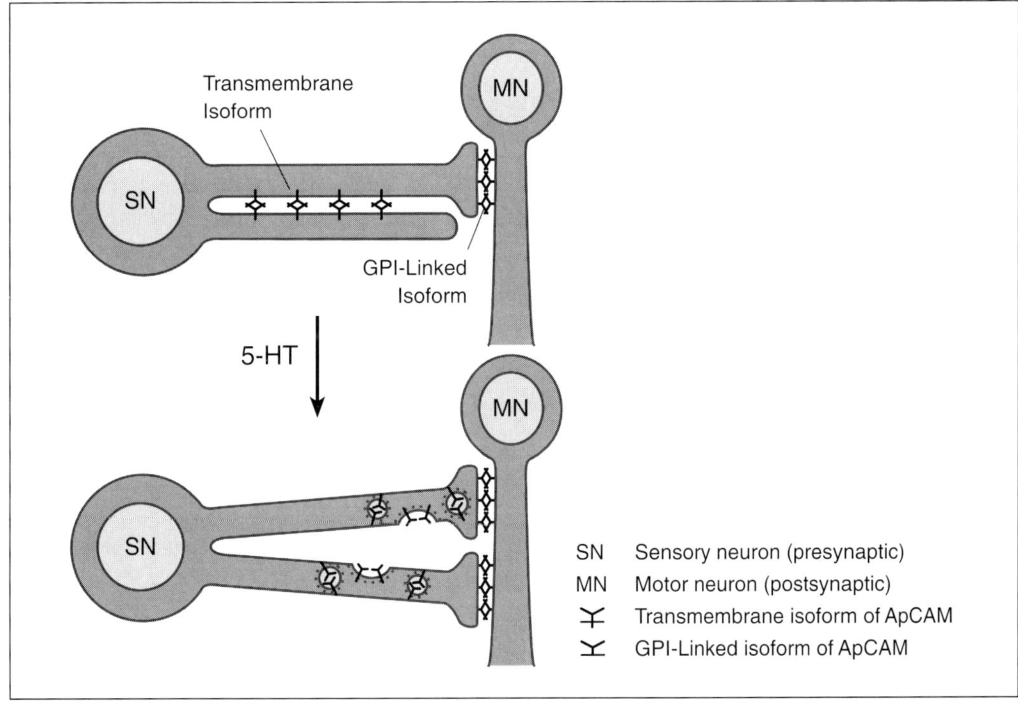

Fig. 1 Involvement of endocytosis in structural changes of *Aplysia* neurons during memory storage. Following stimulation by 5-HT, MAP kinase activation, and endocytosis of transmembrane ApCAMs, inhibitory contacts on extending neurites are relieved allowing the creation of new synapses. This figure has been reproduced with modification from ref. 5 with permission of E. Kandel.

The finding that ApCAM down-regulation by endocytosis may be crucial for the long-term facilitation underlying this form of learning is backed up by many other studies demonstrating the importance of the cell adhesion molecules L1 and NCAM in synaptic plasticity. In the chick, for example, transfer of short-term memories to long-term storage after a period of passive-avoidance learning is correlated with two waves of enhanced glycoprotein synthesis (46). These glycoproteins include the adhesion molecules L1 and NCAM. Antibodies raised to either and injected intracerebrally at defined periods before and after training cause amnesia, as do fragments of their extracellular domains (46). In the songbird brain, NCAM and the related N-cadherin are serially required in the migration of newly born neurons along radial fibres (47). Further evidence emerges from studies of the *Drosophila* nerve-muscle synapse: here, down-regulation of the ApCAM fly homologue, fasciclin II, is both necessary and sufficient for activity-dependent synaptic outgrowth (48, 49). And in cultured mouse sensory neurons, expression of L1 (but not, interestingly, NCAM) was reduced after low-frequency electrical stimulation (0.1 Hz), a treatment that also greatly reduced cell adhesion and neurite fasciculation (50). Higher-frequency stimulation did not have any effect, suggesting that neuronal adhesion and the formation of synaptic contacts is finely regulated by firing rates. Further, this study also showed an involvement of L1 in the myelination of axons by Schwann cells: at higher frequency stimulation, adhesion to these glial cells was increased, suggesting a multi-stage process of neurite outgrowth in which the initially low firing rates initiate reduction in the density of L1 adhesion molecules, encouraging axonal defasciculation, and resulting neurite outgrowth—until synaptic terminals begin forming. Then the firing rate is predicted to increase, leading to a return to higher L1 levels, and increased adhesion between pre- and postsynaptic neurons, as well as myelination by Schwann cells (50).

4.2 Endocytosis proteins, neurite outgrowth, and long-term depression

The studies above suggest that, by modulating the surface expression of adhesion molecules, endocytosis literally has the potential to help shape the very structure and function of the nervous system. In the future, it would be fascinating to test this hypothesis directly, by blocking endocytosis in selected populations of neurons, and observing any effects on patterning of the central nervous system. Glimmerings of the importance of endocytosis on neurite outgrowth can already be seen, from three studies conducted on proteins important in the fission step of synaptic vesicle reuptake, dynamin and amphiphysin. Dynamin is a molecular motor that collars the necks of clathrin-coated pits (51), and upon hydrolysis of bound GTP, either 'pops' (52) or 'pinches' (53) off the nascent vesicle. Amphiphysin is an SH3 domain-containing protein thought to recruit dynamin to the coated pit in the first place, and is implicated in certain neurological autoimmune diseases (54). As long ago as 1987, differences were noticed in the behaviour of neurons cultured from the *Drosophila* mutant *shibire* (in which a point mutation in the dynamin molecule causes a potent temperature-sensitive block in synaptic vesicle recycling). At the restrictive temperature, neurite outgrowth was greatly reduced, and adhesive properties altered (55). At the time, the reasons for this were not clear, but now that the role of dynamin in endocytosis is fully appreciated, it would be interesting to test if there is a higher surface expression of CAMs at the restrictive temperature. Similar effects on neurite outgrowth have been observed in antisense experiments. Incubation of cultured hippocampal neurons with antisense oligonucleotides to dynamin I (56), and more recently amphiphysin I (57), potently blocked neurite outgrowth.

If synaptic vesicle recycling is indeed important in remodelling of synaptic surface membranes during learning, then it should be possible to find a change in expression of key endocytic proteins. A test of this kind was indeed conducted by Napolitano *et al.* (1999), using differential display to isolate genes that are modulated after tetanic long-term depression (LTD) in slices derived from the striatum (60). Of the three genes isolated, one was found to be identical in sequence to dynamin I, and another to be amphiphysin II. The former was up-regulated 3.2-fold, and the latter (2.4 kb transcript only) activated 3.5-fold in LTD-induced slices, as compared to control slices.

Thus, endocytosis, by allowing a rapid, modulated down-regulation of surface molecules, may serve very important and broad roles in the spreading out of axons, and the dynamics of growth cones, in the development of synaptic connections during learning and memory. Development of screens for other molecules whose expression on surface membranes is affected by endocytosis might lead to the discovery of other adhesion molecules important in this process. This is a field of many interesting possibilities. If the differently internalized ApCAM isoforms expressed on the surface of *Aplysia* neurons really do maintain distinct spatial distributions, this might turn out to be a general mechanism by which particular regions of the cell surface, or particular active synapses, are 'marked' for endocytosis, returning membrane components to other, more distant parts of the neuron, or even to growing areas of active synapse formation (see Fig. 1). Many, if not all, synaptic surface proteins exist as multiple isoforms, each isoform being capable of alternative splicing. In theory, there should be good biological reasons for this 'expense'; but few isoform-specific functions have been identified to date. Development of splice form-specific antibodies might lead to fascinating insights as to the extent of spatial (or temporal?) specificity in the cellular localization of these distinct proteins. Such specificity might explain how neurons maintain their identities and become committed to particular fates during development. Do distinct patches of different 'stickiness' exist on neuronal surfaces at different times of development? Endocytosis might serve to maintain and modulate such patches in activity-dependent patterns. Might these patches then determine which other neurons will be sought out, adhered to, and co-opted for synapse formation?

4.3 Postsynaptic endocytosis: dynamin-dependent uptake of AMPA receptors

It has long been suspected that a great deal of synaptic plasticity, in particular LTP, is determined by receptor modulation at the postsynaptic level. However, the view until very recently held that the postsynaptic density (the protein scaffold enriched in neurotransmitter receptors and ion channels such as the NMDA and AMPA-type glutamate receptors) was by and large static, with most of the documented dynamic changes in membrane composition occurring presynaptically. However, new evidence suggests that one class of receptors, the AMPA channels, is continuously being recycled by endocytosis at the postsynaptic density (6). Endocytosis in CA1 pyramidal cells was blocked with a specific peptide predicted to perturb the dynamin–amphiphysin interaction, and excitatory postsynaptic currents monitored (EPSCs). These AMPA receptor postsynaptic currents increased within minutes of administering the peptide (or the GTPase inhibitor GDPβS), whilst a scrambled peptide control had no effect (6). Evidence that it is AMPA receptors that are internalized includes their co-localization with AP-2 adaptors at coated pits, and the fact that their uptake from the membrane is blocked by transfection of dominant negative dynamin mutants (58).

The cycling of AMPA receptors into and out of the postsynaptic membrane is constitutive, and does not seem to be affected by synaptic activity (6). However, the process clearly has ramifications for synaptic plasticity, and in particular LTD. Changes in the postsynaptic levels of AMPA receptor at the membrane surface, through differences in relative rates of exocytosis and endocytosis, might potently affect the excitability of the postsynaptic neuron in response to glutamate, in turn possibly feeding back to the presynaptic neuron. This mechanism may extend to other postsynaptic receptors, although the NMDA class of ion channels was found to be not affected in this study (6). Interestingly, substance P receptor, a molecule implicated in pain, is known to be internalized in a ligand-dependent manner (59). Such molecular changes could explain some of the intricacies of how synaptic weights are changed during development, or during memory storage.

References

1. Alvarez-Buylla, A. and Nottebohm, F. (1988) Migration of young neurons in adult avian brain. *Nature*, **335**, 353.
2. Gould, E., Reeves, A. J., Graziano, M. S., and Gross, C. G. (1999) Neurogenesis in the neocortex of adult primates. *Science*, **286**, 548.
3. Grigliatti, T. A., Hall, L., Rosenbluth, R., and Suzuki, D. T. (1973) Temperature-sensitive mutations in *Drosophila* melanogaster. XIV. A selection of immobile adults. *Mol. Gen. Genet.*, **120**, 107.
4. Koenig, J. H. and Ikeda, K. (1989) Disappearance and reformation of synaptic vesicle membrane upon transmitter release observed under reversible blockage of membrane retrieval. *J. Neurosci.*, **9**, 3844.
5. Bailey, C. H., Kaang, B. K., Chen, M., Martin, K. C., Lim, C. S., Casadio, A., *et al.* (1997) Mutation in the phosphorylation sites of MAP kinase blocks learning- related internalization of apCAM in *Aplysia* sensory neurons. *Neuron*, **18**, 913.
6. Luscher, C., Xia, H., Beattie, E. C., Carroll, R. C., von Zastrow, M., Malenka, R. C., *et al.* (1999) Role of AMPA receptor cycling in synaptic transmission and plasticity. *Neuron*, **24**, 649.

7. Palfrey, H. C. and Artalejo, C. R. (1998) Vesicle recycling revisited: rapid endocytosis may be the first step. *Neuroscience*, **83**, 969.
8. Fesce, R., Grohovaz, F., Valtorta, F., and Meldolesi, J. (1994) Neurotransmitter release: fusion or 'kiss-and-run'? *Trends Cell Biol.*, **4**, 1.
9. Lindau, M. and Almers, W. (1995) Structure and function of fusion pores in exocytosis and ectoplasmic membrane fusion. *Curr. Opin. Cell Biol.*, **7**, 509.
10. Pearse, B. M. (1976) Clathrin: a unique protein associated with intracellular transfer of membrane by coated vesicles. *Proc. Natl. Acad. Sci. USA*, **73**, 1255.
11. Pearse, B. M. F. and Robinson, M. S. (1990) Clathrin, adaptors and sorting. *Annu. Rev. Cell Biol.*, **6**, 151.
12. Heuser, J. E. and Reese, T. S. (1973) Evidence for recycling of synaptic vesicle membrane during transmitter release at the frog neuromuscular junction. *J. Cell Biol.*, **57**, 315.
13. Heuser, J. E. and Reese, T. S. (1981) Structural changes after transmitter release at the frog neuromuscular junction. *J. Cell Biol.*, **88**, 564.
14. Calakos, N. and Scheller, R. H. (1996) Synaptic vesicle biogenesis, docking, and fusion: a molecular description. *Physiol. Rev.*, **76**, 1.
15. Sudhof, T. C. (1995) The synaptic vesicle cycle: a cascade of protein-protein interactions. *Nature*, **375**, 645.
16. Maycox, P. R., Link, E., Reetz, A., Morris, S. A., and Jahn, R. (1992) Clathrin-coated vesicles in nervous tissue are involved primarily in synaptic vesicle recycling. *J. Cell Biol.*, **118**, 1379.
17. Colasante, C. and Pecot-Dechavassine, M. (1995) Cd(2+)-and K(+)-evoked ACh release induce different synaptophysin and synaptobrevin immunolabelling at the frog neuromuscular junction. *J. Neurocytol.*, **24**, 547.
18. Colasante, C. and Pecot-Dechavassine, M. (1996) Ultrastructural distribution of synaptophysin and synaptic vesicle recycling at the frog neuromuscular junction. *J. Neurosci. Res.*, **44**, 272.
19. Koenig, J. H. and Ikeda, K. (1996) Synaptic vesicles have two distinct recycling pathways. *J. Cell Biol.*, **135**, 797.
20. von Gersdorff, H. and Matthews, G. (1994) Dynamics of synaptic vesicle fusion and membrane retrieval in synaptic terminals. *Nature*, **367**, 735.
21. Kosaka, T. and Ikeda, K. (1983) Possible temperature-dependent blockage of synaptic vesicle recycling induced by a single gene mutation in *Drosophila*. *J. Neurobiol.*, **14**, 207.
22. Gonzalez-Gaitan, M. and Jackle, H. (1997) Role of *Drosophila* α-adaptin in presynaptic vesicle recycling. *Cell*, **88**, 767.
23. Morris, S. A. and Schmid, S. L. (1995) The ferrari of endocytosis? *Curr. Biol.*, **5**, 113.
24. Marsh, M. and Helenius, A. (1980) Adsorptive endocytosis of Semliki Forest virus. *J. Mol. Biol.*, **142**, 439.
25. Betz, W. J., Mao, F., and Smith, C. B. (1996) Imaging exocytosis and endocytosis. *Curr. Opin. Neurobiol.*, **6**, 365.
26. Wu, L. G. and Betz, W. J. (1996) Nerve activity but not intracellular calcium determines the time course of endocytosis at the frog neuromuscular junction. *Neuron*, **17**, 769.
27. Ryan, T. A., Smith, S. J., and Reuter, H. (1996) The timing of synaptic vesicle endocytosis. *Proc. Natl. Acad. Sci. USA*, **93**, 5567.
28. Parsons, T. D., Lenzi, D., Almers, W., and Roberts, W. M. (1994) Calcium-triggered exocytosis and endocytosis in an isolated presynaptic cell: capacitance measurements in saccular hair cells. *Neuron*, **13**, 875.
29. DeCamilli, P. and Takei, K. (1996) Molecular mechanisms in synaptic vesicle endocytosis and recycling. *Neuron*, **16**, 481.

30. Takei, K., Mundigl, O., Daniell, L., and DeCamilli, P. (1996) The synaptic vesicle cycle—a single vesicle budding step involving clathrin and dynamin. *J. Cell Biol.*, **133**, 1237.
31. von Gersdorff, H. and Matthews, G. (1994) Inhibition of endocytosis by elevated internal calcium in a synaptic terminal. *Nature*, **370**, 652.
32. Artalejo, C. R., Henley, J. R., McNiven, M. A., and Palfrey, H. C. (1995) Rapid endocytosis coupled to exocytosis in adrenal chromaffin cells involves Ca^{2+}, GTP, and dynamin but not clathrin. *Proc. Natl. Acad. Sci. USA*, **92**, 8328.
33. Neves, G. and Lagnado, L. (1999) The kinetics of exocytosis and endocytosis in the synaptic terminal of goldfish retinal bipolar cells. *J. Physiol. (Lond).*, **515**, 181.
34. Ales, E., Tabares, L., Poyato, J. M., Valero, V., Lindau, M., and Alvarez de Toledo, G. (1999) High calcium concentrations shift the mode of exocytosis to the kiss-and-run mechanism. *Nat. Cell Biol.*, **1**, 40.
35. Klingauf, J., Kavalali, E. T., and Tsien, R. W. (1998) Kinetics and regulation of fast endocytosis at hippocampal synapses. *Nature*, **394**, 581.
36. Murthy, V. N. and Stevens, C. F. (1998) Synaptic vesicles retain their identity through the endocytic cycle. *Nature*, **392**, 497.
37. Marks, B. and McMahon, H. T. (1998) Calcium triggers calcineurin-dependent synaptic vesicle recycling in mammalian nerve terminals. *Curr. Biol.*, **8**, 740.
38. Turner, K. M., Burgoyne, R. D., and Morgan, A. (1999) Protein phosphorylation and the regulation of synaptic membrane traffic. *Trends Neurosci.*, **22**, 459.
39. Slepnev, V. I., Ochoa, G. C., Butler, M. H., Grabs, D., and Camilli, P. D. (1998) Role of phosphorylation in regulation of the assembly of endocytic coat complexes. *Science*, **281**, 821.
40. Thiele, C., Hannah, M. J., Fahrenholz, F., and Huttner, W. B. (2000) Cholesterol binds to synaptophysin and is required for biogenesis of synaptic vesicles. *Nat. Cell Biol.*, **2**, 42.
41. Bailey, C. H. and Kandel, E. R. (1993) Structural changes accompanying memory storage. *Annu. Rev. Physiol.*, **55**, 397.
42. Michael, D., Martin, K. C., Seger, R., Ning, M. M., Baston, R., and Kandel, E. R. (1998) Repeated pulses of serotonin required for long-term facilitation activate mitogen-activated protein kinase in sensory neurons of *Aplysia. Proc. Natl. Acad. Sci. USA*, **95**, 1864.
43. Hu, Y., Barzilai, A., Chen, M., Bailey, C. H., and Kandel, E. R. (1993) 5-HT and cAMP induce the formation of coated pits and vesicles and increase the expression of clathrin light chain in sensory neurons of *Aplysia. Neuron*, **10**, 921.
44. Strous, G. J. and Govers, R. (1999) The ubiquitin-proteasome system and endocytosis. *J. Cell Sci.*, **112**, 1417.
45. Kamiguchi, H., Long, K. E., Pendergast, M., Schaefer, A. W., Rapoport, I., Kirchhausen, T., *et al.* (1998) The neural cell adhesion molecule L1 interacts with the AP-2 adaptor and is endocytosed via the clathrin-mediated pathway. *J. Neurosci.*, **18**, 5311.
46. Rose, S. P. (1995) Cell-adhesion molecules, glucocorticoids and long-term memory formation. *Trends Neurosci.*, **18**, 502.
47. Barami, K., Kirschenbaum, B., Lemmon, V., and Goldman, S. A. (1994) N-cadherin and Ng-CAM/8D9 are involved serially in the migration of newly generated neurons into the adult songbird brain. *Neuron*, **13**, 567.
48. Schuster, C. M., Davis, G. W., Fetter, R. D., and Goodman, C. S. (1996) Genetic dissection of structural and functional components of synaptic plasticity. II. Fasciclin II controls presynaptic structural plasticity. *Neuron*, **17**, 655.
49. Schuster, C. M., Davis, G. W., Fetter, R. D., and Goodman, C. S. (1996) Genetic dissection of structural and functional components of synaptic plasticity. I. Fasciclin II controls synaptic stabilization and growth. *Neuron*, **17**, 641.

50. Itoh, K., Stevens, B., Schachner, M., and Fields, R. D. (1995) Regulated expression of the neural cell adhesion molecule L1 by specific patterns of neural impulses. *Science*, **270**, 1369.
51. Takei, K., McPherson, P. S., Schmid, S. L., and DeCamilli, P. (1995) Tubular membrane invaginations coated by dynamin rings are induced by GTPγS in nerve terminals. *Nature*, **374**, 186.
52. Stowell, M. H. B., Marks, B., Wigge, P., and McMahon, H. T. (1999) Nucleotide-dependent conformational changes in dynamin: Evidence for a mechanochemical molecular spring. *Nat. Cell Biol.*, **1**, 27.
53. Hinshaw, J. E. and Schmid, S. L. (1995) Dynamin self-assembles into rings suggesting a mechanism for coated vesicle budding. *Nature*, **374**, 190.
54. Wigge, P. and McMahon, H. T. (1998) The amphiphysin family of proteins and their role in endocytosis at the synapse. *Trends Neurosci.*, **21**, 339.
55. Kim, Y. T. and Wu, C. F. (1987) Reversible blockage of neurite development and growth cone formation in neuronal cultures of a temperature-sensitive mutant of *Drosophila*. *J. Neurosci.*, **7**, 3245.
56. Torre, E., McNiven, M. A., and Urrutia, R. (1994) Dynamin I antisense oligonucleotide treatment prevents neurite formation in cultured hippocampal neurons. *J. Biol. Chem.*, **269**, 32411.
57. Mundigl, O., Ochoa, G. C., David, C., Slepnev, V. I., Kabanov, A., and DeCamilli, P. (1998) Amphiphysin I antisense oligonucleotides inhibit neurite outgrowth in cultured hippocampal neurons. *J. Neurosci.*, **18**, 93.
58. Carroll, R. C., Beattie, E. C., Xia, H., Luscher, C., Altschuler, Y., Nicoll, R. A., *et al.* (1999) Dynamin-dependent endocytosis of ionotropic glutamate receptors. *Proc. Natl. Acad. Sci. USA*, **96**, 14112.
59. Mantyh, P. W., Allen, C. J., Ghilardi, J. R., Rogers, S. D., Mantyh, C. R., Liu, H., *et al.* (1995) Rapid endocytosis of a G protein-coupled receptor: substance P evoked internalization of its receptor in the rat striatum *in vivo*. *Proc. Natl. Acad. Sci. USA*, **92**, 2622.
60. Napolitano, M., Marfia, G. A., Vacca, A., Centourje, D., Bellavia, D., Di Marcotullio, L., *et al.* (1999) Modulation of gene expression following long-term synaptic depression in the striatum. *Brain Res. Mol. Brain Res.*, **72**, 89.

10 | Endocytosis in *Saccharomyces cerevisiae*: involvement of actin, actin-associated protein complexes, and lipids in the internalization step

RUBEN LOMBARDI, SYLVIE FRIANT, and HOWARD RIEZMAN

1. Introduction

All eukaryotic cells are able to internalize extracellular material together with portions of their plasma membrane through a mechanism called endocytosis. The budding yeast *Saccharomyces cerevisiae* (further referred to as yeast) is an organism well suited for the study of cell biological processes like endocytosis. It is an unicellular eukaryote with a membrane organization and organelles similar to higher eukaryotes (see Fig. 1), and it offers well developed genetic manipulation techniques that enable the identification of conditional mutants defective in the process of interest. Indeed, several studies have led to the isolation of endocytosis-deficient mutants in yeast based on three major approaches, defective accumulation of fluorescent dyes in their vacuole (1, 2), defective pheromone receptor endocytosis (3, 4), and their synthetic lethality with a mutation in the vacuolar H^+-ATPase (5).

Yeast can exist in three different cell types, two haploids with opposite mating types **a** or α and a diploid **a**/α. The haploid **a** or α cells can either grow by mitotic cell division or they can mate with a cell of the opposite mating type thus forming an **a**/α diploid cell. The diploid cell is no longer competent for mating. It also grows by mitotic cell division and upon starvation for nutrients (especially nitrogen) the diploid yeast undergoes meiosis and produces four haploid spores, two of each mating type (see ref. 6 and references therein). The differences between **a**, α, and **a**/α cells are determined by the mating type locus. Two different alleles encode for different regulatory proteins causing cell-specific expression or repression of a subset of genes ultimately

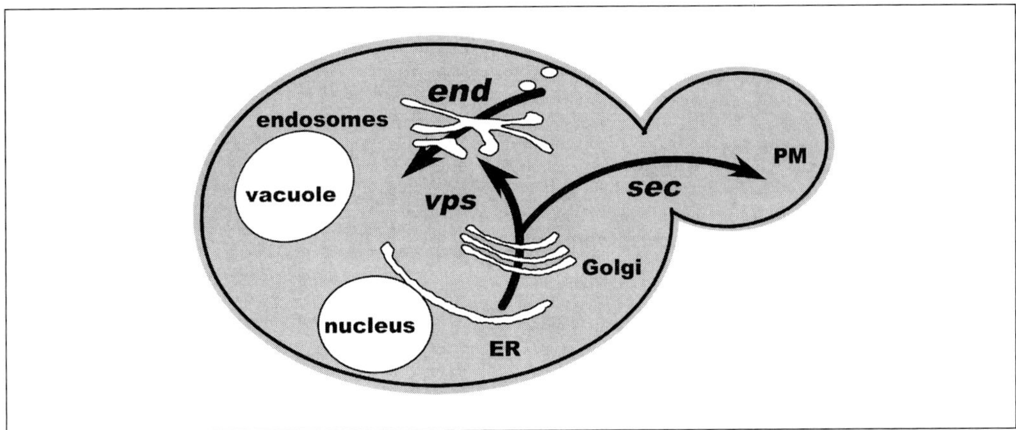

Fig. 1 Membrane trafficking in yeast. A schematic overview of a budding yeast cell with internal organelles is drawn showing the major membrane trafficking pathways in yeast. *end*, endocytic mutants; *sec*, secretory mutants; *vps*, vacuolar protein sorting mutants.

generating the respective cell type (7). In contrast to the diploid cells, both haploid mating types secrete and recognize short peptides called pheromones which bind to receptors on the cell surface of cells of the opposite mating type. Both the α-factor receptor Ste2p and the **a**-factor receptor Ste3p are plasma membrane proteins with seven membrane-spanning segments and are coupled to a heterotrimeric G protein that is involved in the mating-specific signal transduction pathway (8). α cells secrete α-factor but recognize **a**-factor which is secreted by **a** cells and vice versa. Binding of either ligand to its receptor leads to several striking changes, including transcriptional induction of a variety of genes required for mating, arrest in the G1 phase of the cell cycle, clearance of pheromone binding sites, and formation of cell surface projections which are the sites of cell–cell fusion (8, 9). Cells displaying these morphological changes have been termed 'shmoos'. Thus the mating pheromone system functions in cell to cell communication to synchronize the cell cycles of the mating partners and to allow the appropriate fusion events.

Receptor-mediated endocytosis was shown to occur in yeast by using radioactively labelled α-factor pheromone (1, 10). The study of the pheromone receptors, especially the α-factor receptor Ste2p, has shed light on the sequence of events resulting in receptor down-regulation. It was shown that Mat **a** cells are competent to respond to α-factor and internalize it throughout the cell cycle (11). The pheromone receptors undergo constitutive endocytosis at a slow rate, and upon ligand binding the internalization rate is greatly stimulated and hyperphosphorylation and ubiquitination of the cytoplasmic tail of the receptors are induced (for review see ref. 12). Multiple internalization signals have been identified in the cytoplasmic tail of Ste2p. The most membrane proximal signal, SINNDAKSS, was shown to be necessary and sufficient for internalization of a truncated receptor. The crucial residue is the lysine that when mutated to arginine completely blocks receptor internalization (13). The lysine was shown to be an acceptor site for ubiquitination and this ubiquitination signals the internalization of the receptor (14). Several other yeast and mammalian plasma membrane proteins have been shown to be ubiquitinated and internalized, suggesting a general role of ubiquitination not only in yeast but also in higher eukaryotes (for review see refs 15, 16). Another internalization signal in yeast, NPFXD, was found in Kex2p and a similar NPF-sequence was shown to be necessary for the pheromone dependent internalization of a truncated **a**-factor receptor (17).

The endocytic pathway in mammalian cells has been well characterized at the morphological level (see refs 18, 19 and references therein) while, until very recently, the characterization of this pathway in yeast was quite poor. Biochemical evidence for two endocytic intermediates between the plasma membrane and the vacuole in yeast came from studies following the internalization of radioactively labelled α-factor. Internalized α-factor travels successively and transiently through two biochemically distinct membrane-bound compartments to the vacuole (20). Based on the kinetics of α-factor movement through those compartments and by analogy to the mammalian pathway, the two intermediates were termed early endosomes and late endosomes. Immunofluorescence studies following the internalization and delivery of the α-factor receptor Ste2p have revealed a peripheral early endocytic compartment and a late endocytic compartment near the vacuole. The formation of the early endocytic compartment was dependent upon *SEC18* gene function (yeast homologue of the mammalian *N*-ethyl maleimide-sensitive fusion protein) which is essential for multiple vesicular fusion events (see ref. 21 and references therein). Recently, an electron microscopy (EM) study has allowed the yeast cell endocytic pathway to be seen at an ultrastuctural level by following the internalization and delivery of positively charged Nanogold™ (Nanoprobes, Stony Brook, NY, USA) to the vacuole (22). In agreement with previous reports, the first endocytic intermediates seen are small vesicles of approximately 50 nm in diameter which also accumulate at non-permissive temperature in the *sec18* mutant. These vesicles are likely to be the primary endocytic vesicles. In wild-type cells, the Nanogold™ is next found in a peripheral compartment with a tubular–vesicular structure. Later the Nanogold™ is found in a large oval structure with internal membranes located near the vacuole and finally in the vacuole. Taken together, these data suggest a very similar organization of the endocytic pathways in mammalian cells and yeast. In yeast, the first detected intermediates are endocytic vesicles that either generate an early peripheral endocytic intermediate by homotypic fusion or fuse with a pre-existing endocytic compartment. These tubular–vesicular structures at the cell periphery were termed early endosomes. The next clearly defined intermediate is a large oval structure with internal membranes located near the vacuole, called late endosome. Finally, the endocytosed material is delivered to the vacuole.

It has been previously shown that vacuolar and endocytosed proteins accumulate in an aberrant compartment in a subset of vacuolar protein sorting (*vps*) mutants (23). This compartment, termed 'class E compartment', has been proposed to be an exaggerated form of a prevacuolar compartment (see Chapter 6) where the endocytic and vacuole biogenesis pathways intersect (23). The prevacuolar compartment could be an intermediate between the early and late endosomes as defined above.

Transport between the endosomal compartments and the vacuole has been shown to require two small GTP-binding proteins of the Rab/Ypt family. Ypt51p functions in the early to late endosome transport step while Ypt7p is involved in late endosome to vacuole trafficking and in homotypic vacuolar fusion (24–27). Several t-SNARES are involved in the endocytic pathway. Vam3p has been localized to the vacuole and shown to be important for several trafficking pathways leading to the vacuole (28, 29), while Pep12p has been localized to a prevacuolar compartment and functions in traffic from the Golgi apparatus to this prevacuolar compartment (30). Recently, two other members of the yeast syntaxin family of t-SNARES, Tlg1p and Tlg2p, have been identified. Both have been implicated in the TGN/endosomal system but there is controversy as to their localization in the cell and their exact function (31–35). Nevertheless, not only the morphology of the endocytic pathway but also the components involved in regulating and mediating the trafficking steps appear to be conserved from yeast to mammals.

In this review, we will focus on the internalization step of receptor-mediated endocytosis (R-ME). Different studies have led to the identification of a great number of mutants impaired in this step (see Table 1). Analysis of these mutants revealed two fundamental aspects of the internalization process: The requirement for actin and a subset of actin-associated proteins, and the importance of certain lipids. Both aspects

Table 1 Yeast genes required for the uptake step of receptor-mediated endocytosis

Yeast gene	Homologies/comments	Motifs/domains	References
ACT1/END7	Actin		36, 37
AKR1		Ankyrin repeat	38
ARC35/END9	Subunit ARP2/3-complex		5, 147
ARP2	Actin related protein		39
ARP3	Actin related protein		C. Schärer-Brodbeck, unpublished data
CHC1	Clathrin heavy chain		40
CLC1	Clathrin light chain		41, 42
CMD1	Calmodulin	Four EF hands	43
END3	Eps15	EH domain	3,44
ERG2/END11	Ergosterol biosynthesis enzyme		5
LAS17/BEE1	Human WASP		45
LCB1/END8	Ceramide biosynthesis enzyme		5
MYO5	Type I myosin	SH3	46
PAN1/DIM2	Eps15	EH domains	2, 47
RSP5/NPI1	Ubiquitin protein ligase	HECT domain	48, 49
RVS161/END6	Amphiphysin		37
RVS167	Amphiphysin	SH3	37
SAC6	Fimbrin		36
SJL1, SJL2, SJL3	Synaptojanin		50
SLA2/END4	Talin		3
SRV2/END14			51
VRP1/END5		Proline rich	37

will be discussed together with the role of clathrin and an overview of the techniques available to study endocytosis in yeast.

2. Techniques used to study endocytosis in yeast

Several reporter systems have been developed to study endocytosis in yeast. Fluid phase endocytosis can be followed by using fluorescent dyes like lucifer yellow (52) or FM 4-64 (53), or by internalization of electron dense particles and analysis by EM (2, 22). R-ME can be assayed by following **a**-factor receptor Ste3p (4) and α-factor pheromone (1) or by measuring the clearance of transporters from the plasma membrane (54–57). In the present chapter we will present and detail these different methods.

2.1 Fluid phase endocytosis

2.1.1 Fluorescent dye uptake

Two fluorescent dyes are commonly used to study fluid phase endocytosis in yeast, lucifer yellow-carbohydrazide (LY), a fluid phase marker, and FM 4-64, a membrane probe. LY is a small hydrophilic fluorescent molecule that is incapable of diffusion across biological membranes. It is non-toxic, highly fluorescent, and resistant to bleaching. Uptake of LY is time-, energy-, and temperature-dependent and it is non-saturable. The rate of endocytic accumulation has been estimated as 27 nl/mg of cellular protein/h at 30 °C for yeast cells (52), compared to 250 nl/mg/h in murine peritoneal macrophages (58). Internalized LY accumulates in the vacuole and can be visualized by fluorescence microscopy using FITC optics. LY has been used to screen for mutants that are defective in endocytosis, since the assay is simple to perform (1). Unfortunately, due to the limitations in the resolution of yeast organelles by light microscopy, it has not been possible to visualize internalized LY in any intermediate compartment.

Recently, the lipophilic styryl dye FM 4-64 (*N*-(3-thiethylammoniumpropyl)-4-(*p*-diethyl-aminophenylhexatrienyl) pyridinium dibromide) has been shown to enter yeast cells by an endocytic mechanism (53). This dye selectively labels the membrane of the intracellular organelles along the endocytic pathway since it is fluorescent only when inserted into membranes. During a time course of FM 4-64 staining, the dye initially stains the plasma membrane, then the cytoplasmic intermediate endosomal compartments, and finally the vacuolar membrane (53). FM 4-64 has the advantage that it can be used to visualize intermediates between the plasma membrane and the vacuole during endocytosis.

2.1.2 Electron dense endocytic markers

Positively charged Nanogold™ (Nanoprobes, Stony Brook, NY) is a new marker used to follow the endocytic pathway in yeast (22). Nanogold™ binds to the plasma membrane of yeast spheroplasts and its internalization and intracellular targeting can be followed by EM. Nanogold™ cannot be degraded, and can be used to visualize all compartments along the endocytic pathway.

Cationized ferritin was also used to follow endocytosis in yeast spheroplasts (2). This marker is also electron dense and can be visualized by EM. Cationized ferritin has been used to identify structures that accumulate in endocytic mutants.

2.2 Receptor-mediated endocytosis

2.2.1 α-factor pheromone uptake

Receptor-mediated endocytosis can be followed using the yeast pheromone α-factor. α-factor binds to its specific cell surface receptor, the *STE2* gene product (59). Internalized α-factor is transported through two intermediate compartments, the early and late endosomes (20, 21), on its way to the vacuole where it is degraded by resident vacuolar proteases (60, 61). The best way to quantitatively assess the earliest stages of endocytosis in yeast is to follow the uptake of radioactively labelled α-factor by Mat **a** cells. The α-factor pheromone is radioactively labelled with $^{35}SO_4$ *in vivo*, recovered from the culture supernatant, and purified to obtain ^{35}S-labelled α-factor (62). [^{35}S]α-factor uptake assays are performed on mid-log phase cells using either the continuous incubation or pulse-chase protocols (62). The percentage of internalized α-factor at each time point is calculated by dividing the internalized counts (pH 1-resistant) by the total cell-associated counts (pH 6-resistant) (see Fig. 2).

2.2.2 a-factor receptor internalization

Receptor-mediated endocytosis can also be assayed by following the yeast **a**-factor receptor, Ste3p. The **a**-factor receptor, like the Ste2p, is subjected to two modes of

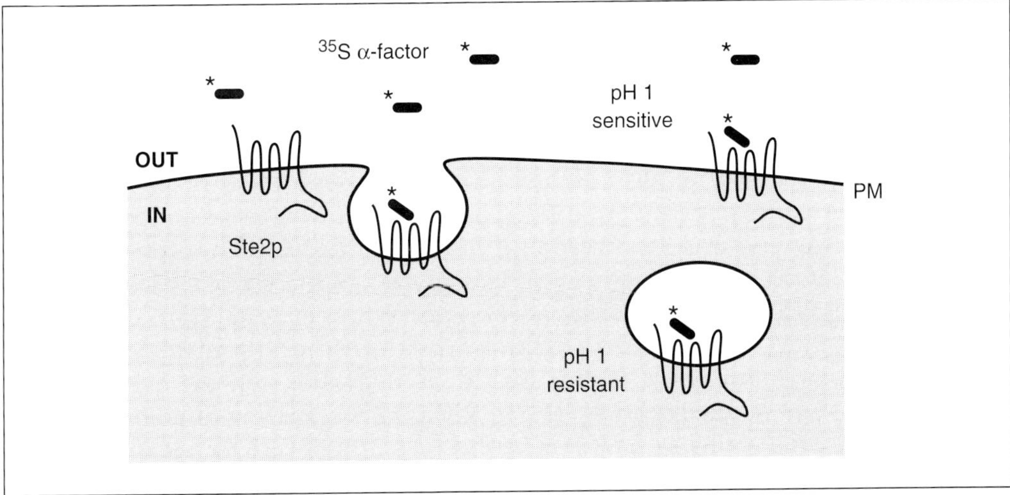

Fig. 2 α-factor uptake assay. [^{35}S]α-factor binds the α-factor receptor Ste2p and is internalized. The internalization rate is determined by dividing the fraction of radiolabelled α-factor that is internalized and therefore cannot be removed from the cells by an acid wash protocol (pH 1-resistant), by the total cell-associated counts (pH 1-sensitive).

endocytosis, a constitutive, ligand-independent mechanism and a regulated, ligand-dependent mechanism (4). Both mechanisms result in delivery of the receptor to the vacuole for subsequent degradation. The endocytic assay was developed based on the down-regulation of the receptor by determining the rate of degradation of Ste3p. The ligand-independent turnover of the Ste3p receptor is rapid, with a $t_{1/2}$ estimated to be ~ 15 min. An endocytic defect leads to a turnover defect of Ste3p and to accumulation of the receptor at the cell surface. Ste3p is labelled with [^{35}S]methionine in a pulse-chase protocol, at various time points after the initiation of the non-radioactive chase, aliquots of the labelled intact cells are taken and subjected to external protease treatment. This assay distinguishes surface-localized receptors (susceptible to external proteases), from receptors that localize to compartments inside the cell (resistant to external proteases). The Ste3p protein is immunoprecipitated from cell extracts and subjected to polyacrylamide gel electrophoresis (4).

2.2.3 Permeases and transporters uptake

Another rapid and sensitive way to follow endocytosis is to measure clearance of permeases and transporters from the plasma membrane. A number of stimuli appear to trigger rapid endocytosis and subsequent vacuolar degradation of various permeases when the uptake of their respective substrate is no longer needed by the cell. Such a control of permease stability in response to nutrients has been demonstrated for inositol, tryptophan, and uracil permeases (54, 56, 63, 64). Similar mechanisms were also reported for the Ste6p **a**-factor pheromone transporter, the Pdr5p ATP-binding cassette (ABC) transporter, and the maltose transporter (55, 57, 65, 66).

The best studied among the yeast permeases is the uracil permease, encoded by the *FUR4* gene (67). Uracil permease is phosphorylated and ubiquitinated at the plasma membrane and undergoes rapid internalization followed by vacuolar degradation in cells submitted to various stress conditions, such as inhibition of protein synthesis (48, 68). The fate of plasma membrane uracil permease can be followed in exponentially growing cells after inhibition of protein synthesis by cycloheximide. [^{14}C]Uracil uptake is measured at various time points after addition of cycloheximide, and protein extracts are prepared and analysed for uracil permease on immunoblots. In wild-type yeast cells, inhibition of protein synthesis triggers rapid loss of uracil uptake, concurrent with permease degradation, whereas in endocytic mutants a protection against cycloheximide-induced loss of uracil uptake is observed (54).

The yeast maltose transporter is degraded in the vacuole after internalization by endocytosis (55). This internalization occurs under certain physiological conditions such as impaired protein synthesis or presence of a fermentable substrate in the medium. Endocytosis of this protein is dependent on the actin network but independent of microtubules (69). In addition, the binding of ubiquitin is required for the internalization step (70). By using yeast mutants defective in the heavy chain of clathrin and in several subunits of the COPI and the COPII complexes, it has recently been shown that clathrin and the two cytosolic subunits of COPII, Sec23p and Sec24p, are involved in endocytosis of the maltose transporter (71).

3. Actin and actin-associated proteins

3.1 The actin cytoskeleton

Initial studies in yeast revealed a fundamental role for the actin cytoskeleton in the internalization step of endocytosis (36). Using two conditional mutations in both actin and β-tubulin, a requirement for actin in the internalization step but not for post-internalization trafficking was demonstrated. Microtubules were not required at all. An *act1-1* strain showed a rapid onset of endocytic defect at 37 °C, even without pre-incubation, and internalized α-factor at less than 10% of the rate detected in wild-type cells suggesting a direct role of actin in the internalization step.

The actin cytoskeleton in yeast consists of cortical patches and actin cables both composed of F-actin. Patches show a polarized distribution that changes during the cell cycle and the cables generally run along the mother-bud axis (see ref. 72 and references therein). At an ultrastuctural level, the actin patches consist of a finger-like invagination of plasma membrane around which actin filaments and actin binding proteins like Abp1p and cofilin are organized (73). These data led to the notion that actin patches might be the sites of endocytosis. However, a recent study following Ste2p internalization by immuno-EM provided evidence that the cortical actin patches are not the sites of receptor internalization (74). The authors showed that Ste2p is not randomly distributed over the plasma membrane but is concentrated in furrow-like invaginations. This localization of Ste2p agrees with previous immunofluorescence studies showing a spotty distribution of Ste2p on the cell surface (21).

Using the drug latrunculin A, the actin filaments in yeast were shown to undergo rapid cycles of assembly and disassembly *in vivo* (75) suggesting a very dynamic actin cytoskeleton even in this non-motile organism. Using conditional mutations in the yeast cofilin gene *COF*, rapid turnover of cortical actin structures was shown to be essential for endocytosis (76). Cells lacking the yeast homologue of the actin filament bundling protein fimbrin, Sac6p, are also defective for internalization (36). Taken together, these data strongly support the central role of an organized dynamic actin cytoskeleton in the internalization step of endocytosis in yeast. Further support comes from the analysis of isolated yeast mutants defective for endocytosis. One mutant allele, *end7-1*, has been shown to be allelic to *ACT1* (37) and several endocytic mutants also exhibit defects in the actin cytoskeleton (77). However, it is important to point out that not all mutants with defects in the actin cytoskeleton are affected in endocytosis. Mutations in *myo2*, *pfy1*, and *tpm1* cause actin cytoskeleton defects similar to those observed in several endocytic mutants but they endocytose as well as wild-type cells (37). The involvement of the actin cytoskeleton in mammalian cells has been a matter of discussion for a long time (see ref. 78 and references therein). Ambiguous results have been obtained depending on the cell line and the actin depolymerizing agent used. A recent study using latrunculin A provided evidence that the actin cytoskeleton is required for R-ME in mammalian cells (79), though other studies have suggested it is not essential (146).

3.2 Actin-associated proteins

As previously mentioned, several endocytic mutants exhibit defects in the actin cytoskeleton and some of the proteins encoded by these genes have been shown to bind actin or are associated with actin-binding proteins. Several protein complexes have been implicated in the internalization step of endocytosis and will be described below (see Fig. 3).

3.2.1 Myo5p-complex

Type I myosins have been implicated in actin-dependent membrane motility processes such as membrane trafficking, organelle movement, phagocytosis, pinocytosis, and

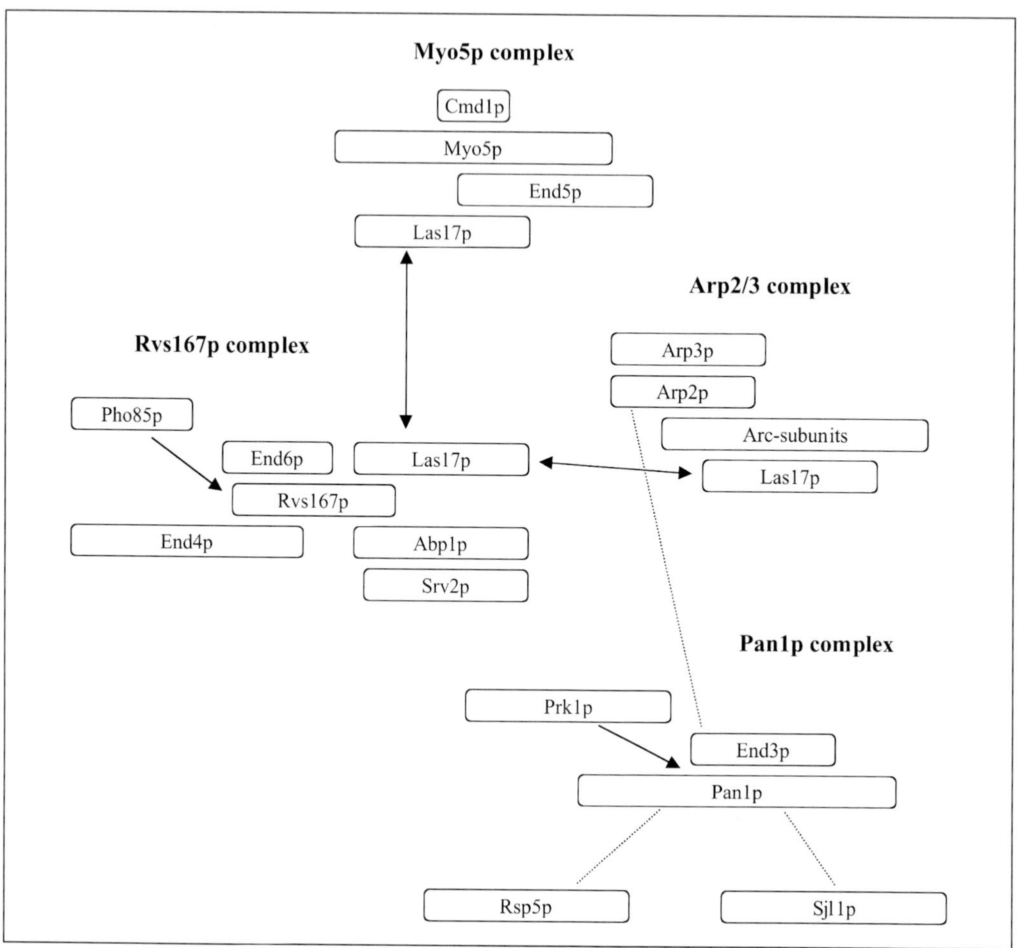

Fig. 3 Actin-associated protein complexes. Schematic drawing of the four protein complexes described in detail in the text. Proteins that have been shown to interact biochemically and/or in the two-hybrid system are drawn near each other. Synthetic lethal interactions are shown by a dashed line and an arrow marks phosphorylation of a protein by a kinase. Las17p might be part of three complexes and therefore connections are drawn by double-headed arrows.

cellular locomotion (see ref. 80 and references therein). The yeast genome encodes two type I myosins, *MYO3* and *MYO5* (46, 81). Single deletions do not lead to any growth phenotype but deletion of both genes results in a severe growth defect or lethality (see ref. 46 and references therein). In contrast to a *myo3Δ* strain, a *myo5Δ* strain is impaired for α-factor internalization at 37 °C, suggesting a more direct role of Myo5p in the internalization step (46). Myo5p contains two IQ motifs that constitute binding sites for the small EF-hand containing protein calmodulin (Cmd1p) and this interaction is required for endocytosis (82). A previous study had already implicated Cmd1p in the internalization step of endocytosis and this function of Cmd1p appears to be Ca^{2+}-independent because a calmodulin allele defective in high-affinity calcium binding (*cmd1-3*) shows no endocytic defect (43). In addition, evidence suggests that at least two distinct calmodulin functions are required for the internalization step. One target of Cmd1p is Myo5p and this function is impaired in the *cmd1-247* mutant. Another function is impaired in the *cmd1-228* mutant for which the target is not known (82). The actual function of Myo5p in the internalization step is unknown. Since none of the dynamin-homologues in yeast have a role in the internalization step of endocytosis, Myo5p was suggested to replace dynamin in yeast endocytosis if one assumes that dynamin works as a mechanochemical enzyme (78). Myo5p has been shown to interact via its SH3-domain with End5p/Vrp1p and this interaction is required for polarized localization of Myo5p (83). Interestingly, a mutant allele of *END5*, *end5-1*, has been isolated in a previous screen for endocytic mutants (37). End5p is very rich in proline residues and contains several putative SH3-binding sites. End5p interacts with actin in the two-hybrid system and an actin-binding domain was mapped to the first 70 amino acids of End5p (84).

In summary, the endocytic function of Myo5p is regulated at least partially by Cmd1p and there is a second function of Cmd1p in the internalization step whose target is unknown. Polarized localization of Myo5p depends on End5p and therefore this interaction might concentrate or activate Myo5p at sites of function.

3.2.2 Arp2/3 complex

As already mentioned, a dynamic actin cytoskeleton is required for the internalization step of endocytosis. Recently, a protein complex, the Arp2/3 complex, has been implicated in the regulation of the actin cytoskeleton and identified in several organisms. The Arp2/3 complex consists of seven subunits and has been shown to stimulate actin filament nucleation and to bind both to the pointed-ends and sides of actin filaments (see ref. 85 and references therein). Interestingly, several subunits of the complex are required for the internalization step of endocytosis. Mutations in genes encoding the actin-related proteins *ARP2* (39) and *ARP3* (C. Schärer-Brodbeck and H. Riezman, unpublished data) block R-ME. A conditional mutant, *end9-1*, has been isolated in a screen for endocytic mutants (5) and subsequently shown to be allelic to *ARC35*, the 35 kDa subunit of the Arp2/3 complex (147). The *end9-1* strain is defective for both fluid phase and R-ME at the restrictive temperature (5). Taken together these data strongly implicate the Arp2/3 complex in the internalization step. To our knowledge no data have been published concerning the endocytic phenotypes of mutations in other subunits of this complex.

Studies from two different laboratories have implicated the yeast homologue of the human Wiskott-Aldrich Syndrome protein (WASP), *LAS17/BEE1*, in Arp2/3 complex function (45, 86). Las17p has been shown to interact with the Arp2/3 complex by co-immunoprecipitation. *LAS17* is an allele-specific multicopy suppressor of *ARP2* and *ARP3* mutations, and overexpression restores the endocytic defect of the *arp2-2* mutant allele. Furthermore, *las17Δ* is synthetically lethal with several *ARP2* mutant alleles and with *arp3-14*. In addition, Las17p stimulated the actin nucleation activity of the Arp2/3 complex *in vitro*. Taken together, these data support an important functional interaction of Las17p and the Arp2/3 complex. Las17p interacts with the Arp2/3 complex via its carboxy terminal WA-domain. Unexpectedly, deletion of this domain, as opposed to a *las17Δ*, caused relatively minor defects suggesting that Las17p does not function solely via the Arp2/3 complex and that other cellular factors act redundantly with Las17p to activate the Arp2/3 complex (86). Interestingly, a *las17Δ* strain exhibits a strong α-factor internalization defect indicating a role in endocytosis (45).

LAS17 has been isolated as a multicopy suppressor of the *end5-1* temperature-sensitive growth defect and overexpression of Las17p also restores the endocytic defect of this mutant allele. Furthermore, *las17Δ* is synthetically lethal with *end5Δ* and the two proteins interact in the two-hybrid system (87). These findings suggest a functional relationship between Las17p and End5p. A similar interaction has been detected in human cells between WASP and WIP (WASP interacting protein), whose yeast homologue probably is *END5* (88).

In summary, the data presented in the previous chapter (Myo5p complex) together with this data suggest that End5p can associate with both Myo5p and Las17p and thus might functionally couple the Arp2/3 complex to the Myo5p complex. However, there is no evidence that End5p interacts with both proteins at the same time leaving the possibility that End5p exerts its function in two separate complexes.

3.2.3 Rvs167p complex

The *end6-1* mutant was isolated in a screen for endocytic mutants and shown to be allelic to *RVS161* (37). Rvs161p shows homologies to a second yeast protein, Rvs167p, and to the mammalian protein amphiphysin. In mammalian cells, amphiphysin interacts with dynamin, synaptojanin, AP-2, and clathrin, all proteins implicated in clathrin-mediated endocytosis, and has been suggested to act as a scaffold protein (see Chapter 1 and 2; ref. 89 and references therein). The entire Rvs161p is homologous to the N-terminal part of Rvs167p and amphiphysin but lacks the SH3-domain present in Rvs167p and amphiphysin. Mutations in either *RVS161* or *RVS167* lead to similar phenotypes except for a cell fusion defect detected in *rvs161Δ* and not in *rvs167Δ* strains (see ref. 90 and references therein; 91). Both proteins are required for the internalization step of endocytosis (37) and they interact with each other (90) further supporting a joint function. Surprisingly, their localization in the cell is different. Rvs161p was shown to be mainly cytosolic in unbudded cells and upon appearance of the bud to localize mainly to the mother-bud-neck (91). In contrast, in unbudded cells Rvs167p is localized mainly in small cortical patches throughout the cell which polarize at the bud emergence site and in the small buds (92). Taken together, these data support both overlapping and unique functions of the proteins in the cell.

Rvs167p interacts with the Pho85 cyclin-dependent kinase complexes and is a substrate for this kinase *in vitro*. The similarities of the phenotypes associated with the deletion of *PHO85* and *RVS167* as well as the reduced phosphorylation of Rvs167p in a *pho85Δ* strain *in vivo* suggest a regulatory function of Pho85 kinases on Rvs167p activity (93). The SH3-domain of Rvs167p interacts with Act1p (94) and Las17p (45, 95) in the two-hybrid system and with Abp1p (96) in a binding assay. Another protein interacting with Rvs167p is End4p/Sla2p (51), a protein isolated both in a screen for endocytic mutants (3) and in a synthetic lethal screen with *abp1Δ* (97). The SH3-domain of Abp1p interacts with Srv2p (98). Deletion of either *ABP1* or *SRV2* show no endocytic defect (36, 51) but interestingly, a mutant allele (*srv2-14*) of *SRV2* has been isolated based on its endocytic defect (51). Furthermore, deletion of the central coiled-coil domain of End4p creates a synthetic endocytic defect in the absence of Abp1p or Srv2p suggesting a redundant endocytic function of both Abp1p and Srv2p with End4p (51). Several lethal double mutant combinations have been shown among disruptions in *ABP1*, *END4*, *RVS167*, and *SRV2* (96) further supporting a functional relationship of these actin-associated proteins.

As mentioned previously, Rvs167p interacts with Las17p in the two-hybrid system. This interaction could link the Rvs167p complex to the Arp2/3 complex and/or the Myo5p complex via Las17p. However, Las17p probably does not interact with all these complexes at the same time. Rather a dynamic model can be envisioned in which Las17p interactions are tightly regulated and possibly mutually exclusive to allow a controlled interplay between the different complexes.

3.2.4 Pan1 complex

The *end3-1* allele was isolated in a screen for endocytic mutants and shown to be required for the internalization of α-factor and for fluid phase endocytosis (3, 44). End3p contains an N-terminal EH-domain (Eps15 homology domain) first identified in the mammalian protein Eps15 (see Chapter 1; 99). Several EH-domain containing proteins are present in yeast, and of these Pan1p has also been shown to be required for endocytosis (2, 47). Interestingly, *END3* was identified as a multicopy suppressor of *pan1-4*, and *end3Δ* is synthetically lethal with *pan1-4*. In addition, both proteins have been shown to interact by co-immunoprecipitation and in the two-hybrid system (47). Loss of function mutations in *PRK1* suppress the growth and actin defect of *pan1-4* (100). Prk1p is a serine/threonine kinase that was shown to phosphorylate Pan1p *in vitro*. This phosphorylation occurs in the Pan1p domain implicated in End3p binding and pre-incubation of this Pan1p-domain with End3p prior to the *in vitro* kinase assay reduced the phosphorylation of the Pan1p-domain (100). *PRK1* shows homologies to another putative yeast kinase, *ARK1*. Single deletions of either kinase gene are viable but an *ark1Δ prk1Δ* strain exhibits large cytoplasmic actin clumps and severe defects in cell growth implicating both kinases in regulating the actin cytoskeleton (101). Interestingly, Ark1p was isolated in a two-hybrid screen using parts of the End4p as bait, and *prk1Δ* is synthetically lethal with *end4Δ* providing a link to the previously described Rvs167p complex. So far, no function for either Prk1p or Ark1p in regulating End4p or the Rvs167p complex has been shown.

Pan1p also interacts with yAP180A and yAP180B, the yeast homologues of the mammalian AP180, via its EH-domain. As with to their mammalian counterparts, the two yeast proteins interact with clathrin and might therefore function as adaptor proteins (102). However, this interaction might not be required for endocytosis because the two yAP180 are not required for endocytosis (103) (see Section 4 below).

END3 and *PAN1* show genetic interactions with several proteins involved in endocytosis. *end3-1* is synthetically lethal with *arp2-1*, a component of the previously described Arp2/3 complex (39) and *pan1-20* is synthetically lethal with *sjl1Δ* but not with *sjl2Δ* or *sjl3Δ*, the yeast homologues of the mammalian synaptojanin (102). A previous study has shown that a strain lacking *SJL1* and *SJL2* (*sjl1Δ sjl2Δ*) exhibits a defect in both R-ME and fluid phase endocytosis, while a *sjl2Δ sjl3Δ* mutant had only a minor defect, and a *sjl1Δ sjl3Δ* had no defect, implicating these proteins in endocytosis like their mammalian counterpart, synaptojanin (50). *PAN1* shows allele-specific synthetic lethality with *RSP5* (104), an ubiquitin-ligase involved in the ubiquitination of several permeases and therefore in their internalization (see ref. 49 and references therein). *RSP5* is an essential gene, and the conserved cysteine in the HECT domain is required for both yeast cell viability and ubiquitination of permeases like Gap1p. Interestingly, deletion of the N-terminal C_2-domain of Rsp5p does not affect viability but it impairs internalization of Gap1p without affecting ubiquitination of the permease indicating a role of Rsp5p in internalization in addition to its function in ubiquitination (49).

Recently, two yeast homologues of the mammalian epsin, Ent1p and Ent2p have been identified and shown to be required for FM4-64 internalization. Ent1p interacts with clathrin via its final eight amino acids (105). A previous study had reported a weak interaction of Ent1p and the EH-domain of Pan1p in the two-hybrid system (102) providing a possible link between the yeast epsins and the Pan1p complex.

Taken together, these data suggest that the two protein kinases Ark1p and Prk1p are involved in regulating the actin cytoskeleton and may work, at least partially, via Pan1p. The genetic interactions detected link the yeast homologues of synaptojanin, *SJL1-3*, and the ubiquitin ligase *RSP5* to the Pan1p complex.

4. Involvement of clathrin in endocytosis

In mammalian cells, the most prominent and best studied mechanism of R-ME occurs via clathrin-coated pits and vesicles (see Chapter 1) (106, 107, and references therein). In addition to the clathrin-mediated endocytic pathway, other so-called clathrin-independent endocytic pathways are present and function as the major internalization routes for some receptors (see Chapter 2) (108, and references therein). Selective inhibition of clathrin-dependent endocytosis causes the up-regulation of these clathrin-independent pathways (109).

The clathrin molecule is a triskelion formed by three clathrin heavy chain molecules, each associated with a clathrin light chain. The yeast genome contains a single clathrin heavy chain gene (*CHC1*) and also a single clathrin light chain gene (*CLC1*). Strains with a disruption of either *CHC1* or *CLC1*, as well as a strain harbouring a *chc1-ts* allele, show a defect in α-factor internalization. However, these strains still internalize radioactive α-factor at about 35–50% of the level detected in wild-type cells (40–42, 110). Compared to the strict requirement of actin and several actin-associated proteins for α-factor internalization (see Section 3 above) this partial effect of clathrin mutations points to a non-essential role of clathrin in R-ME in yeast. Several models have been proposed to explain this partial effect of clathrin mutations:

(a) Clathrin is only required to concentrate receptors at the sites of internalization but not for the actual budding of the vesicles.

(b) At least two endocytic pathways exist in yeast that are actin-dependent but only one of them is clathrin-dependent.

(c) Another protein can partially substitute for mutations in clathrin and take over some of its functions.

(d) The defect of clathrin mutations is indirect.

The findings that no other genes with extended homologies to *CHC1* are found in the yeast genome and the rapid onset of the endocytic defect in a *chc1-ts* strain (40) would argue against the later two models. An interesting recent finding is that the internalization of the uracil permease Fur4p is unaffected at restrictive temperature in a *chc1-ts* strain (A. Gratias and R. Haguenauer-Tsapis, unpublished data). Therefore, the requirement for clathrin in endocytosis seems to depend on the protein to be internalized. According to the second model, the two proteins might be internalized by two different endocytic pathways. According to the first model, a cell might need to remove an actively signalling receptor (Ste2p with bound α-factor) from the plasma membrane faster than a pyrimidine base permease (Fur4p) and thus only Ste2p would normally be concentrated for internalization.

In mammalian cells, the heterotetrameric adaptor complexes have been implicated both in recruiting clathrin to membranes and in concentrating receptors in clathrin-coated pits via interactions with their cytoplasmic tails (107). A recent study demonstrated that clathrin can function in the absence of both heterotetrameric adaptors and AP180-related proteins in yeast (103). A yeast strain with disruptions of all six heterotetrameric AP large chain genes (all β chains, α/γ/δ chains removed) as well as a strain with disruption of the three AP large chain β-subunit genes and the two *YAP180* genes did not display the phenotypes of clathrin-deficient cells. Endocytosis was not affected in these strains and the authors were able to isolate CCVs from them. Taken together, these data suggest that clathrin can be recruited to membranes in the absence of functional adaptors and that clathrin can function in the absence of adaptors. There is still the possibility that other proteins can function as adaptor molecules and therefore mediate the membrane association of clathrin and the concentration of receptors. In mammalian cells, β-arrestin was shown to act as a clathrin adaptor in the endocytosis of the β2-adrenergic receptor and some other seven transmembrane domain G protein-coupled receptors (111). Taken together, these data suggest that a re-evaluation of the requirements to form a CCV must take place.

Additional support for a non-essential role for clathrin in the internalization step in yeast comes from the finding that none of the dynamin-homologues in yeast are involved in endocytosis (see ref. 78 and references therein). It is well established in mammalian cells that dynamin is required for clathrin-mediated endocytosis (see Chapter 1) (112). Nevertheless, several proteins involved in endocytosis in yeast have homologues in mammalian cells and vice versa pointing to at least some functional homologies of endocytosis in both cell types (for a review see ref. 78).

5. Role of lipids in the endocytic pathway

5.1 Lipid requirement in membrane trafficking in yeast

Recently, not only proteins but also specific lipids were shown to be required for the endocytic pathway. Lipids are responsible for the structural integrity of biological membranes and confer specific dynamic properties to the bilayer. Major lipid components of eukaryotic membranes are phospholipids, sterols, sphingolipids, and glycerolipids. Phosphatidic acid, diacylglycerol (DAG), sphingolipids, and phosphoinositides (PI) have been implicated in several stages of membrane trafficking in yeast.

The vacuolar protein sorting (*VPS*) pathway of yeast mediates transport of vacuolar protein precursors from the late Golgi to the lysosome-like vacuole (see Fig. 1). Sorting of some vacuolar proteins occurs via the prevacuolar endosomal compartment and mutations in a subset of *VPS* genes interfere with the Golgi-to-endosome transport step. The *VPS34* gene encodes a PI 3-kinase and this enzyme is required for protein sorting to the vacuole (113). Inactivation of *VPS34* results in impaired fusion of endosomal transport intermediates with the vacuole. These data implicate PI(3)P as a regulator of membrane traffic (see Chapter 5) (114).

The *SEC14* gene product encodes a phosphatidylinositol/phosphatidylcholine transfer protein that is required for the production of secretory vesicles from the Golgi (115). This requirement can be relieved by inactivation of the cytosine 5′-diphosphate (CDP)- choline pathway for phosphatidylcholine biosynthesis (116) or by increasing the supply of DAG to the Golgi (117). Recently, a *sec14* mutant that was inactivated for phosphatidylinositol (PI), but not phosphatidylcholine (PC) transfer activity, was shown to be able to rescue the lethality and the Golgi secretory defects associated with *sec14-ts* or *sec14Δ* mutations. These findings indicate that PI binding/transfer seems to be dispensable for Sec14p function *in vivo* (118).

GPI-anchored proteins are attached to the membrane via a glycosylphosphatidyl inositol-(GPI) anchor whose carbohydrate core is conserved in all eukaryotes. Apart from membrane attachment, the precise role of the GPI-anchor is not known, but it has been proposed to play a role in protein sorting. *In vitro* and *in vivo* data suggest that ceramides are required for trafficking of GPI-anchored proteins from the endoplasmic reticulum (ER) to the Golgi apparatus in yeast (119–121).

Members of the synaptobrevin/VAMP family of v-SNAREs are involved in vesicle docking and have been shown to be essential for exocytosis in yeast. Recessive mutations in either *ELO2* or *ELO3*, two genes that mediate the elongation of very long chain fatty acids, allow yeast to grow normally and secrete in the absence of v-SNAREs. Thus, the v-SNARE requirement in constitutive exocytosis can be abrogated by mutations in genes involved in lipid synthesis (122).

5.2 Lipid requirement in endocytosis

The *VMA2* gene encodes the 60 kDa vacuolar H^+-ATPase (V-ATPase) regulatory subunit (subunit B). In cells bearing a disruption of *VMA2* (*vma2Δ*), the V-ATPase does not assemble on the vacuolar membrane and consequently the lumen of the vacuole fails to become acidified. Yeast *vma2Δ* mutants are able to grow if the external medium is around pH 5.5 but not at pH 7 or higher pH (123), because the cells are able to take up protons from the medium by fluid phase endocytosis. Based on this result a screen for isolation of endocytosis-deficient (*end*) mutants was developed, taking advantage of the synthetic lethality of endocytic mutants with *vma2Δ*. The principle of this screen is presented in Fig. 4 and resulted in isolation of the *end8-1* and *end11-1* mutants among others (5).

The cloning and sequencing of these two genes revealed that *END8* is allelic to *LCB1*, an enzyme required in sphingolipid synthesis and that *END11* encodes *ERG2* an enzyme involved in ergosterol synthesis (124). These findings showed that mutants affected in lipid biosynthesis are defective in endocytosis.

5.2.1 Sphingosine bases requirement for endocytosis

The *LCB1* gene is essential and encodes a subunit of the serine palmitoyl-transferase enzyme (SPT) (125). SPT catalyses the first step in *de novo* sphingolipid synthesis, the condensation of serine and palmitoyl-CoA to yield the 3-ketosphinganine (KDS) (see Fig. 5) (126). In yeast, sphingosine bases (KDS, PHS, and DHS) are generated by *de novo* synthesis in the ER.

The *end8-1* = *lcb1-100* mutant cells have a temperature-sensitive growth defect. At 24 °C this mutant exhibits an α-factor uptake kinetics that are almost like wild-type cells, but at 37 °C the mutant is clearly defective. Moreover, the *lcb1-100* cells are defective for accumulation of LY in the vacuole at 24 °C and 37 °C and showed a few small vacuoles when viewed by Nomarski optics. In summary, the *lcb1-100* mutant is defective in the internalization step of both R-ME and fluid phase endocytosis at non-permissive temperature (5).

These findings suggest that sphingosine bases might play an important role in the internalization step of endocytosis. To test this hypothesis, sphingosine bases (PHS and DHS) were added externally to the *lcb1-100* strain and the rate of R-ME was determined. Both phytosphingosine (PHS) and dihydrosphingosine (DHS) were able

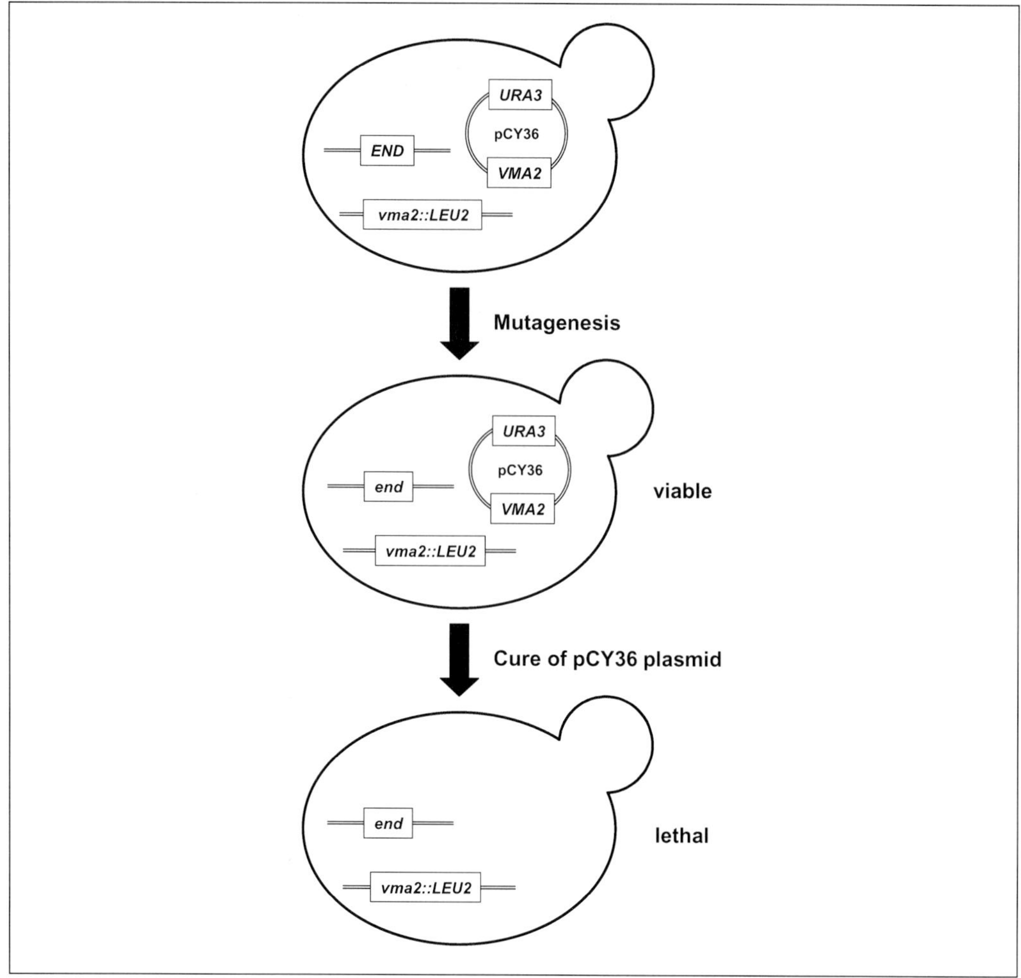

Fig. 4 The *vma2* synthetic lethality mutant screen. The *VMA2* chromosomal gene is disrupted by *LEU2* in a yeast strain carrying the wild-type *VMA2* gene on a *URA3* low copy number plasmid (pCY36). After mutagenesis some of the cells will become endocytosis-deficient (*end*) mutants. These *end* mutants cannot grow without a wild-type copy of *VMA2*, thus cannot lose the pCY36 plasmid. 5-fluoro-orotic acid (5-FOA) is used to screen for *end* mutants, because only the Ura^- cells, which are END^+ can grow on this selective medium.

to suppress the endocytic defect observed in *lcb1-100* cells (148). The use of genetic approaches should help to identify the sphingosine base compound (KDS, DHS, or PHS) that is required for endocytosis in yeast. The Sur2p activity is required to interconvert DHS and PHS in yeast (see Fig. 5) (127). To check whether one of these two compounds is specifically required for the suppression, the double mutant *lcb1 sur2* cells can be tested for α-factor uptake in the presence of DHS or PHS. The recent identification of two *S. cerevisiae* genes encoding sphingosine kinases, *LCB4* and *LCB5* (128), should indicate whether phosphorylated sphingosine bases are required to trigger endocytosis in yeast.

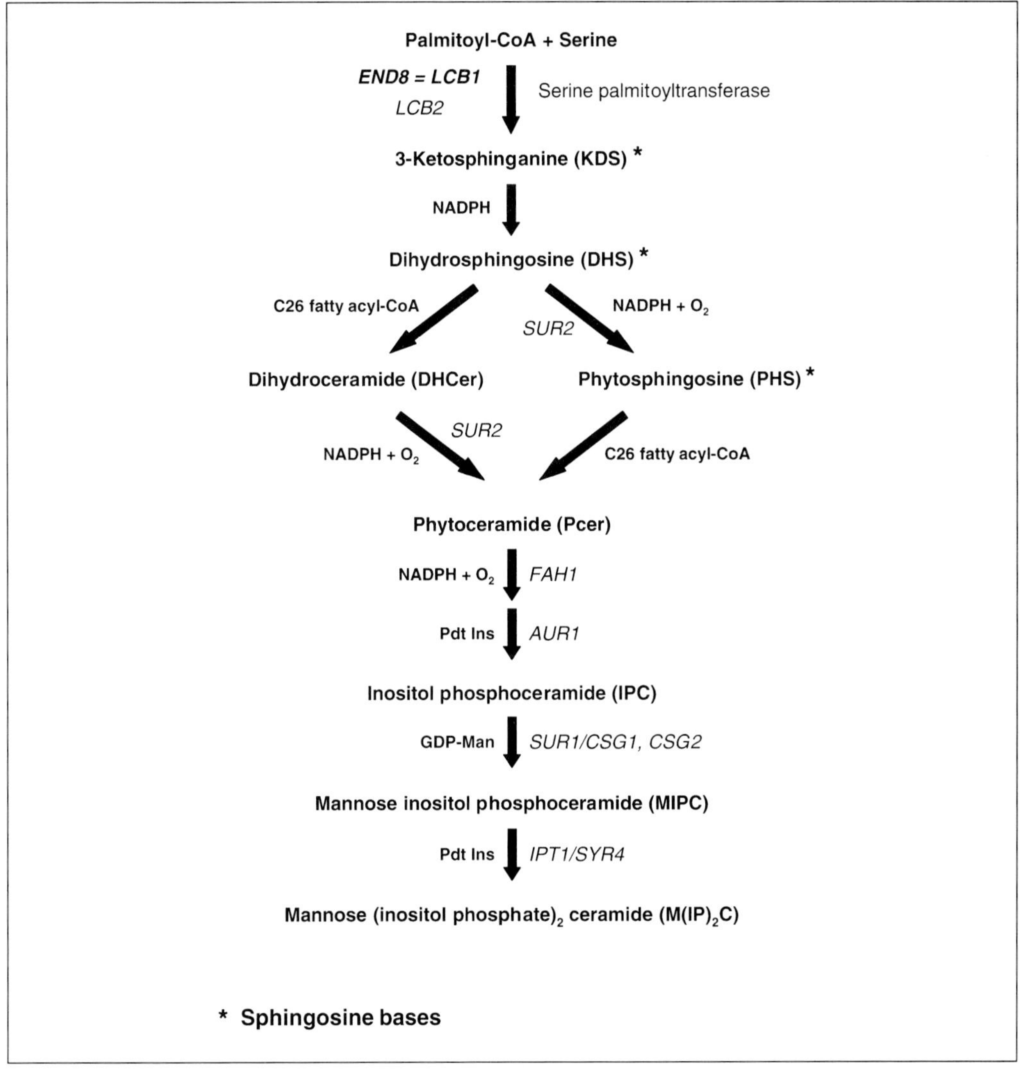

Fig. 5 Sphingolipid biosynthetic pathway in yeast. Known pathway intermediates, substrates, and genes implicated in sphingolipid biosynthesis are indicated. Sphingolipids are characterized by the presence of a long chain fatty acid that is amide-linked to a sphingosine long chain base moiety. Sphingosine long chain bases are indicated.

Sphingosine bases and their phosphorylated derivatives (DHS-1P and PHS-1P) are thought to be signalling molecules for regulating a variety of mammalian cellular processes including cell growth, motility, and apoptosis (129, 130). Sphingolipids and phosphorylated sphingosine bases have also been implicated in the yeast stress response (131–133). Sphingosine bases or ceramides have been proposed to activate protein phosphatases. The best candidate for a ceramide-activated protein phosphatase (CAPP) in yeast is the protein phosphatase 2A (PP2A), which has two regulatory subunits, Cdc55p and Tpd3p (134, 135), and a catalytic subunit. The catalytic subunit of CAPP has been postulated to be Sit4p (136), but three other genes that are functionally overlapping, *PPH21*, *PPH22*, and *PPH3*, encode the major yeast PP2A catalytic activity (137, 138). In a *lcb1 cdc55Δ* double mutant and in a *lcb1 pph21Δ pph22Δ pph3Δ pph21-ts* strain, endocytosis is restored in the absence of sphingosine synthesis (149). Therefore, the sphingosine requirement for endocytosis can be suppressed by PP2A mutations. Furthermore, overexpression of two kinases can suppress the sphingosine requirement for both R-ME and fluid phase endocytosis (149). In summary, the sphingosine requirement for endocytosis can be suppressed by the loss of PP2A activity or by overexpression of two kinases, suggesting a signalling function of sphingosine in the activation of a protein kinase and a protein phosphatase acting sequentially in endocytosis.

5.2.2 Specific sterol requirement for endocytosis

The *end11-1* mutant is defective for the internalization step of endocytosis and has been shown to be allelic to *ERG2* (124). The *ERG2* gene encodes the sterol C-8 isomerase, an enzyme required for one of the late steps in ergosterol synthesis (see Fig. 6) (139). Ergosterol is the principal sterol in yeast and is an essential lipid component for proper function of the membranes (140). The *end11(erg2)-1* mutant is defective for LY accumulation in the vacuole and for α-factor uptake at both 24 °C and 37 °C (5). These results suggest an important role of ergosterol in the internalization step of endocytosis. To test this hypothesis some other *erg* mutants were analysed for endocytosis (124). Yeast cells require sterols for viability. Only the last steps of ergosterol biosynthesis (formation of zymosterol to ergosterol) are not essential and therefore enzymes that are involved in these steps can be mutated (see Fig. 6). The *erg6Δ* and the double mutant *erg2Δ erg6Δ* strains were analysed for α-factor uptake and LY accumulation at both 24 °C and 37 °C. The *erg6Δ* cells exhibited reduced internalization of α-factor at 24 °C and 37 °C and a wild-type LY accumulation in the vacuole, whereas in the *erg2Δ erg6Δ* mutant cells both LY and α-factor uptake were completely defective (124). The *ERG6* gene encodes the sterol C-24 methyl transferase and catalyses the formation of fecosterol that serves as substrate for the Erg2p enzyme (see Fig. 6) (141). One characteristic of the ergosterol biosynthesis pathway is that mutations in the Erg2p and/or Erg6p do not prevent subsequent enzymes from altering the improperly modified substrates (142, 143). Thus these mutations lead to accumulation of sterol structures that are not normal intermediates of the pathway. To correlate the endocytic defects that are observed in *erg2*, *erg6*, and *erg2 erg6* mutants with the sterol composition, the sterols accumulating in these different *erg* mutants were analysed. The results suggested that the state of desaturation of the B-ring may be critical for the internalization step of endocytosis in yeast (124).

Recently, plasma membrane cholesterol was shown to play a critical role in clathrin-coated pit internalization in mammalian cells (144, 145). Cholesterol depletion from the plasma membrane inhibited transferrin and EGF endocytosis, demonstrating an essential role of cholesterol in the formation of CCVs. In summary, sterols were identified as a novel requirement for endocytosis in yeast and in animal

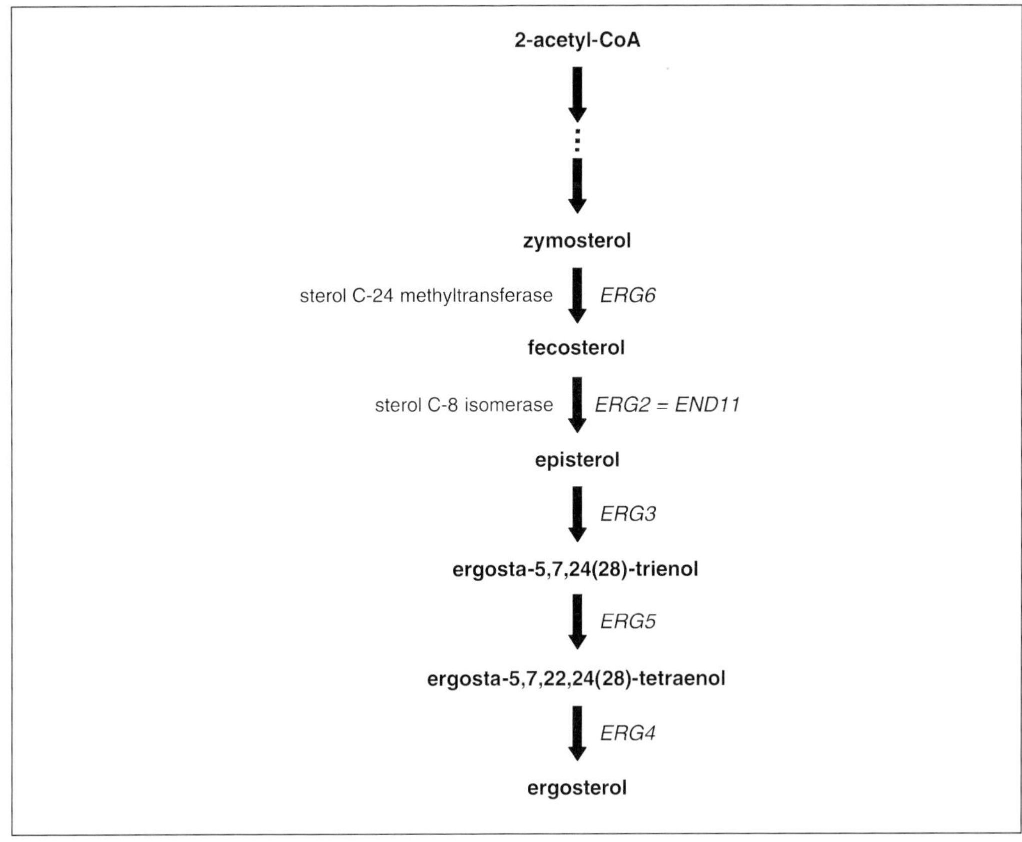

Fig. 6 Ergosterol biosynthesis in yeast. The ergosterol biosynthetic pathway is described from zymosterol to ergosterol. The first steps of this pathway to the mevalonate pathway and the farnesyl pyrophosphate to zymosterol pathway are not described. The last steps of ergosterol synthesis represented here are non-essential.

cells, but the exact role(s) of sterols are unknown up to now. The use of genetic approaches should help to clarify the ergosterol requirement for endocytosis in yeast.

6. Outlook

The studies in yeast have revealed several crucial aspects of the internalization step of R-ME. The actin cytoskeleton plays a major role and several actin-associated protein complexes are involved in the process. A great number of proteins have been isolated based on their role in the internalization step of endocytosis and surely more proteins will be discovered in the future. However, very little is known about the function of these proteins in endocytosis. The Arp2/3 complex has been implicated in actin-nucleation and pointed-end capping but whether these functions are required for endocytosis or if there is an additional endocytic function of the complex is unknown. The type I myosin, Myo5p, is a motor protein and thus may be involved in force generation in endocytosis. However, nothing is known about which step(s) in the endocytic pathway require force and how this force is generated. The major goals for the future are to understand the precise role of the different protein complexes involved in the internalization step, how their functions are regulated, and how they co-operate to drive the internalization step of endocytosis.

The lipid requirement for endocytosis in yeast is still conceptually at an early stage. The link between a sphingosine signalling pathway and the internalization step of endocytosis remains to be defined. The near future should provide significant advances in the identification and characterization of downstream effectors and upstream regulators of the sphingosine activated signalling pathway. Likewise, there is a gap in our understanding of the ergosterol requirement for the internalization step. Analysis of other viable *erg* single and double mutants should permit the development of theories to explain how sterols function in endocytosis.

The combination of genetic and biochemical approaches should help to answer the remaining questions about the protein and lipid requirements in yeast endocytosis.

Acknowledgements

We thank Antje Heese-Peck for critical comments on this manuscript and the other members of the Riezman laboratory for stimulating discussions. This work was funded by the Canton of Basel Stadt and by a grant to H. R. from the Swiss National Science Foundation, and a long-term fellowship from HFSPO to S. F.

References

1. Chvatchko, Y., Howald, I., and Riezman, H. (1986). Two yeast mutants defective in endocytosis are defective in pheromone response. *Cell*, **46**, 355.
2. Wendland, B., McCaffery, J. M., Xiao, Q., and Emr, S. D. (1996) A novel fluorescence-activated cell sorter-based screen for yeast endocytosis mutants identifies a yeast homologue of mammalian eps15. *J. Cell Biol.*, **135**, 1485.
3. Raths, S., Rohrer, J., Crausaz, F., and Riezman, H. (1993) end3 and end4: two mutants defective in receptor-mediated and fluid- phase endocytosis in Saccharomyces cerevisiae. *J. Cell Biol.*, **120**, 55.
4. Davis, N. G., Horecka, J. L., and Sprague, G. F., Jr. (1993) Cis- and trans-acting functions required for endocytosis of the yeast pheromone receptors. *J. Cell Biol.*, **122**, 53.
5. Munn, A. L. and Riezman, H. (1994) Endocytosis is required for the growth of vacuolar H(+)-ATPase- defective yeast: identification of six new END genes. *J. Cell Biol.*, **127**, 373.
6. Herskowitz, I. (1988) Life cycle of the budding yeast Saccharomyces cerevisiae. *Microbiol. Rev.*, **52**, 536.
7. Herskowitz, I. (1989) A regulatory hierarchy for cell specialization in yeast. *Nature*, **342**, 749.
8. Marsh, L., Neiman, A. M., and Herskowitz, I. (1991) Signal transduction during pheromone response in yeast. *Annu. Rev. Cell Biol.*, **7**, 699.

9. Kurjan, J. (1992) Pheromone response in yeast. *Annu. Rev. Biochem.*, **61**, 1097.
10. Jenness, D. D. and Spatrick, P. (1986) Down regulation of the alpha-factor pheromone receptor in S. cerevisiae. *Cell*, **46**, 345.
11. Zanolari, B. and Riezman, H. (1991) Quantitation of alpha-factor internalization and response during the Saccharomyces cerevisiae cell cycle. *Mol. Cell. Biol.*, **11**, 5251.
12. Riezman, H. (1998) Down regulation of yeast G protein-coupled receptors. *Semin. Cell. Dev. Biol.*, **9**, 129.
13. Rohrer, J., Benedetti, H., Zanolari, B., and Riezman, H. (1993) Identification of a novel sequence mediating regulated endocytosis of the G protein-coupled alpha-pheromone receptor in yeast. *Mol. Biol. Cell*, **4**, 511.
14. Hicke, L. and Riezman, H. (1996) Ubiquitination of a yeast plasma membrane receptor signals its ligand- stimulated endocytosis. *Cell*, **84**, 277.
15. Hicke, L. (1997) Ubiquitin-dependent internalization and down-regulation of plasma membrane proteins. *FASEB J.*, **11**, 1215.
16. Hicke, L. (1999) Gettin' down with ubiquitin: turning off cell-surface receptors, transporters and channels. *Trends Cell Biol.*, **9**, 107.
17. Tan, P. K., Howard, J. P., and Payne, G. S. (1996) The sequence NPFXD defines a new class of endocytosis signal in Saccharomyces cerevisiae. *J. Cell Biol.*, **135**, 1789.
18. Gruenberg, J. and Maxfield, F. R. (1995) Membrane transport in the endocytic pathway. *Curr. Opin. Cell Biol.*, **7**, 552.
19. Mellman, I. (1996) Endocytosis and molecular sorting. *Annu. Rev. Cell Dev. Biol.*, **12**, 575.
20. Singer-Krüger, B., Frank, R., Crausaz, F., and Riezman, H. (1993) Partial purification and characterization of early and late endosomes from yeast. Identification of four novel proteins. *J. Biol. Chem.*, **268**, 14376.
21. Hicke, L., Zanolari, B., Pypaert, M., Rohrer, J., and Riezman, H. (1997) Transport through the yeast endocytic pathway occurs through morphologically distinct compartments and requires an active secretory pathway and Sec18p/N-ethylmaleimide-sensitive fusion protein. *Mol. Biol. Cell*, **8**, 13.
22. Prescianotto-Baschong, C. and Riezman, H. (1998) Morphology of the yeast endocytic pathway. *Mol. Biol. Cell*, **9**, 173.
23. Piper, R. C., Cooper, A. A., Yang, H., and Stevens, T. H. (1995) VPS27 controls vacuolar and endocytic traffic through a prevacuolar compartment in Saccharomyces cerevisiae. *J. Cell Biol.*, **131**, 603.
24. Wichmann, H., Hengst, L., and Gallwitz, D. (1992) Endocytosis in yeast: evidence for the involvement of a small GTP- binding protein (Ypt7p). *Cell*, **71**, 1131.
25. Schimmöller, F., and Riezman, H. (1993) Involvement of Ypt7p, a small GTPase, in traffic from late endosome to the vacuole in yeast. *J. Cell Sci.*, **106**, 823.
26. Singer-Krüger, B., Stenmark, H., Dusterhoft, A., Philippsen, P., Yoo, J. S., Gallwitz, D., *et al.* (1994) Role of three rab5-like GTPases, Ypt51p, Ypt52p, and Ypt53p, in the endocytic and vacuolar protein sorting pathways of yeast. *J. Cell Biol.*, **125**, 283.
27. Singer-Krüger, B., Stenmark, H., and Zerial, M. (1995) Yeast Ypt51p and mammalian Rab5: counterparts with similar function in the early endocytic pathway. *J. Cell Sci.*, **108**, 3509.
28. Nichols, B. J., Ungermann, C., Pelham, H. R., Wickner, W. T., and Haas, A. (1997) Homotypic vacuolar fusion mediated by t- and v-SNAREs. *Nature*, **387**, 199.
29. Wada, Y., Nakamura, N., Ohsumi, Y., and Hirata, A. (1997) Vam3p, a new member of syntaxin related protein, is required for vacuolar assembly in the yeast Saccharomyces cerevisiae. *J. Cell Sci.*, **110**, 1299.

30. Becherer, K. A., Rieder, S. E., Emr, S. D., and Jones, E. W. (1996) Novel syntaxin homologue, Pep12p, required for the sorting of lumenal hydrolases to the lysosome-like vacuole in yeast. *Mol. Biol. Cell*, **7**, 579.
31. Abeliovich, H., Grote, E., Novick, P., and Ferro-Novick, S. (1998) Tlg2p, a yeast syntaxin homolog that resides on the Golgi and endocytic structures. *J. Biol. Chem.*, **273**, 11719.
32. Holthuis, J. C., Nichols, B. J., Dhruvakumar, S., and Pelham, H. R. (1998) Two syntaxin homologues in the TGN/endosomal system of yeast. *EMBO J.*, **17**, 113.
33. Holthuis, J. C. M., Nichols, B. J., and Pelham, H. R. B. (1998) The syntaxin Tlg1p mediates trafficking of chitin synthase III to polarized growth sites in yeast. *Mol. Biol. Cell*, **9**, 3383.
34. Seron, K., Tieaho, V., Prescianotto-Baschong, C., Aust, T., Blondel, M. O., Guillaud, P., *et al.* (1998) A yeast t-SNARE involved in endocytosis. *Mol. Biol. Cell*, **9**, 2873.
35. Coe, J. G., Lim, A. C., Xu, J., and Hong, W. (1999) A role for Tlg1p in the transport of proteins within the Golgi apparatus of Saccharomyces cerevisiae. *Mol. Biol. Cell*, **10**, 2407.
36. Kübler, E. and Riezman, H. (1993) Actin and fimbrin are required for the internalization step of endocytosis in yeast. *EMBO J.*, **12**, 2855.
37. Munn, A. L., Stevenson, B. J., Geli, M. I., and Riezman, H. (1995) end5, end6, and end7: mutations that cause actin delocalization and block the internalization step of endocytosis in Saccharomyces cerevisiae. *Mol. Biol. Cell*, **6**, 1721.
38. Givan, S. A. and Sprague, G. F., Jr. (1997) The ankyrin repeat-containing protein Akr1p is required for the endocytosis of yeast pheromone receptors. *Mol. Biol. Cell*, **8**, 1317.
39. Moreau, V., Galan, J. M., Devilliers, G., Haguenauer-Tsapis, R., and Winsor, B. (1997) The yeast actin-related protein Arp2p is required for the internalization step of endocytosis. *Mol. Biol. Cell*, **8**, 1361.
40. Tan, P. K., Davis, N. G., Sprague, G. F., and Payne, G. S. (1993) Clathrin facilitates the internalization of seven transmembrane segment receptors for mating pheromones in yeast. *J. Cell Biol.*, **123**, 1707.
41. Chu, D. S., Pishvaee, B., and Payne, G. S. (1996) The light chain subunit is required for clathrin function in Saccharomyces cerevisiae. *J. Biol. Chem.*, **271**, 33123.
42. Huang, K. M., Gullberg, L., Nelson, K. K., Stefan, C. J., Blumer, K., and Lemmon, S. K. (1997) Novel functions of clathrin light chains: clathrin heavy chain trimerization is defective in light chain-deficient yeast. *J. Cell Sci.*, **110**, 899.
43. Kübler, E., Schimmöller, F., and Riezman, H. (1994) Calcium-independent calmodulin requirement for endocytosis in yeast. *EMBO J.*, **13**, 5539.
44. Benedetti, H., Raths, S., Crausaz, F., and Riezman, H. (1994) The END3 gene encodes a protein that is required for the internalization step of endocytosis and for actin cytoskeleton organization in yeast. *Mol. Biol. Cell*, **5**, 1023.
45. Madania, A., Dumoulin, P., Grava, S., Kitamoto, H., Scharer-Brodbeck, C., Soulard, A., *et al.* (1999) The saccharomyces cerevisiae homologue of human wiskott-aldrich syndrome protein las17p interacts with the Arp2/3 complex. *Mol. Biol. Cell*, **10**, 3521.
46. Geli, M. I. and Riezman, H. (1996) Role of type I myosins in receptor-mediated endocytosis in yeast. *Science*, **272**, 533.
47. Tang, H. Y., Munn, A., and Cai, M. (1997) EH domain proteins Pan1p and End3p are components of a complex that plays a dual role in organization of the cortical actin cytoskeleton and endocytosis in Saccharomyces cerevisiae. *Mol. Cell. Biol.*, **17**, 4294.
48. Galan, J. M., Moreau, V., Andre, B., Volland, C., and Haguenauer-Tsapis, R. (1996) Ubiquitination mediated by the Npi1p/Rsp5p ubiquitin-protein ligase is required for endocytosis of the yeast uracil permease. *J. Biol. Chem.*, **271**, 10946.
49. Springael, J. Y., De Craene, J. O., and Andre, B. (1999) The yeast Npi1/Rsp5 ubiquitin

ligase lacking its N-terminal C2 domain is competent for ubiquitination but not for subsequent endocytosis of the gap1 permease. *Biochem. Biophys. Res. Commun.*, **257**, 561.
50. Singer-Krüger, B., Nemoto, Y., Daniell, L., Ferro-Novick, S., and De Camilli, P. (1998) Synaptojanin family members are implicated in endocytic membrane traffic in yeast. *J. Cell Sci.*, **111**, 3347.
51. Wesp, A., Hicke, L., Palecek, J., Lombardi, R., Aust, T., Munn, A. L., *et al.* (1997) End4p/Sla2p interacts with actin-associated proteins for endocytosis in Saccharomyces cerevisiae. *Mol. Biol. Cell*, **8**, 2291.
52. Riezman, H. (1985) Endocytosis in yeast: several of the yeast secretory mutants are defective in endocytosis. *Cell*, **40**, 1001.
53. Vida, T. A. and Emr, S. D. (1995) A new vital stain for visualizing vacuolar membrane dynamics and endocytosis in yeast. *J. Cell Biol.*, **128**, 779.
54. Volland, C., Urban-Grimal, D., Geraud, G., and Haguenauer-Tsapis, R. (1994) Endocytosis and degradation of the yeast uracil permease under adverse conditions. *J. Biol. Chem.*, **269**, 9833.
55. Riballo, E., Herweijer, M., Wolf, D. H., and Lagunas, R. (1995) Catabolite inactivation of the yeast maltose transporter occurs in the vacuole after internalization by endocytosis. *J. Bacteriol.*, **177**, 5622.
56. Lai, K., Bolognese, C. P., Swift, S., and McGraw, P. (1995) Regulation of inositol transport in Saccharomyces cerevisiae involves inositol-induced changes in permease stability and endocytic degradation in the vacuole. *J. Biol. Chem.*, **270**, 2525.
57. Berkower, C., Loayza, D., and Michaelis, S. (1994) Metabolic instability and constitutive endocytosis of STE6, the a- factor transporter of Saccharomyces cerevisiae. *Mol. Biol. Cell*, **5**, 1185.
58. Swanson, J. A., Yirinec, B. D., and Silverstein, S. C. (1985) Phorbol esters and horseradish peroxidase stimulate pinocytosis and redirect the flow of pinocytosed fluid in macrophages. *J. Cell Biol.*, **100**, 851.
59. Jenness, D. D., Burkholder, A. C., and Hartwell, L. H. (1983) Binding of alpha-factor pheromone to yeast a cells: chemical and genetic evidence for an alpha-factor receptor. *Cell*, **35**, 521.
60. Singer, B. and Riezman, H. (1990) Detection of an intermediate compartment involved in transport of alpha- factor from the plasma membrane to the vacuole in yeast. *J. Cell Biol.*, **110**, 1911.
61. Schandel, K. A. and Jenness, D. D. (1994) Direct evidence for ligand-induced internalization of the yeast alpha- factor pheromone receptor. *Mol. Cell. Biol.*, **14**, 7245.
62. Dulic, V., Egerton, M., Elguindi, I., Raths, S., Singer, B., and Riezman, H. (1991) Yeast endocytosis assays. *Methods Enzymol.*, **194**, 697.
63. Beck, T., Schmidt, A., and Hall, M. N. (1999) Starvation induces vacuolar targeting and degradation of the tryptophan permease in yeast. *J. Cell Biol.*, **146**, 1227.
64. Seron, K., Blondel, M. O., Haguenauer-Tsapis, R., and Volland, C. (1999) Uracil-induced down-regulation of the yeast uracil permease. *J. Bacteriol.*, **181**, 1793.
65. Kolling, R. and Hollenberg, C. P. (1994) The ABC-transporter Ste6 accumulates in the plasma membrane in a ubiquitinated form in endocytosis mutants. *EMBO J.*, **13**, 3261.
66. Egner, R., Mahe, Y., Pandjaitan, R., and Kuchler, K. (1995) Endocytosis and vacuolar degradation of the plasma membrane-localized Pdr5 ATP-binding cassette multidrug transporter in Saccharomyces cerevisiae. *Mol. Cell. Biol.*, **15**, 5879.
67. Weber, E., Jund, R., and Chevallier, M. R. (1986) Chromosomal mapping of the uracil permease gene of Saccharomyces cerevisiae. *Curr. Genet.*, **11**, 93.

68. Volland, C., Garnier, C., and Haguenauer-Tsapis, R. (1992) In vivo phosphorylation of the yeast uracil permease. *J. Biol. Chem.*, **267**, 23767.
69. Penalver, E., Ojeda, L., Moreno, E., and Lagunas, R. (1997) Role of the cytoskeleton in endocytosis of the yeast maltose transporter. *Yeast*, **13**, 541.
70. Lucero, P. and Lagunas, R. (1997) Catabolite inactivation of the yeast maltose transporter requires ubiquitin-ligase npi1/rsp5 and ubiquitin-hydrolase npi2/doa4. *FEMS Microbiol. Lett.*, **147**, 273.
71. Penalver, E., Lucero, P., Moreno, E., and Lagunas, R. (1999) Clathrin and two components of the COPII complex, Sec23p and Sec24p, could be involved in endocytosis of the Saccharomyces cerevisiae maltose transporter. *J. Bacteriol.*, **181**, 2555.
72. Amberg, D. C. (1998) Three-dimensional imaging of the yeast actin cytoskeleton through the budding cell cycle. *Mol. Biol. Cell*, **9**, 3259.
73. Mulholland, J., Preuss, D., Moon, A., Wong, A., Drubin, D., and Botstein, D. (1994) Ultrastructure of the yeast actin cytoskeleton and its association with the plasma membrane. *J. Cell Biol.*, **125**, 381.
74. Mulholland, J., Konopka, J., Singer-Kruger, B., Zerial, M., and Botstein, D. (1999) Visualization of receptor-mediated endocytosis in yeast. *Mol. Biol. Cell*, **10**, 799.
75. Ayscough, K. R., Stryker, J., Pokala, N., Sanders, M., Crews, P., and Drubin, D. G. (1997) High rates of actin filament turnover in budding yeast and roles for actin in establishment and maintenance of cell polarity revealed using the actin inhibitor latrunculin-A. *J. Cell Biol.*, **137**, 399.
76. Lappalainen, P. and Drubin, D. G. (1997) Cofilin promotes rapid actin filament turnover in vivo. *Nature*, **388**, 78.
77. Riezman, H., Munn, A., Geli, M. I., and Hicke, L. (1996) Actin-, myosin- and ubiquitin-dependent endocytosis. *Experientia*, **52**, 1033.
78. Geli, M. I. and Riezman, H. (1998) Endocytic internalization in yeast and animal cells: similar and different. *J. Cell Sci.*, **111**, 1031.
79. Lamaze, C., Fujimoto, L. M., Yin, H. L., and Schmid, S. L. (1997) The actin cytoskeleton is required for receptor-mediated endocytosis in mammalian cells. *J. Biol. Chem.*, **272**, 20332.
80. Mermall, V., Post, P. L., and Mooseker, M. S. (1998) Unconventional myosins in cell movement, membrane traffic, and signal transduction. *Science*, **279**, 527.
81. Goodson, H. V., Anderson, B. L., Warrick, H. M., Pon, L. A., and Spudich, J. A. (1996) Synthetic lethality screen identifies a novel yeast myosin I gene (MYO5): myosin I proteins are required for polarization of the actin cytoskeleton. *J. Cell Biol.*, **133**, 1277.
82. Geli, M. I., Wesp, A., and Riezman, H. (1998) Distinct functions of calmodulin are required for the uptake step of receptor-mediated endocytosis in yeast: the type I myosin Myo5p is one of the calmodulin targets. *EMBO J.*, **17**, 635.
83. Anderson, B. L., Boldogh, I., Evangelista, M., Boone, C., Greene, L. A., and Pon, L. A. (1998) The Src homology domain 3 (SH3) of a yeast type I myosin, Myo5p, binds to verprolin and is required for targeting to sites of actin polarization. *J. Cell Biol.*, **141**, 1357.
84. Vaduva, G., Martin, N. C., and Hopper, A. K. (1997) Actin-binding verprolin is a polarity development protein required for the morphogenesis and function of the yeast actin cytoskeleton. *J. Cell Biol.*, **139**, 1821.
85. Machesky, L. M. and Gould, K. L. (1999) The Arp2/3 complex: a multifunctional actin organizer. *Curr. Opin. Cell Biol.*, **11**, 117.
86. Winter, D., Lechler, T., and Li, R. (1999) Activation of the yeast Arp2/3 complex by Bee1p, a WASP-family protein. *Curr. Biol.*, **9**, 501.

87. Naqvi, S. N., Zahn, R., Mitchell, D. A., Stevenson, B. J., and Munn, A. L. (1998) The WASp homologue Las17p functions with the WIP homologue End5p / verprolin and is essential for endocytosis in yeast. *Curr. Biol.*, **8**, 959.
88. Ramesh, N., Anton, I. M., Hartwig, J. H., and Geha, R. S. (1997) WIP, a protein associated with wiskott-aldrich syndrome protein, induces actin polymerization and redistribution in lymphoid cells. *Proc. Natl. Acad. Sci. USA*, **94**, 14671.
89. Wigge, P. and McMahon, H. T. (1998) The amphiphysin family of proteins and their role in endocytosis at the synapse. *Trends Neurosci.*, **21**, 339.
90. Navarro, P., Durrens, P., and Aigle, M. (1997) Protein-protein interaction between the RVS161 and RVS167 gene products of Saccharomyces cerevisiae. *Biochim. Biophys. Acta*, **1343**, 187.
91. Brizzio, V., Gammie, A. E., and Rose, M. D. (1998) Rvs161p interacts with Fus2p to promote cell fusion in Saccharomyces cerevisiae. *J. Cell Biol.*, **141**, 567.
92. Balguerie, A., Sivadon, P., Bonneu, M., and Aigle, M. (1999) Rvs167p, the budding yeast homolog of amphiphysin, colocalizes with actin patches. *J. Cell Sci.*, **112**, 2529.
93. Lee, J., Colwill, K., Aneliunas, V., Tennyson, C., Moore, L., Ho, Y., *et al.* (1998) Interaction of yeast Rvs167 and Pho85 cyclin-dependent kinase complexes may link the cell cycle to the actin cytoskeleton. *Curr. Biol.*, **8**, 1310.
94. Amberg, D. C., Basart, E., and Botstein, D. (1995) Defining protein interactions with yeast actin in vivo. *Nat. Struct. Biol.*, **2**, 28.
95. Colwill, K., Field, D., Moore, L., Friesen, J., and Andrews, B. (1999) In vivo analysis of the domains of yeast Rvs167p suggests Rvs167p function is mediated through multiple protein interactions. *Genetics*, **152**, 881.
96. Lila, T. and Drubin, D. G. (1997) Evidence for physical and functional interactions among two Saccharomyces cerevisiae SH3 domain proteins, an adenylyl cyclase- associated protein and the actin cytoskeleton. *Mol. Biol. Cell*, **8**, 367.
97. Holtzman, D. A., Yang, S., and Drubin, D. G. (1993) Synthetic-lethal interactions identify two novel genes, SLA1 and SLA2, that control membrane cytoskeleton assembly in Saccharomyces cerevisiae. *J. Cell Biol.*, **122**, 635.
98. Freeman, N. L., Lila, T., Mintzer, K. A., Chen, Z., Pahk, A. J., Ren, R., *et al.* (1996) A conserved proline-rich region of the Saccharomyces cerevisiae cyclase- associated protein binds SH3 domains and modulates cytoskeletal localization. *Mol. Cell. Biol.*, **16**, 548.
99. Wong, W. T., Schumacher, C., Salcini, A. E., Romano, A., Castagnino, P., Pelicci, P. G., *et al.* (1995) A protein-binding domain, EH, identified in the receptor tyrosine kinase substrate Eps15 and conserved in evolution. *Proc. Natl. Acad. Sci. USA*, **92**, 9530.
100. Zeng, G. and Cai, M. (1999) Regulation of the actin cytoskeleton organization in yeast by a novel serine / threonine kinase Prk1p. *J. Cell Biol.*, **144**, 71.
101. Cope, M. J., Yang, S., Shang, C., and Drubin, D. G. (1999) Novel protein kinases Ark1p and Prk1p associate with and regulate the cortical actin cytoskeleton in budding yeast. *J. Cell Biol.*, **144**, 1203.
102. Wendland, B. and Emr, S. D. (1998) Pan1p, yeast eps15, functions as a multivalent adaptor that coordinates protein-protein interactions essential for endocytosis. *J. Cell Biol.*, **141**, 71.
103. Huang, K. M., D'Hondt, K., Riezman, H., and Lemmon, S. K. (1999) Clathrin functions in the absence of heterotetrameric adaptors and AP180-related proteins in yeast. *EMBO J.*, **18**, 3897.
104. Zolladek, T., Tobiasz, A., Vaduva, G., Boguta, M., Martin, N. C., and Hopper, A. K. (1997) MDP1, a Saccharomyces cerevisiae gene involved in mitochondrial / cytoplasmic protein distribution, is identical to the ubiquitin-protein ligase gene RSP5. *Genetics*, **145**, 595.

105. Wendland, B., Steece, K. E., and Emr, S. D. (1999) Yeast epsins contain an essential N-terminal ENTH domain, bind clathrin and are required for endocytosis. *EMBO J.*, **18**, 4383.
106. Schmid, S. L. (1997) Clathrin-coated vesicle formation and protein sorting: an integrated process. *Annu. Rev. Biochem.*, **66**, 511.
107. Hirst, J. and Robinson, M. S. (1998) Clathrin and adaptors. *Biochim. Biophys. Acta*, **1404**, 173.
108. Lamaze, C. and Schmid, S. L. (1995) The emergence of clathrin-independent pinocytic pathways. *Curr. Opin. Cell Biol.*, **7**, 573.
109. Damke, H., Baba, T., van der Bliek, A. M., and Schmid, S. L. (1995) Clathrin-independent pinocytosis is induced in cells overexpressing a temperature-sensitive mutant of dynamin. *J. Cell Biol.*, **131**, 69.
110. Payne, G. S., Baker, D., van Tuinen, E., and Schekman, R. (1988) Protein transport to the vacuole and receptor-mediated endocytosis by clathrin heavy chain-deficient yeast. *J. Cell Biol.*, **106**, 1453.
111. Goodman, O. B., Jr., Krupnick, J. G., Santini, F., Gurevich, V. V., Penn, R. B., Gagnon, A. W., *et al.* (1996) Beta-arrestin acts as a clathrin adaptor in endocytosis of the beta2- adrenergic receptor. *Nature*, **383**, 447.
112. Schmid, S. L., McNiven, M. A., and De Camilli, P. (1998) Dynamin and its partners: a progress report. *Curr. Opin. Cell Biol.*, **10**, 504.
113. Schu, P. V., Takegawa, K., Fry, M. J., Stack, J. H., Waterfield, M. D., and Emr, S. D. (1993) Phosphatidylinositol 3-kinase encoded by yeast VPS34 gene essential for protein sorting. *Science*, **260**, 88.
114. Wurmser, A. E. and Emr, S. D. (1998) Phosphoinositide signaling and turnover: PtdIns(3)P, a regulator of membrane traffic, is transported to the vacuole and degraded by a process that requires lumenal vacuolar hydrolase activities. *EMBO J.*, **17**, 4930.
115. Bankaitis, V. A., Aitken, J. R., Cleves, A. E., and Dowhan, W. (1990) An essential role for a phospholipid transfer protein in yeast Golgi function. *Nature*, **347**, 561.
116. Cleves, A. E., McGee, T. P., Whitters, E. A., Champion, K. M., Aitken, J. R., Dowhan, W., *et al.* (1991) Mutations in the CDP-choline pathway for phospholipid biosynthesis bypass the requirement for an essential phospholipid transfer protein. *Cell*, **64**, 789.
117. Kearns, B. G., McGee, T. P., Mayinger, P., Gedvilaite, A., Phillips, S. E., Kagiwada, S., *et al.* (1997) Essential role for diacylglycerol in protein transport from the yeast Golgi complex. *Nature*, **387**, 101.
118. Phillips, S. E., Sha, B., Topalof, L., Xie, Z., Alb, J. G., Klenchin, V. A., *et al.* (1999) Yeast Sec14p deficient in phosphatidylinositol transfer activity is functional in vivo. *Mol. Cell*, **4**, 187.
119. Horvath, A., Sutterlin, C., Manning-Krieg, U., Movva, N. R., and Riezman, H. (1994) Ceramide synthesis enhances transport of GPI-anchored proteins to the Golgi apparatus in yeast. *EMBO J.*, **13**, 3687.
120. Sutterlin, C., Doering, T. L., Schimmoller, F., Schroder, S., and Riezman, H. (1997) Specific requirements for the ER to Golgi transport of GPI-anchored proteins in yeast. *J. Cell Sci.*, **110**, 2703.
121. Skrzypek, M., Lester, R. L., and Dickson, R. C. (1997) Suppressor gene analysis reveals an essential role for sphingolipids in transport of glycosylphosphatidylinositol-anchored proteins in Saccharomyces cerevisiae. *J. Bacteriol.*, **179**, 1513.
122. David, D., Sundarababu, S., and Gerst, J. E. (1998) Involvement of long chain fatty acid elongation in the trafficking of secretory vesicles in yeast. *J. Cell Biol.*, **143**, 1167.

123. Yamashiro, C. T., Kane, P. M., Wolczyk, D. F., Preston, R. A., and Stevens, T. H. (1990) Role of vacuolar acidification in protein sorting and zymogen activation: a genetic analysis of the yeast vacuolar proton- translocating ATPase. *Mol. Cell. Biol.*, **10**, 3737.
124. Munn, A. L., Heese-Peck, A., Stevenson, B. J., Pichler, H., and Riezman, H. (1999) Specific sterols required for the internalization step of endocytosis in yeast. *Mol. Biol. Cell*, **10**, 3943.
125. Buede, R., Rinker-Schaffer, C., Pinto, W. J., Lester, R. L., and Dickson, R. C. (1991) Cloning and characterization of LCB1, a Saccharomyces gene required for biosynthesis of the long-chain base component of sphingolipids. *J. Bacteriol.*, **173**, 4325.
126. Nagiec, M. M., Baltisberger, J. A., Wells, G. B., Lester, R. L., and Dickson, R. C. (1994) The LCB2 gene of Saccharomyces and the related LCB1 gene encode subunits of serine palmitoyltransferase, the initial enzyme in sphingolipid synthesis. *Proc. Natl. Acad. Sci. USA*, **91**, 7899.
127. Haak, D., Gable, K., Beeler, T., and Dunn, T. (1997) Hydroxylation of Saccharomyces cerevisiae ceramides requires Sur2p and Scs7p. *J. Biol. Chem.*, **272**, 29704.
128. Nagiec, M. M., Skrzypek, M., Nagiec, E. E., Lester, R. L., and Dickson, R. C. (1998) The LCB4 (YOR171c) and LCB5 (YLR260w) genes of Saccharomyces encode sphingoid long chain base kinases. *J. Biol. Chem.*, **273**, 19437.
129. Hannun, Y. A. (1996) Functions of ceramide in coordinating cellular responses to stress. *Science*, **274**, 1855.
130. Perry, D. K. and Hannun, Y. A. (1998) The role of ceramide in cell signaling. *Biochim. Biophys. Acta*, 1**436**, 233.
131. Jenkins, G. M., Richards, A., Wahl, T., Mao, C., Obeid, L., and Hannun, Y. (1997) Involvement of yeast sphingolipids in the heat stress response of Saccharomyces cerevisiae. *J. Biol. Chem.*, **272**, 32566.
132. Mandala, S. M., Thornton, R., Tu, Z., Kurtz, M. B., Nickels, J., Broach, J., *et al.* (1998) Sphingoid base 1-phosphate phosphatase: a key regulator of sphingolipid metabolism and stress response. *Proc. Natl. Acad. Sci. USA*, **95**, 150.
133. Skrzypek, M. S., Nagiec, M. M., Lester, R. L., and Dickson, R. C. (1999) Analysis of phosphorylated sphingolipid long-chain bases reveals potential roles in heat stress and growth control in Saccharomyces. *J. Bacteriol.*, **181**, 1134.
134. Healy, A. M., Zolnierowicz, S., Stapleton, A. E., Goebl, M., DePaoli-Roach, A. A., and Pringle, J. R. (1991) CDC55, a Saccharomyces cerevisiae gene involved in cellular morphogenesis: identification, characterization, and homology to the B subunit of mammalian type 2A protein phosphatase. *Mol. Cell. Biol.*, **11**, 5767.
135. van Zyl, W., Huang, W., Sneddon, A. A., Stark, M., Camier, S., Werner, M., *et al.* (1992) Inactivation of the protein phosphatase 2A regulatory subunit A results in morphological and transcriptional defects in Saccharomyces cerevisiae. *Mol. Cell. Biol.*, **12**, 4946.
136. Nickels, J. T. and Broach, J. R. (1996) A ceramide-activated protein phosphatase mediates ceramide-induced G1 arrest of Saccharomyces cerevisiae. *Genes Dev.*, **10**, 382.
137. Sneddon, A. A., Cohen, P. T., and Stark, M. J. (1990) Saccharomyces cerevisiae protein phosphatase 2A performs an essential cellular function and is encoded by two genes. *EMBO J.*, **9**, 4339.
138. Ronne, H., Carlberg, M., Hu, G. Z., and Nehlin, J. O. (1991) Protein phosphatase 2A in Saccharomyces cerevisiae: effects on cell growth and bud morphogenesis. *Mol. Cell. Biol.*, **11**, 4876.
139. Arthington, B. A., Hoskins, J., Skatrud, P. L., and Bard, M. (1991) Nucleotide sequence of the gene encoding yeast C-8 sterol isomerase. *Gene*, **107**, 173.

140. Daum, G., Lees, N. D., Bard, M., and Dickson, R. (1998) Biochemistry, cell biology and molecular biology of lipids of Saccharomyces cerevisiae. *Yeast*, **14**, 1471.
141. McCammon, M. T., Hartmann, M. A., Bottema, C. D., and Parks, L. W. (1984) Sterol methylation in Saccharomyces cerevisiae. *J. Bacteriol.*, **157**, 475.
142. Parks, L. W. and Casey, W. M. (1995) Physiological implications of sterol biosynthesis in yeast. *Annu. Rev. Microbiol.*, **49**, 95.
143. Lees, N. D., Skaggs, B., Kirsch, D. R., and Bard, M. (1995) Cloning of the late genes in the ergosterol biosynthetic pathway of Saccharomyces cerevisiae—a review. *Lipids*, **30**, 221.
144. Subtil, A., Gaidarov, I., Kobylarz, K., Lampson, M. A., Keen, J. H., and McGraw, T. E. (1999) Acute cholesterol depletion inhibits clathrin-coated pit budding. *Proc. Natl. Acad. Sci. USA*, **96**, 6775.
145. Rodal, S. K., Skretting, G., Garred, O., Vilhardt, F., van Deurs, B., and Sandvig, K. (1999) Extraction of cholesterol with methyl-beta-cyclodextrin perturbs formation of clathrin-coated endocytic vesicles. *Mol. Biol. Cell*, **10**, 961.
146. Fujimoto, L. M., Roth, R., Heuser, J. E., and Schmid, S. L. (2000) Actin assembly plays a variable, but not obligatory role in receptor-mediated endocytosis. *Traffic*, **1**, 161.
147. Schaerer-Brodbeck, C. and Riezman, H. (2000) Functional interaction between the p35 subunit of the Arp2/3 complex and calmodulin in yeast. *Mol. Cell. Biol.*, **11**, 1113.
148. Zanolari, B., Friant, S., Funato, K., Sütterlin, C., Stevenson, B. J., and Riezman, H. (2000) Sphingoid base synthesis requirement for endocytosis in *Saccharomyces cerevisiae*. *EMBO J.*, **19**, 2824.
149. Friant, S., Zanolari, B., and Riezman, H. (2000) Increased protein kinase or decreased PP2A activity bypasses sphingoid base requirement in endocytosis. *EMBO J.*, **19**, 2834.

11 | Endocytosis in pathogen entry and replication

DAVID G. RUSSELL and MARK MARSH

1. Introduction

The endocytic network in eukaryotic cells has evolved to fulfil many diverse functions dealing predominantly with either the acquisition of nutrients or communication with the extracellular environment. To potential pathogens, however, the endocytic network represents a virtually irresistible route into the cell. Uptake of infectious agents by endocytosis or phagocytosis is a frequent pathway of infection. While pathogens have evolved to exploit this route of entry, higher eukaryotes have in turn developed a complex interplay between the endocytic network and their immune system to control infection. Infectious microbes are once again proving to be extremely interesting probes to elucidate the 'checks and balances' that regulate function during passage from endosomes to lysosomes. In this review we will consider the mechanisms through which viral, bacterial, and protozoan intracellular pathogens exploit endocytic pathways to enter cells, and how these agents are providing new insights to the interface between endocytosis and the immune system.

2. Viral interactions with the endocytic pathway

Cellular membranes pose a major barrier to virus infection. All viruses must deliver their genome to the cytoplasm or nucleus of a potential host without damaging the integrity of the plasma membrane or compromising the viability of the target cell. To this end viruses have evolved multiple mechanisms to penetrate their targets. Enveloped (membrane-containing) viruses use membrane fusion, and non-enveloped viruses use various mechanisms, most of which are still poorly understood. These reactions can occur either at the cell surface or following endocytosis. Virus uptake into endocytic vesicles has been known since electron microscopy first allowed detailed observations of cell ultrastructure (1). However, a clear role for endocytosis in viral infection was not established until the early 1980s, when work with several enveloped viruses showed that viral fusion could be activated by the acidic conditions found in endocytic organelles. Subsequently, it became clear that endocytosis and related cellular functions, such as transcytosis, are involved not only in infection but in virus assembly, viral modulation of cell surface antigen expression, the host's immune response to infection, viral evasion of this response, and virus transport across epithelial boundaries.

2.1 Virus endocytosis

The link between endocytosis and infection was first established for the enveloped alphavirus Semliki Forest virus (SFV). SFV was shown to internalize in clathrin-coated vesicles (CCVs) and to be delivered to early endosomes. The low pH within endosomes triggers SFV fusion, resulting in merger of the viral membrane with the endosome limiting membrane and delivery of the viral nucleocapsid to the cytoplasmic compartment of the target cell. Significantly, SFV fusion and infection could be inhibited by weak bases and carboxylic ionophores that raise the pH in acidic organelles (2). Subsequently, these agents were used to show that other enveloped viruses, including influenza and vesicular stomatitis virus (VSV), as well as non-enveloped viruses, including reovirus, adenovirus, and some picornaviruses, also use pH-dependent mechanisms to trigger fusion or penetration (reviewed in ref. 3). These experiments also demonstrated that a number of viruses, including paramyxoviruses and most retroviruses, do not require exposure to low pH for entry. Thus the concept emerged that viruses use either the low pH within intracellular acidic organelles to facilitate fusion or penetration (pH-dependent viruses), or they undergo fusion or penetration at the cell surface (pH-independent viruses).

Endocytosis, however, is not without risk to a virus. The endocytic pathway plays a major role in delivering cell surface components and extracellular ligands to lysosomes for degradation. Thus for pH-dependent viruses, entry via the endocytic route must confer some significant advantages for infection and / or replication. In addition, though pH-independent viruses can potentially penetrate at the cell surface, this property does not preclude them undergoing fusion or penetration from intracellular vesicles following endocytosis.

2.1.1 Endocytosis in virus entry

i. Penetration

The delivery of pH-dependent viruses to acid organelles is the best characterized mechanism through which endocytosis facilitates virus entry. For SFV and influenza, exposure of the viral fusion proteins (the E1E2E3 complex and haemagglutinin (HA), respectively) to pHs within the range found in endosomes (pH 5.0–6.5) triggers the release of energy held in the metastable neutral pH form of the envelope protein to drive membrane fusion. For HA these conformational changes lead to exposure of a hydrophobic region of the protein, the so-called fusion peptide. Current models suggest that individual fusion peptides within a trimeric HA complex are able to interact with either the target or the viral membrane to bring about close apposition of the external leaflets of the membrane surfaces and subsequent fusion (4, 5). Indeed, structural analyses of influenza and other viral fusion proteins suggest that common mechanisms may underlie both viral and cellular SNARE-mediated membrane fusion (see Chapter 5). Under physiological conditions low pH is the only trigger for the flu fusion reaction and, as low pH is normally only encountered within intracellular organelles, endocytosis is essential for entry and infection by these viruses. Similarly, for pH-dependent non-enveloped viruses, low pH also triggers conformation changes in the surface proteins that are required for penetration. With adenoviruses, for example, acidification generates a form of the virus that can lyse the membrane of endosomes. The viral particle is then uncoated in the cytoplasm of the new host cell by a succession of proteolytic and reductive changes (6).

Thus one function of exposure to acid pH is to initiate the conformational changes required for viral fusion or penetration. However, the fact that in many cases viruses with very similar surface proteins are pH-independent for entry indicates that acid-induced reactions are not the only way of getting into a cell. Indeed, low pH-induced penetration might be regarded as a mechanism to ensure that fusion or penetration only occur after endocytosis. As only viable cells are capable of endocytosis, this could provide one means to ensure that a virus enters a cell in which it can replicate. Alternatively, endocytosis might confer other advantages.

ii. Post-penetration

For influenza A viruses, exposure of virions to low pH does have a further role. In addition to HA and the neuraminidase, the membranes of these viruses contain a small membrane protein, M2. M2 is a pH-dependent pore that opens in acidic conditions to allow protons to access the viral core. This acidification has important consequences for the virus. During virus assembly the viral RNA is exported from the infected cell nucleus through association with the M1 protein. When the virus subsequently infects a new cell, this interaction must be disrupted to allow the viral RNA to enter the target cell nucleus. The RNA-M1 association is disrupted when the viral ribonuclear particles are acidified. Thus endocytosis and exposure to acid pH are crucial for post-fusion targeting of influenza RNA to the nucleus (for references see 7).

iii. A cytoskeletal barrier?

Experiments with SFV, and to a lesser extent with other viruses, indicate that in certain cell lines endocytosis is important for events other than delivery of the virus to acidic organelles. Fusion of SFV at the surface of CHO cells (by transient acidification of the medium) does not generate a productive infection. Why infection fails is unclear. One proposal is that the cortical cytoskeleton underlying the plasma membrane imposes a post-fusion barrier to infection. The web of actin filaments and associated proteins that comprise the cortical cytoskeleton is present in most cell types, though it is more prominent in some cells (e.g. leukocytes) than in others (e.g. baby hamster kidney cells) (8). The cortical cytoskeleton excludes virtually all organelles, including ribosomes, from areas of the cell adjacent to the plasma membrane and has the capacity to prevent incoming viral cores from translocating into the body of the cell. Viruses such as the human immunodeficiency viruses (HIV types 1 and 2) carry a protease that can digest actin and actin-associated proteins and may be able to penetrate this barrier by localized proteolytic digestion. Other viruses such as Herpes simplex can capture motor proteins and use cytoskeletal elements to pull themselves through the web (9). Viruses lacking these options may capitalize on the inherent, though poorly understood, ability of endocytic vesicles to pass through the cortical cytoskeleton. Dependence on exposure to endosomal pH provides a mechanism that ensures fusion or penetration only occurs after endocytic uptake.

iv. Distributing virus within the cell

Endocytosis may also play a role in bringing viruses to the correct location in cells for replication to occur. For example, replication of alphaviruses occurs in association with endocytic membranes. Complexes containing viral RNA, capsid protein, and the virally encoded RNA-dependent RNA polymerase are associated with membranes of late endosomes, lysosomes, and the endoplasmic reticulum (10). Fusion following endocytosis may help to locate viral RNA on endosomal membranes for replication. Although not essential in tissue culture (SFV can replicate in certain tissue culture cells after fusion at the cell surface) (2, 8), this positioning of incoming RNA may have more importance for infection *in vivo*. Entry of alphaviruses into neurons may be facilitated by endocytosis. A delay in fusion, retrograde transport of the viruses in endosomes, and fusion in the cell body, would help to ensure that viruses are moved from the cell periphery to regions of the cell where replication can proceed efficiently. Movement of endosomes along microtubule tracks is well documented (see for example ref. 11) and microtubule-mediated movement of endosomes containing internalized reoviruses has been observed in living cells (12).

The need to penetrate the cortical cytoskeleton or reach an appropriate site for replication applies equally to pH-dependent and pH-independent viruses. A number of viruses that are pH-independent for entry may use endocytic vesicles to achieve these aims. Epstein–Barr virus (EBV) and duck hepatitis B virus, for example, undergo endocytosis but are pH-independent for fusion (13, 14). Similarly, the pH-independent picornavirus, human rhinovirus (HRV) 14, appears to transfer its RNA across the membrane of endosome carrier vesicles (ECVs) following uptake in CCVs (15). Further studies using inhibitors other than just weak bases or carboxylic ionophores may help to build a more complete picture of the role of endocytosis in virus entry.

2.1.2 Mechanisms of virus endocytosis

i. Do pH-dependent viruses use endocytic receptors?

The initial interaction with a potential host cell occurs when a virus binds to receptors on the target cell surface. For viruses such as HIV, docking to specific receptors is key for initiating conformational changes in the envelope protein that lead to fusion and infection. For others, the interaction may be crucial for recruiting viruses to endocytic vesicles. Though many cellular virus receptors have now been identified, and signals that mediate endocytosis via clathrin-coated pits have been defined (see Chapter 7), there are few examples of pH- or endocytosis-dependent viruses using receptors with well-established endocytosis signals. HRV2 uses a member of the low density lipoprotein receptor family (16) that contains an endocytic NPxY type signal (where x = any amino acid) in its cytoplasmic domain. The integrin subunits used for entry by adenoviruses also have NPxY motifs. These motifs have been implicated in the internalization of some bacterial pathogens (17), but it is unclear whether they are required for viral endocytosis. Although it is likely that novel endocytosis signals will be discovered (18), the lack of a clear preference for receptors containing known endocytosis signals may indicate that rapid endocytosis is not essential for viruses that use endocytic routes. Alternatively, virus binding might modify the endocytic properties of cell surface molecules. The clustering of receptors, for example, through multipoint virus attachment may be sufficient to bring about endocytosis. Indeed differences in the mode of virus binding may account for the observation that both SV40 and SFV can use MHC I to enter cells through different endocytic pathways (19–23).

ii. The vesicular carriers

It is now clear that multiple mechanisms for endocytosis operate from the cell surface (see Chapters 1–4). Morphologically viruses have been seen associated with a number of these different endocytic vesicles, including CCVs, non-coated vesicles, and caveolae (Table 1). However, in many cases the role of this uptake in virus infection is unclear. Morphological studies with viruses can only be made using high numbers of particles. Virus preparations usually carry inactive particles, sometimes in vast excess, and it is impossible to determine by morphology alone whether individual endocytic events are part of a pathway leading to productive infection. Images of pH-independent viruses within endocytic organelles have often been considered to represent the uptake of non-infectious particles or uptake that fails to result in productive infection. Thus methods for interfering with specific endocytic mechanisms have been sought. As discussed above, inhibitors of endosomal/lysosomal acidification (weak bases, carboxylic ionophores, H^+ATPase) have been useful in identifying viruses that require exposure to acid pH for fusion or penetration. However, these agents do not discriminate between the different endocytic pathways leading to acidic compartments, nor do they identify viruses that are pH-independent for entry but still require endocytosis. Recent molecular analysis has enabled the development of dominant negative reagents that allow more detailed analyses of the endocytic mechanisms used by specific viruses.

iii. Clathrin-coated vesicles (CCVs)

Clathrin-mediated endocytosis has been studied extensively at the molecular level and a number of reagents have been developed that will inhibit CCV formation (24). The temperature-sensitive and K44E forms of dynamin (25) have been used most extensively in studies of virus entry. Expression of these proteins inhibits CCV-mediated endocytosis and entry of the related SFV and Sindbis virus, supporting previous data indicating that alphaviruses are taken up via CCVs (2, 23). Dynamin mutants also inhibit the entry of adeno- and adeno-associated viruses (26, 27) and HRV-14 (28).

Table 1 Entry properties for viruses that use endocytic mechanisms

Family	Species	Inhibitors	pH dependence	Endocytosis	References
Enveloped viruses					
Arenaviridae	Junin	Weak bases	Acid dependent. Infection inhibited by weak bases. Fusion at cell surface and viral-dependent cell–cell fusion induced at pH 5.5.		177, 178
Bunyaviridae	LaCrosse		Acid dependent. Infection inhibited by weak bases and ionophores. Viral induced cell–cell fusion induced at low pH. Acid induced conformational changes in envelope protein.		179–182
Iridoviridae	African swine fever	Weak bases	Acid dependent. Infection inhibited by weak bases.	Viruses associated with coated pits (EM).	183, 184
Orthomyxoviridae	Influenza A	Weak bases, carboxylic ionophores, vATPase inhibitors.	Acid dependent (fusion pH 5.5–6.5 depending on isolate)—established with inhibitors, cell-based and liposome fusion assays, biochemical analysis of envelope protein.	Clathrin and non-clathrin mediated pathways (EM).	1, 38, 185–189
Retroviridae	MMTV		Acid dependent. Fusion activity triggered at low pH.		190
Rhabdoviridae	Vesicular stomatitis virus (VSV), Rabies	Weak bases, carboxylic ionophores, vATPase inhibitors.	Acid dependent (fusion pH ~ 6.0). Infection inhibited by weak bases and vATPase inhibitors.	Clathrin and non-clathrin mediated pathways (EM).	1, 191–193
Togaviridae	Alphaviruses Semliki Forest (SFV), Sindbis	Weak bases, carboxylic ionophores, vATPase inhibitors. Dominant negative dynamin (K44E), anti-clathrin antibodies.	Acid dependent (fusion pH ~ 6.0) Established with inhibitors, cell-based and liposome fusion assays, biochemical analysis of envelope protein.	CCV— established by EM and inhibition of infection with dominant negative dynamin.	3, 23, 28, 194, 195
	Flaviviruses West Nile virus, Tick borne encephalitis virus	Weak bases.	Acid dependent (fusion pH ~ 6.5). Infection inhibited by weak bases, fusion with liposomes induced at low pH. Low pH induced conformational changes in envelope protein.		196–200
Non-enveloped viruses					
Adenoviridae	Adenovirus Various strains	Weak bases, carboxylic ionophores, vATPase inhibitors.	Acid dependent. Infection inhibited by weakbases, ionophores, and vATPase inhibitors.	Entry inhibited by dominant negative dynamin, also by dominant negative Rab5	26, 52, 53, 201, 202

		Dominant negative dynamin (K44E), Rab5 and Cdc42. Phosphatidyl inositol-3 kinase inhibitors.		and Cdc42 and by PI3 kinase inhibitor.	
Calciviridae	Feline calcivirus	Weak bases.	Acid dependent. Infection inhibited by weak bases.		203
Hepadnaviridae	Duck hepatitis B virus	Sodium azide, 2-deoxyglucose	Acid independent. Infection insensitive to weak bases.	Infection sensitive to ATP depletion. Receptor trafficking experiments suggest endocytosis is required.	14, 204
Herpesviridae	Epstein–Barr virus		Acid independent.	Non-coated vesicles (EM).	13, 205
	Human herpes virus 6	Weak bases	Acid dependent.	Non-coated vesicles.	39
Papovaviridae	SV40	Nystatin, methyl-β-cyclodextrin, dominant negative caveolin 1, tyrosine kinase inhibitors		Caveolae (EM). Inhibited by reagents that deplete membrane cholesterol and disrupt caveolae, and by dominant negative caveolin.	19, 20, 206
	JC virus	Chloropromazine, but not nystatin.		Infection not inhibited by agents that block SV40 entry. Clathrin dependent?	37
Parvoviridae	Canine parvovirus	vATPase inhibitors, dominant negative dynamin.	Acid dependent?	Clathrin-mediated (EM), inhibited by dominant negative dynamin.	207
	Adeno-associated virus	Weak bases, vATPase inhibitors, dominant negative dynamin (K44E).	Acid dependent— established with weak bases and vATPase inhibitors.	Clathrin-mediated – inhibited by dominant negative dynamin.	27, 208
Picornaviridae	Human rhinovirus (HRV) -14	Dominant negative dynamin and hypertonic medium.	pH independent. Not inhibited by vATPase inhibitors.	Delayed entry in the presenceof vATPase inhibitors suggests penetration from endocytic vesicles. Infection inhibited by dominant negative dynamin.	15, 28
	HRV-2	vATPase inhibitors.	Acid dependent. Infection inhibited by vATPase inhibitors. Conformational changes at $pH < 5.6$.		51, 209
Totiviridae	Giardiavirus	Weak bases, sodium azide	Acid dependent.	Seen in intracellular vesicles (EM).	210

Although widely used to block clathrin-mediated endocytosis, dynamin activity may not be restricted to CCV formation. Recent results suggest that dynamin is also involved in endocytosis through macropinocytosis and caveolae (29). In addition, cells can compensate for the loss of clathrin-mediated endocytosis by up-regulating an alternative, poorly understood, endocytic pathway (30). Although experiments using dynamin mutants should be interpreted with caution, these results together with morphological studies make a strong case that some viruses absolutely require clathrin-mediated endocytosis for productive infection. The use of other dominant negative inhibitors of proteins involved in clathrin-mediated endocytosis should further clarify these findings (for references see 24).

iv. Caveolae

Caveolae, or plasmalemmal vesicles, are seen in many, though not all, cells (see Chapter 2). The contribution of caveolae to cellular endocytic activity has still to be fully established (see Chapter 2). Nevertheless, there is evidence that these structures internalize and deliver their content to endosomes and / or the ER (19). Caveolae have been implicated in the internalization, and infection, of SV40 virus. In part this association is based on experiments in which cholesterol depleting drugs that disrupt caveolae (e.g. nystatin) have been found to inhibit SV40 infection (20, 21). Similar experiments have suggested that dendritic cells acquire respiratory syncytial virus antigens via caveolae (31). Although cholesterol depletion was initially thought to influence primarily caveolae, recent studies with methyl-β-cyclodextrin have indicated that cholesterol depletion can also influence CCV formation and sorting in the endocytic pathway (32–34). Nonetheless, morphological studies together with recent experiments with dominant negative forms of caveolin, the major structural protein of caveolae, support the notion that caveolae do indeed play a role in SV40 entry (19, 35, 36). However, the properties of this pathway that are of benefit to SV40 during its entry remain to be established. Interestingly, infection by another polyomavirus, the human JC virus, appears not to be inhibited by cholesterol depleting drugs suggesting that this virus does not exhibit the same requirement for caveolae (37).

v. Other pathways

The role of macropinocytosis in virus entry is unclear. In higher organisms this endocytic mechanism may be restricted mainly to cells involved in antigen presentation and may therefore have a role in the entry of viruses into macrophages and dendritic cells (see Chapter 4).

A number of studies, particularly those using dominant negative dynamin constructs that potentially block CCV, caveola, and macropinosome formation have indicated that an alternative endocytic pathway can compensate for the loss of dynamin-dependent endocytosis at least in tissue culture cells (25). Little is known of the molecular mechanisms underlying this form of endocytosis and its contribution to virus uptake under normal conditions remains to be established. However, morphological studies have frequently reported viruses in non-coated endocytic vesicles. In some cases these profiles may be CCVs that have lost their coats or endosomes that have received viruses. However, there is evidence that influenza viruses, for example, can be internalized through a non-coated vesicle-mediated pathway. In these cases, the membrane of the endocytic vesicle is tightly apposed to the virus particles suggesting that membrane might wrap around the virus in a process similar to the 'zippering' mechanism proposed for phagocytosis (see Chapter 3) (38). The herpes viruses EBV and HHV6 are reported to enter B and T cells, respectively, in non-coated vesicles (13, 39). Whether the mechanisms used by these viruses are dynamin-independent or not remains to be established.

The failure of dominant negative dynamin to inhibit infection by some viruses might indicate the use of an alternative endocytic pathway. However, it might also provide support for penetration of a virus at the cell surface as appears to be the case with poliovirus (40, 41).

vi. Endocytosis in virus transfer across tissue barriers

Epithelial and endothelial cell layers form crucial barriers that line the external body surfaces, blood vessels, and lymphatics. Despite the continuous epithelium lining the gut, many viruses use oral routes of transmission and exploit the endocytic properties of epithelial cells to enter a host. In particular, epithelial M cells that use endocytic mechanisms to continually sample the antigenic content of the gut provide a prominent route for transfer (42). Reo- and picornaviruses are able to exploit these cells in the small intestine to cross the epithelial barrier. HIV may use a similar route to cross the rectal epithelium (43). In this case the proximity of M cells to dendritic and lymphoid cells on the ad-lumenal surface of the epithelium is likely to provide the virus with direct access to vulnerable host cell populations. This provides one example of how viruses can exploit endocytic properties that are primarily designed to protect the host. Although a conduit for virus entry, these same routes may provide opportunities for delivering recombinant viral-based vaccines (42).

2.1.3 Sites of fusion and penetration

With the possible exception of caveolae, endocytic vesicles deliver internalized ligands to organelles of the endocytic pathway, primarily early endosomes. Subsequent trafficking events can transfer these ligands to other endocytic compartments (recycling endosomes, late endosomes, and lysosomes) or to a retrograde pathway through the Golgi apparatus to the ER (44). The exact locations along these pathways that are used by viruses for penetration are determined to a large extent by the pH dependence of fusion or penetration. SFV strains that are activated at pH 6 fuse rapidly after endocytosis, and have been proposed to fuse from early endosomes. Viruses that are activated at more acidic pH fuse with delayed kinetics from later endocytic compartments (45). Similarly, influenza viruses that are activated at $<$ pH 5.5 (e.g. X31) are likely to fuse from late endosomes, whereas strains that are activated at pH values closer to neutral (e.g. Rostok) may fuse from early endosomes.

The pH-dependent viruses have been used to probe the acidity of endocytic compartments. Previously, early endosomes were found to be $\approx$ pH 6.0 (45, 46). However, recent studies have indicated that VSV and SFV entry is blocked in cells lacking a functional COP1 complex (47, 48). In the absence of COP1 transport from early to late endosomes is inhibited (49). These results suggest that fusion of VSV and SFV may occur from endosomal carrier vesicles (ECVs) or late endosomes, rather than from early endosomes and that early endosomes do not become more acidic than pH 6.2. The vATPase inhibitor Bafilomycin A also blocks transport from early to late endosomes (50). Thus the ability of Bafilomycin A to inhibit infection of pH-dependent viruses such as HRV2 may be partly attributable to a block in transfer of virus to late endosomes and not solely an effect on intra endosomal pH (51). Conversely, HRV14 can infect Bafilomycin A-treated cells suggesting that this virus undergoes pH-independent penetration in early endosomes (15), although this does not exclude the possibility that HRV14 may uncoat in later endocytic compartments under normal circumstances.

Rab5, an early endosome-associated Rab protein (see Chapter 5), PI3 kinase, and the small GTPases Rac and CDC42 functions are required for adenovirus entry (52, 53). These proteins have been implicated in regulating traffic through the endocytic pathway, though their precise mode of action in relation to virus entry remains to be determined. Similarly, activation of a tyrosine kinase signalling pathway appears to be required for SV40 infection via caveolae, but the precise role of this signalling is unclear (54).

2.2 Endocytosis in virus assembly

Although the endocytic pathway is usually associated with virus entry there is evidence that some viruses may also use endocytic organelles for assembly. All enveloped viruses must assemble on and/or bud through a cellular membrane system to acquire an envelope. Many viruses such as influenza, SFV, VSV, and most retroviruses bud from the plasma membrane. Other viruses including bunyaviruses, coronaviruses, herpesviruses, reoviruses, and poxviruses assemble on intracellular membranes. By restricting their membrane proteins to internal membranes, these viruses avoid drawing the attentions of the immune system to the infected cell. For the most part, intracellular assembly has been linked to compartments of the exocytic pathway, the ER, Golgi apparatus, and intermediate compartments (55). However, morphological studies have suggested that the final envelopment steps of the human cytomegalovirus occur on endosomal membranes (56, 57).

The assembly of HIV also raises interesting questions. For most enveloped viruses, the site of virus budding and assembly is determined by sorting signals that direct viral membrane proteins to specific compartments of the cell (55). The envelope protein (Env) of HIVs and the closely related SIV, contain highly conserved signals that mediate endocytosis and sorting of Env to endocytic organelles (58–61). Thus, the possibility exists that some HIV particles assemble on membranes of late endocytic compartments. Alternatively, these viruses must contain mechanisms to redistribute Env to the cell surface during virus assembly (62). HIV particles have frequently been seen to bud into intracellular membrane-bound compartments (see ref. 63), but the nature of these compartments and the role of this internal assembly in the biology of the virus has yet to be addressed.

3. Bacterial and protozoan pathogens

The problems that beset protist and bacterial pathogens have strong parallels with viral pathogens with respect to initial cell entry. However, once inside the cell there are major differences in how these pathogens interface with the host's endocytic machinery. Even between bacterial and protozoan pathogens there are only limited parallels, the majority of species having evolved their own unique means of subverting host cell function. One would have imagined, given the promiscuity of bacterial plasmids and phages that more commonality would exist between at least related bacterial species. However, while there is clear evidence that related pathogenicity islands provide similar mechanisms of delivery of effector molecules into the host cell, these molecules are themselves highly divergent and induce many different responses in the host (64, 65). The infectious agents discussed in this section are restricted to intracellular pathogens. Although several extracellular bacteria exploit the host's endocytic system for delivery of toxins to their sites of action this subject is too broad to be dealt with in this chapter (see ref. 44).

Unlike viruses that suborn the host's synthetic and endocytic machinery, protozoal and bacterial pathogens remain entities discrete from their hosts, yet they may require the host's membrane trafficking systems to maintain and nurture their infection. For many pathogens the endosomal/lysosomal continuum represents either a haven or a graveyard. The outcome depends on multiple factors including the route of entry, the successful modulation of normal phagosome maturation, or escape into the host cell cytosol, and finally, in phagocytes at least, the activation status of the host cell. This portion of the chapter will deal with these issues sequentially using individual pathogens to exemplify the ways in which they deal with normal host cell function.

3.1 Host cell entry

Invasion and infection must be regarded as a continuum because each step can have consequences on the subsequent processes. Obviously this varies from pathogen to pathogen and, for ease of discussion, the following sections are organized into themes rather than along phylogenetic lines. The infectious agents discussed are shown in Table 2.

The mechanisms of entry into mammalian cells by prokaryote and lower eukaryote pathogens fall into three discrete categories depending on which cell is actively involved in the process. The most practised mode of entry, phagocytosis, is dependent wholly on the host cell for its execution. During this process the pathogen does little more than ensure that it is coated with, or displays, an appropriate ligand for recognition by receptors on the surface of the phagocyte. Many investigators have suggested that the nature of the host cell receptor(s) ligated may influence the subsequent survival of the pathogens. However, the only persuasive data argue that, beyond the

Table 2 Cellular pathogens discussed in this chapter

Pathogen/disease	Protist or bacteria	Mode of invasion	Intracellular location
Trypanosoma cruzi (Chagas' disease)	Protozoan	Active invasion	Cytosolic
Shigella flexneri (bacillary dysentery)	Bacterium	Phagocytosis	Cytosolic
Listeria monocytogenes (listeriosis)	Bacterium	Phagocytosis	Cytosolic
Microsporidia (diarrhea, anorexia—*Enterocytozoon bieneusi*)	Protozoan	Active invasion	Cytosolic
***Mycobacterium* spp.** (tuberculosis—*M. tuberculosis*)	Bacterium	Phagocytosis	Early endosomal compartment
***Ehrlichia* spp.** (monocytic ehrlichiosis—*E. chaffeensis*)	Bacterium	Phagocytosis	Early endosomal compartment
***Plasmodium* spp.** (malaria)	Protozoan	Active invasion	Sequestered vacuole
Toxoplasma gondii (toxoplasmosis)	Protozoan	Active invasion	Sequestered vacuole
***Chlamydia* spp.** (psittacosis—*C. psittaci*; trachoma, lymphogranuloma venerium—*C. trachomatis*)	Bacterium	Phagocytosis	Sequestered vacuole
Legionella pneumophila (legionnaire's disease)	Bacterium	Phagocytosis	Sequestered vacuole that becomes fusigenic
Brucella abortus (abortion in cattle)	Bacterium	Phagocytosis	Autophagic vacuole associated with ER
Salmonella typhi (enteritis, typhoid)	Bacterium	Induced endocytosis	Modified endosomal compartment
Coxiella burnetii (Q fever)	Bacterium	Phagocytosis	Lysosomal compartment
***Leishmania* spp.** (Kala azar—*L. donovani*, tropical ulcers—*L. major*, *L. mexicana*)	Protozoan	Phagocytosis	Lysosomal compartment

use of complement receptors versus Fc receptors that trigger the release of reactive oxygen intermediates, routes of uptake via different phagocytic receptors have at most only subtle effects on the efficiency of infection. The signalling pathways, activated by receptor ligation, involved in phagocytosis are multiple and complex and are dealt with in Chapter 3.

Induced endocytosis is the means of entry into non-professional phagocytes and usually requires energy expenditure by both the host cell and the invader. Pathogens, such as *Yersinia* or *Salmonella* bind to receptors not normally associated with uptake or endocytosis. However, the pathogen induces the host cell to internalize these receptors either through the aggregation of multiple receptors (66), or though delivery of effector molecules into the host cell through the use of a type III secretion system (67). Despite the commonality of the delivery systems (which resemble bacteriophage genes in structure and organization and were probably acquired by horizontal gene transfer), the effector molecules delivered into the cytoplasm fulfil different functions. *Salmonella* and *Yersinia* are prime examples of this divergence. Although *Yersinia* gains entry to some cell types through binding to integrins using the invasin protein, the bacterium has a strong anti-phagocytic response that blocks its uptake (68, 69). This is mediated through molecules such as YopT that interfere with the host cell's cytoskeletal elements (68). In contrast, *Salmonella* uses the needle-like SPI-1 type III secretion system to inject at least five proteins into the host cell (67). One of these proteins, SopE, leads to enhanced ruffling, macropinocytosis, and nuclear activation through stimulation of Cdc42 and JNK (70) and mediates uptake of the bacterium into the cell. Our appreciation of the means of manipulation of host cell cytoskeletal elements through these effector molecules is expanding rapidly (see ref. 71).

The third means of host cell entry is the least subtle but is highly effective. Active invasion of the host cell is practised by three, phylogenetically-distinct groups of protozoa. All members of the phyla *Apicomplexa* and *Microsporidia*, and the trypomastigote stage of the kinetoplastid flagellate *Trypanosoma cruzi*, are capable of active entry into host cells that remain virtually passive throughout the process. *Apicomplexans* include both *Plasmodium* and *Toxoplasma gondii* and invasion is restricted to the motile, or 'zoite', stages of these parasites. Zoites are long, sausage-shaped cells that exhibit a gliding motility that requires binding to substrata and is actin-based. These parasites bind to the surface of a target cell forming a circular-junction that they then cap down the parasite body, thus inoculating themselves into the host cell (72–75). The host cell membrane is not ruptured during the process, though all host cell plasma membrane proteins with cytosolic domains are excluded from the forming parasitophorous vacuole by the circular junction (76). Once inside, the parasite remodels the vacuole through incorporation of its own proteins into the vacuolar membrane (77). *T. cruzi* trypomastigotes also invade actively but the process is enhanced by recruitment and fusion of lysosomes to the forming entry vacuole (78). This process is parasite-induced but is reliant on a transient calcium flux mediated by signalling pathways in the host cell (79, 80).

3.2 Post entry

3.2.1 The established infection: Where to go from here

Once inside a potential host cell, pathogens have to deal with the inevitable progression of their phagosome through the endosomal continuum to the lysosome were they would encounter both low pH and hydrolytic lysosomal enzymes. Maturation through this pathway is achieved through multiple vesicle fission and fusion events and the protein trafficking pathways that are involved are the subject of intense study. The GTPases Rab5 and Rab7, that are involved in the early to late endosome transition (see Chapter 5), also appear active in phagosome maturation. Data from recent studies on several pathogens have provided fresh insights into some of the host proteins that affect this process. The metal ion transporter Nramp1 was originally identified as a genetic locus that segregated with resistance to infection with *Leish-*

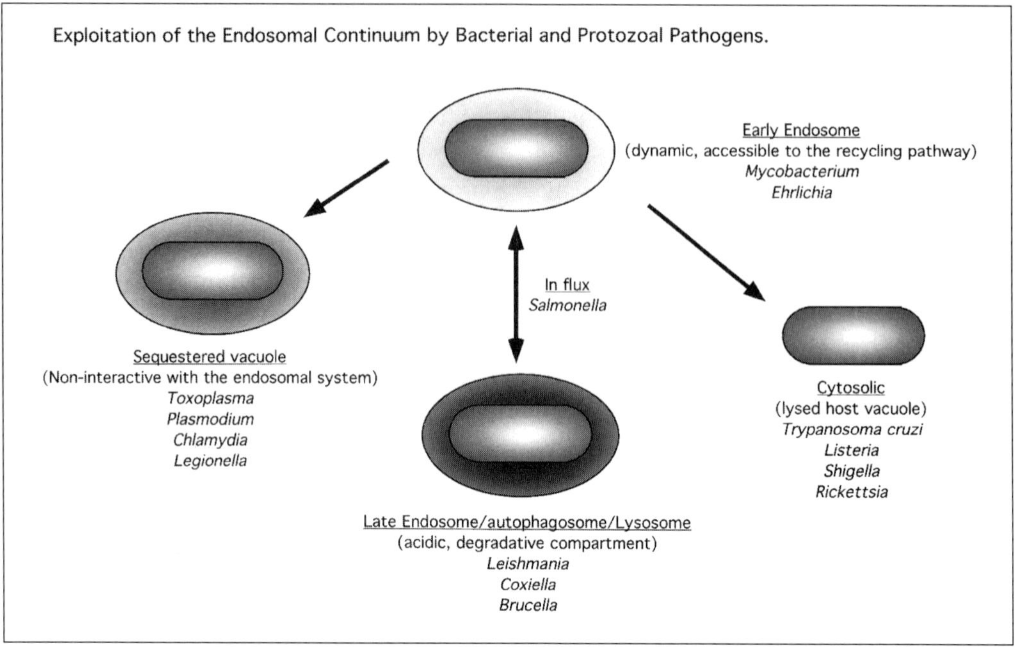

Fig. 1 A cartoon summarizing the different intracellular niches exploited by bacterial and protozoal pathogens.

mania donovani, Salmonella typhimurium, and *Mycobacterium bovis* (81, 82). However, this effect may be mediated through the profound influence that this transporter has on the rate and extent of phagosome acidification (83). Another example to have emerged from studies of pathogens is the Toll-like receptor family. These molecules sample the lumenal contents of phagosomes and alert the cell to the contents. Toll-like receptors can discriminate between Gram positive and Gram negative bacteria and transduce cellular responses that culminate in release of proinflammatory cytokines that modulate macrophage function (84).

As with viruses, bacterial and protozoal pathogens have evolved to exploit different niches within their host cell (Fig. 1). These 'choices' divide readily into three discrete groups.

(a) Pathogens that rely on rapid exit from the endocytic network and invasion of the host cell cytosol for survival.

(b) Parasites that modulate the normal maturation of their phagosome to prevent its development into a lysosomal vacuole.

(c) Pathogens that are resistant to the digestive capacity of the lysosome and go on to reside in acidic, hydrolytically-competent lysosomes.

3.2.2 Intracytosolic parasites

The bacteria *Listeria, Shigella,* and *Rickettsia* and the protozoan *T. cruzi* all deal with the increasing hostility of the endosomal/lysosomal continuum by lysing the limiting vacuolar membrane and escaping into the host cell cytosol. It has been shown in *T. cruzi*, *Listeria*, and *Shigella* that the lytic activities generated by the pathogens have acid pH optima, and that raising intravacuolar pH blocks escape and reduces the parasites' viability. In the case of *Listeria* the proteins responsible for this activity have been identified and the genes sequenced. The bacteria encode three different proteins with differing lytic activity. Listeriolysin is a 58 kDa pore-forming protein that is capable, on its own, of allowing bacteria to escape from phagosomes (85, 86). Two additional phospholipases, one a broad spectrum PLC and the other a phosphotidylinositol-specific PLC, appear to be required for maximal activity and cell/cell spread in culture (87–90).

3.2.3 Pathogens that modulate vacuolar function

This is the largest and most diverse group of parasites and is subdivided into two further groups for ease of discussion. The first group consists of pathogens that reside in vacuoles that are modified yet continue to be accessible to host membrane trafficking pathways, and includes *Mycobacterium*, *Ehrlichia*, and *Legionella*. The second group consists of pathogens that sequester their vacuole in apparent isolation from the host's membrane trafficking pathways, and includes *Chlamydia*, *Toxoplasma*, and *Plasmodium*.

Mycobacterium spp., including *M. avium* and *M. tuberculosis*, are the most studied first group. These cells reside in vacuoles inside macrophages in their host. Early work indicated that *Mycobacterium*-containing vacuoles failed to fuse with lysosomes (91, 92), did not acidify below pH 6.3, and yet the vacuoles were positive for lysosomal proteins such as LAMP 1 and cathepsin D (93–95). Closer analysis of this apparently anomalous set of characteristics revealed that these vacuoles were highly dynamic and readily accessible to transferrin or GM1 ganglioside (complexed with cholera toxin B subunit) internalized at the cell surface (95–97). Autoradiography performed on infected macrophages incubated with ^{59}Fe-loaded transferrin indicates that the bacterium can exploit the transferrin as an iron source. These data demonstrated that the vacuoles were within the sorting/recycling endosome system of the host cell and that it was an arrested maturation that was responsible for the apparently anomalous distribution of host proteins. The cathepsin D in the *Mycobacteria*-containing vacuoles has since been shown to derive from the TGN of the host cell (95, 98). Some groups contended that *M. tuberculosis* is capable of mediating escape from its vacuole (99). However, this is not a widely held belief. Recent work indicates that the *Mycobacterium*-containing vacuole may be 'porous' facilitating access of cytosolic tracers into the vacuole (100). The authors suggest this data may explain tap transporter-dependent presentation of mycobacterial antigens on MHC I. However, even inert particles coated with antigen are capable of delivering antigen to this pathway, so it does not constitute proof that *Mycobacteria*-containing vacuoles are leaky (101, 102).

Although the mechanism by which *Mycobacterium* arrests the maturation process is unknown, it has been shown that *Mycobacterium*-containing vacuoles retain proteins associated with regulation of phagosomal differentiation such as Rab5 (103) and the coat protein Coronin I (called TACO in the original paper) (104). The restricted maturation limits the accumulation of vATPases, underlining the intimate association between pH and passage along the endosomal/lysosomal continuum, and suggests that limiting acidification could influence maturation directly. In a related way, expression of the metal ion transporter NrampI, the product of the mouse locus bcg^r that confers resistance to infection with less virulent species of *Mycobacterium* (e.g. *M. avium* and BCG [*M. bovis*]), correlates with vacuolar acidification (83).

Ehrlichia is retained in a vacuole that bears similarities with that of *Mycobacterium spp*. These vacuoles appear to be early endosomes as limited accumulation of the weak base DAMP indicates they are mildly acidic (105), and they are positive for the transferrin receptor, EEA1 and Rab5 (105, 106). *Ehrlichia* go one step further than *Mycobacterium* and up-regulate expression of the transferrin receptor through activation of the host's iron-responsive protein 1 (IRP-1) (107). This enhancement of iron-uptake capabilities in the host cell appears important for sustaining bacterial growth because treatment of infected cells with iron chelating agents leads to death of the *Ehrlichia*.

Legionella is more complex because the vacuole in which it resides, the replicative phagosome, is initially a non-acidic vacuole that accumulates host organelles around its periphery (108–111). Nevertheless, at later time points it appears to acidify and become interactive with the endosomal network (Sturgill-Koszycki, personal communication). The *Legionella dot/icm* gene family encodes a secretion system (similar to the conjugal transposon delivery system found in *Agrobacterium*) that may translocate bacterial products into the host cell cytoplasm (112). Mutants in these genes are unable to recruit rough ER to the vacuolar membrane and fail to establish a productive infection (113). It has been suggested that a cell treats the replicative phagosome like a nascent autophagic vacuole. However, these vacuoles acquire rough ER rather than the smooth ER normally associated with autophagosome formation. The bacteria lyse out of the host cell 16–24 hours after infection. Prior to this, the bacteria-containing vacuoles acidify and re-establish fusigenicity with endosomal/lysosomal vesicles. The significance of this differentiation to the life cycle of *Legionella* has yet to be established but it suggests that the replicative phagosome is a dynamic compartment that needs to be actively maintained as a sequestered entity.

Brucella abortus also shows an association with a nascent autophagosome-like compartment. A recent study demonstrated that vacuoles containing viable bacteria were negative for mannose-6-phosphate receptor and cathepsin D, yet were studded with ribosomes and stained positive for the ER proteins sec61β and protein disulfide isomerase and the lysosomal protein LAMP1 (114, 115). Moreover mutants in the two component sensor system BvrS-BvrR, which is known to regulate bacterial virulence, were incapable of arresting delivery of the bacilli to lysosomes suggesting that an intracellular stimulus was required to induce the bacteria to modulate maturation of its vacuole.

The above examples contrast sharply with the sequestered vacuoles formed by the second group of obligate intracellular parasites *Toxoplasma*, *Plasmodium*, and *Chlamydia*, all of which exhibit a marked dependence on their host cell for survival. Although *Chlamydia* uses phagocytosis as a means of entry into the host cell it very rapidly remodels its compartment into a vacuole that shows no interaction with the host cell's endocytic network (116–118). The ability to sequester the vacuole is independent of the receptors activated during the entry process (119). Shortly after uptake, the bacteria synthesize a range of proteins that are known to incorporate into the membrane of the vacuole which is now termed the 'inclusion body' (119, 120). Some of these proteins, such as IncA, are transmembrane proteins in which the cytosolic domain is a substrate for host cell kinases (121). Once fully formed, the inclusion body membrane appears inaccessible to all host trafficking pathways with the exception of NBD-ceramide that, in the form of sphingomyelin, is incorporated into the inclusion body membrane (122). Whether this indicates an active intersection with the glycosphingolipid biogenesis pathway remains to be determined. An interesting recent study showed that uptake and infection of HeLa cells overexpressing a dominant negative dynamin (see Chapter 2) was unaffected during initial stages. Nevertheless, development of the inclusion body was inhibited, presumably through disruption of the pathway from the TGN to inclusion bodies (123). In contrast to *Toxoplasma*, discussed below, *Chlamydia* does not gain access to cytosolic nutrients through the use of passive transport through pores in the inclusion body membrane. This suggests that membrane proteins of the inclusion body must actively transport these nutrients into the bacteria-containing vacuole.

Toxoplasma is an active invader that excludes host proteins from its entry vacuole during invasion. Unlike *Chlamydia*, ligation of receptors such as the Fc receptor during entry will culminate in death of the parasite because the vacuole will fuse with lysosomes (124). Once the parasitophorous vacuole of *Toxoplasma* is established the parasite can be killed by drugs such as pyrimethamine and the vacuole remains non-fusogenic with lysosomes (125) (Fig. 2). During the infection process the parasites discharge proteins from three discrete organelles; the micronemes during early invasion, the rhoptries during initial vacuole formation, and the dense granules during remodelling of the parasitophorous vacuoles (126). The dense granules contain an extensive family of proteins (seven to date) the majority of which intercalate into either the tubular network released into the vacuole or the limiting membrane of the vacuole (77, 127, 128). Once remodelled the vacuole associates strongly with host organelles, most notably mitochondria, and has pores in its membrane that allow passive transfer of material up to 3 kDa (129). In contrast to *Chlamydia*, the vacuoles containing *Toxoplasma* are not accessible to NBD-ceramide-derived sphingomyelin exiting from the host cell Golgi (130).

3.2.4 Intralysosomal parasites

Coxiella burnetii reside and multiply in large, fluid-filled vacuoles inside their host cells. These vacuoles are acidic and positive for all lysosomal proteins examined (122). Indeed the bacilli appear to be exquisitely evolved for survival in this environment because treatment of infected cells with Bafilomycin A reduces the viability of the bacteria. *Coxiella* increase the fusigenicity of this vacuole inside the cell and appear to cause a collapse in the boundary between lysosomes and late endosomes (131). The infection also causes an increased association between the MHC II proteins HLA-DM and HLA-DR, though the steady state expression levels of these proteins remain

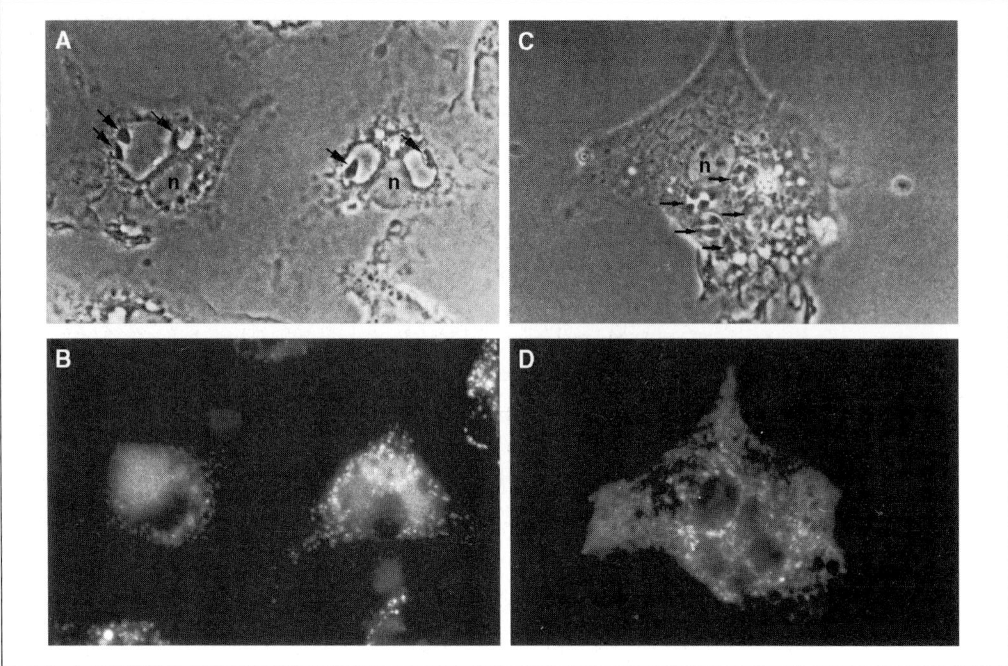

Fig. 2 Bone marrow-derived macrophages infected with *Leishmania mexicana* (A, B) and *Toxoplasma gondii* (C, D), arrowed. The cells were incubated with FITC-dextran (10 kDa) for 60 min pulse and 60 min chase prior to fixation and examination. The fluorescent tracer is present in the parasitophorous vacuoles containing *Leishmania* amastigotes and absent from those containing *Toxoplasma*. n = nucleus.

unchanged (132). It is thought that this alteration may affect the efficiency of peptide-loading into MHC II molecules and subsequent antigen presentation.

Parasites of the genus *Leishmania* also persist in acidic compartments. The most studied of these are members of the *L. mexicana* complex that form large vacuous compartments in host macrophages (133). The amastigotes of *Leishmania* are acidophilic stages that grow only at low pH. The parasitophorous vacuoles containing amastigotes of *Leishmania spp.* all intersect freely with the host cell's endocytic network and fluid phase markers internalized by the cell accumulate in the parasite-containing vacuoles (134–136) (Fig. 2). However, recent data indicate that following phagocytosis of the promastigote, or insect form of the parasite, the parasite's surface lipidoglycan (LPG, or lipophosphoglycan) retards phagosome maturation, buying time for the promastigote to transform to the amastigote stage (137, 138). The parasitophorous vacuoles of *Leishmania spp.* are also highly fusigenic and interact with nascent autophagosomes facilitating access of the parasite to potential nutrients from the host cell cytosol (139) (Fig. 3).

Although we have included *Salmonella* within this group of pathogens the data on this organism are confusing and may reflect a variety of strategies activated by different cell types. In non-professional phagocytes *Salmonella* resides within a tubular vesicular network that is positive for LAMP 1, yet appears segregated from the late

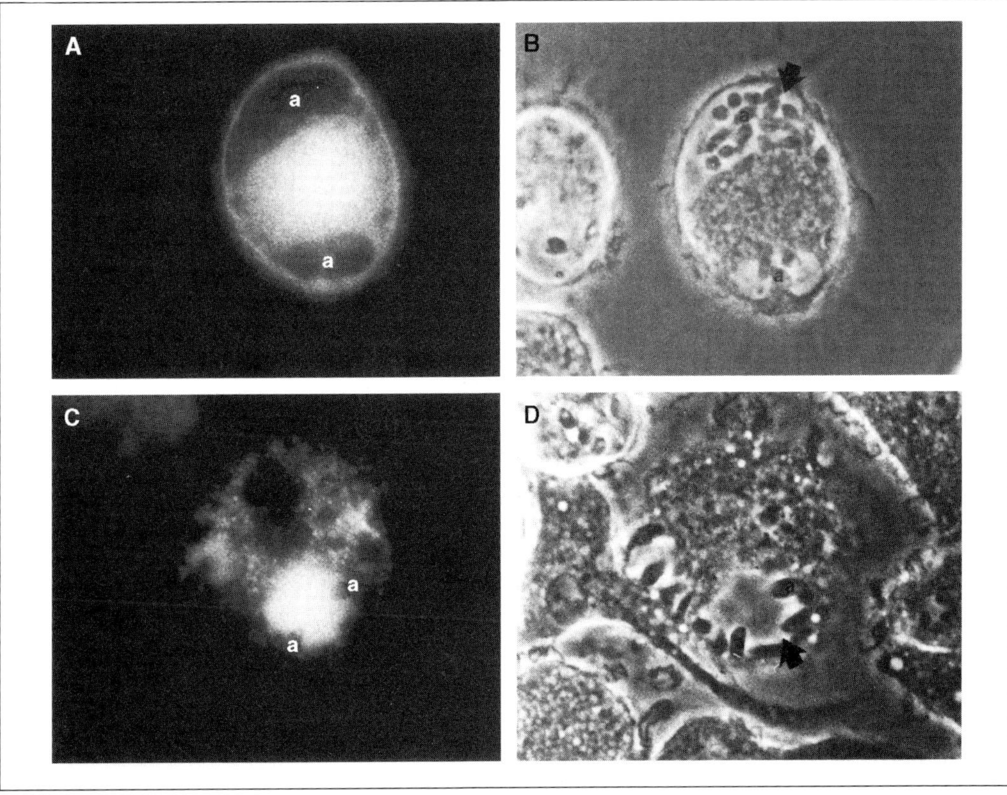

Fig. 3 Bone marrow-derived macrophages infected with *Leishmania mexicana* and bead-loaded with Texas-red dextran (10 kDa). The bead-loading procedure introduces the tracer into the cytosol of the infected macrophage. After 4 h (A, B) the tracer is visible in the cytosol of the cell. However, after 24 h (C, D) much of the tracer has been sequestered into vesicles and delivered into the parasitophorous vacuoles, arrowed. Transfer from the cytosol appears to be mediated through an autophagous process (139). a = amastigote.

endosomes/lysosomes of the host cell (140–142). Nonetheless, treatment of infected macrophages with Bafilomycin A causes a reduction in bacterial survival indicating that some acidification is necessary to support growth or to trigger an appropriate intracellular response from the bacterium (143, 144). A recent study of infected HeLa cells expressing dominant negative forms of Rab7 showed that acquisition of lysosomal glycoproteins (Lgps) was Rab7-dependent. Moreover, these Lgps were derived from a Rab7 +ve, Lgp +ve vesicular pool that was not accessible to the endosomal network (145). These data imply a subtle and highly directional inhibition of vesicular trafficking or fusion. Interestingly SipC, a substrate of the SPI-2 pathogenicity island secretion system that is known to activate at low pH inside the host cell (144), is capable of suppressing vesicular fusion in an *in vitro* fusion assay (146). Whether this reflects the steady state arrest of vacuolar fusion in infected cells, or is a similar retardation strategy to enable the bacterium to respond to a changing environment, has yet to be explored.

4. Modulating the immune response

The data generated from tissue culture models give only one facet of the complex interactions between viral and microbial pathogens and their host organisms. The cells in which they reside are capable of both alerting the immune system to the presence of an infection and generating anti-microbial behaviour in response to cytokines released by stimulated T cells. The majority of pathogens have strategies to avoid the induction of the immune response. Some bacterial and protozoan pathogens down-regulate surface expression of MHC I or II molecules, but none are as effective as viruses in this regard (132, 147, 148). Other pathogens, like *Leishmania*, minimize the induction of an immune response by avoiding release of proteinaceous antigens (149, 150).

4.1 Viral regulation of cell surface antigen expression

The ability of some viruses to down-modulate MHC I expression through retention of the newly synthesized protein in the ER, and subsequent degradation, is well established (151). However, down-modulation of MHC antigens can also occur through perturbation of endocytic mechanisms. The best characterized scenario for these effects is the HIV/SIV Nef protein (152, 153). Nef is a small, myristoylated protein that is synthesized early after virus infection. *In vivo* Nef has been shown to be crucial for SIV pathogenicity. Animals infected with Nef-deficient viruses have low levels of virus production and progress to disease much more slowly. Moreover, these animals can be protected from infection by more pathogenic virus strains. Nef has multiple actions in cells. Two well-documented effects are its ability to down-modulate cell surface CD4 and MHC I. CD4 down-modulation is proposed to facilitate the release of progeny viruses from infected cells, while MHC I clearance may reduce the exposure of infected cells to the immune system.

The mechanisms of down-regulation have been linked to the ability of Nef to couple to sorting proteins associated with the endocytic pathway. For CD4, Nef is believed to bind to the cytoplasmic domain of CD4 and to the clathrin AP-2 complex. Interaction with AP-2 appears to occur through a di-leucine-containing loop and, in SIV Nef, by an additional tyrosine-based motif. Coupling to the adaptors increases the rate of CD4 endocytosis from the cell surface. Nef also appears to bind endosomal COP proteins and, in so doing diverts internalized CD4 away from a recycling pathway towards lysosomes where it is degraded (154). Nef also increases the rate of MHC I endocytosis through a mechanism that is not understood and is clearly different to that described for CD4. In Nef expressing cells, MHC I molecules are seen in a perinuclear compartment overlapping the Golgi. Recent data have suggested that Nef can cross-link MHC I to a protein called PACS1 that has been implicated in recycling between endosomes and the Golgi apparatus (see Chapter 7) (155). The association with PACS1 promotes the delivery of cell surface MHC I to the Golgi region of the cell.

Nef may exert other effects on trafficking through the endocytic pathway. A screen for interacting proteins identified a subunit of the vATPase as a prominent Nef binding partner (156). Nef-induced swelling of endocytic organelles (157) and changes in the sorting of specific membrane proteins could both be a consequence of Nef perturbation of vATPase function. Proteins expressed by other viruses have also been found to modify the activity of the vATPase. The E5 oncoprotein of bovine papillomavirus binds to the 16 K subunit of the vATPase. E5 expressed in cells is seen predominantly in the Golgi region of the cell and raises the pH in Golgi cisternae. The p12I protein of the human T cell lymphotrophic virus (HTLV-1) also interacts with this same vATPase subunit (158). Whether the pH in other acidic compartments is changed by these interactions is unclear. A second transforming protein of BPV binds the AP-1 adaptor complex of TGN CCVs (159). Presumably the combined activities of these proteins modify sorting in the exocytic pathway, though the effects of these modifications remain to be determined.

By contrast to Nef, the *m06* gene product of murine cytomegalovirus (gp48) is a type 1 transmembrane protein that binds newly synthesized MHC I molecules and carries them to lysosomes where they are degraded (160). Lysosomal targeting is dependent on a di-leucine motif in the gp48 cytoplasmic domain similar to the sorting signals identified in tyrosinase and other integral membrane proteins of late endocytic organelles (Chapter 7).

Overall a pattern emerges of viral proteins that are able to subvert the machineries that regulate cellular membrane trafficking to redistribute specific molecules or classes of proteins. In several of these cases the viral proteins carry motifs that mimic the sorting signals found in normal cellular proteins allowing them to interact directly with the cellular sorting machineries. The viral proteins essentially act as adaptors to couple cellular proteins to sorting pathways with which they would not normally interact.

4.2 Modification of the immune response by bacterial and protozoal pathogens

Bacterial and protozoal pathogens do not have the same access to a host cell's antigen processing and presentation machinery as viral pathogens. Nonetheless, they can influence both the induction and consequences of an immune response. There are numerous reports of intramacrophage pathogens such as *Mycobacterium* and *Leishmania* down-regulating surface expression of MHC I and II through either reduced expression, or degradation (133, 147). However, the sequestration of parasite-derived antigens is an alternate and demonstrable means of achieving the same effect (149, 161).

However, such strategies merely delay the inevitable and the immune system will mount a response that usually leads to activation of the host cell. Killing of pathogens by macrophages is mediated predominantly through the arginine-dependent production of nitric oxide products by the inducible-nitric oxide synthase (iNOS). Activation of macrophages induces a complex range of physiological responses including the apparent condensation of the endosomal/lysosomal network. As a consequence pathogens like *Mycobacterium*, that normally reside in early endosomes, are delivered

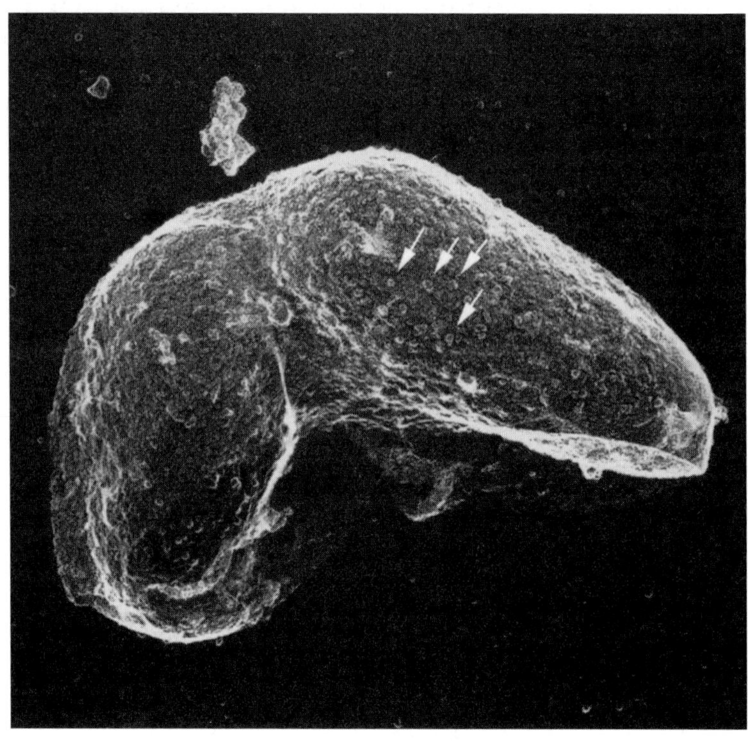

Fig. 4 A replica of a rapid-freeze phagosome isolated from bone marrow-derived macrophages activated with interferon-γ and LPS. The macrophages were incubated with *M. avium* for 3 h prior to lysis and isolation of the bacteria-containing phagosomes. The vATPase complexes are visible as 'doughnut-shaped' bumps on the phagosome surface, arrowed (94, 162).

to acidic, Rab 7-positive vacuoles capable of restricting their growth (162, 163). The accumulation of vATPase complexes that accompanies the decreased pH can be visualized on replicas of rapid-frozen phagosomes isolated from activated macrophages infected with *M. avium* (Fig. 4). To avoid this outcome, *M. tuberculosis* can block the macrophage's response to activating cytokines by suppressing GAS (interferon-γ activation sequences) transcription (164). The mediator(s) of this effect has not been identified but it has long been known that mycobacterial lipids show immunosuppressive activities (165–167) and certain of these lipids are released in copious amounts in infected cells (168). Not surprisingly the ability to suppress T cell responses and host cell activation is a frequently recurring theme in microbial infections (169–176).

5. Conclusions

The molecular mechanisms underlying infection by viral and cellular pathogens are becoming increasingly well understood. Not surprisingly, these agents make effective use of cellular processes to facilitate their interactions with hosts and they continue to generate novel and unexpected insights into basic cell functions. Growing concern over new pathogens, the re-emergence of old pathogens, and drug resistance, requires that increased attention is directed towards understanding the cellular mechanisms involved in pathogen entry and replication. This will undoubtedly have an impact on our understanding of endocytic mechanisms. Moreover, the possibilities to manipulate endocytic processes to prevent pathogen entry or divert pathogens from productive to non-productive routes of infection will emerge from basic studies of the molecular basis of endocytosis. There is much to learn and there will be interesting times ahead.

Acknowledgements

The NIH and the UK Medical Research Council support work in the authors' laboratories. We thank our colleagues Marie-José Bijlmakers and Annegret Pelchen-Matthews for critical comments on the paper. We also apologize to colleagues whose work has not been cited due to space restrictions.

References

1. Dales, S. (1973) Early events in cell-animal virus interactions. *Bacteriol. Rev.*, **37**, 103.
2. Helenius, A., Kartenbeck, J., Simons, K., and Fries, E. (1980) On the entry of Semliki Forest virus into BHK-21 cells. *J. Cell Biol.*, **84**, 404.
3. Marsh, M. and Helenius, A. (1989) Virus entry into animal cells. *Adv. Virus Res.*, **36**, 107.
4. Weissenhorn, W., Dessen, A., Calder, L. J., Harrison, S. C., Skehel, J. J., and Wiley, D. C. (1999) Structural basis for membrane fusion by enveloped viruses. *Mol. Membr. Biol.*, **16**, 3.
5. Stegmann, T. (2000) Membrane fusion mechanisms: the influenza hemagglutinin paradigm and its implications for intracellular fusion. *Traffic*, **1**, 598.
6. Greber, U. F. (1998) Virus assembly and disassembly: the adenovirus cysteine protease as a trigger factor. *Rev. Med. Virol.*, **8**, 213.
7. Marsh, M. (1992) Keeping the viral coat on. *Curr. Biol.*, **2**, 379.
8. Marsh, M. and Bron, R. (1997) SFV infection in CHO cells: Cell-type specific restrictions to productive virus entry at the cell surface. *J. Cell Sci.*, **110**, 95.
9. Sodeik, B., Ebersold, M. W., and Helenius, A. (1997) Microtubule-mediated transport of incoming herpes simplex virus 1 capsids to the nucleus. *J. Cell Biol.*, **136**, 1007.
10. Froshauer, S., Kartenbeck, J., and Helenius, A. (1988) Alphavirus RNA replicase is located on the cytoplasmic surface of endosomes and lysosomes. *J. Cell Biol.*, **107**, 2075.
11. Matteoni, R. and Kreis, T. E. (1987) Translocation and clustering of endosomes and lysosomes depends on microtubules. *J. Cell Biol.*, **105**, 1253.
12. Georgi, A., Mottola-Hartshorn, C., Warner, A., Fields, B., and Chen, L. B. (1990) Detection of individual fluorescently labeled reovirions in living cells. *Proc. Natl. Acad. Sci. USA*, **87**, 6579.
13. Miller, N. and Hutt-Fletcher, L. M. (1992) Epstein-Barr virus enters B cells and epithelial cells by different routes. *J. Virol.*, **66**, 3409.
14. Kock, J., Borst, E. M., and Schlicht, H. J. (1996) Uptake of duck hepatitis B virus into hepatocytes occurs by endocytosis but does not require passage of the virus through an acidic intracellular compartment. *J. Virol.*, **70**, 5827.

15. Bayer, N., Prchla, E., Schwab, M., Blaas, D., and Fuchs, R. (1999) Human rhinovirus HRV14 uncoats from early endosomes in the presence of bafilomycin. *FEBS Lett.*, **463**, 175.
16. Hofer, F., Gruenberger, M., Kowalski, H., Machat, H., Huettinger, M., Kuechler, E., *et al.* (1994) Members of the low density lipoprotein receptor family mediate cell entry of a minor-group common cold virus. *Proc. Natl. Acad. Sci. USA*, **91**, 1839.
17. Van Nhieu, G. T., Krukonis, E. S., Reszka, A. A., Horwitz, A. F., and Isberg, R. R. (1996) Mutations in the cytoplasmic domain of the integrin beta1 chain indicate a role for endocytosis factors in bacterial internalization. *J. Biol. Chem.*, **271**, 7665.
18. Lewis, C. M., Latham, K., and Roth, M. G. (2000) A screen of random sequences for those that alter the trafficking of the influenza virus hemagglutinin *in vivo*. *Traffic*, **1**, 282.
19. Parton, R. G. and Lindsay, M. (1999) Exploitation of major histocompatibility complex class I molecules and caveolae by simian virus 40. *Immunol. Rev.*, **168**, 23.
20. Anderson, H. A., Chen, Y., and Norkin, L. C. (1996) Bound simian virus 40 translocates to caveolin-enriched membrane domains, and its entry is inhibited by drugs that selectively disrupt caveolae. *Mol. Biol. Cell*, **7**, 1825.
21. Stang, E., Kartenbeck, J., and Parton, R. G. (1997) Major histocompatibility complex class I molecules mediate association of SV40 with caveolae. *Mol. Biol. Cell*, **8**, 47.
22. Helenius, A., Morein, B., Fries, E., Simons, K., Robinson, P., Schirrmacher, V., *et al.* (1978) Human (HLA-A and HLA-B) and murine (H-2K and H-2D) histocompatibility antigens are cell surface receptors for Semliki Forest virus. *Proc. Natl. Acad. Sci. USA*, **75**, 3846.
23. Marsh, M. and Helenius, A. (1980) Adsorptive endocytosis of Semliki Forest virus. *J. Mol. Biol.*, **142**, 439.
24. Marsh, M. and McMahon, H. (1999) The structural era of endocytosis. *Science*, **285**, 215.
25. Damke, H., Gossen, M., Freundlieb, S., Bujard, H., and Schmid, S. L. (1995) Tightly regulated and inducible expression of dominant interfering dynamin mutant in stably transformed HeLa cells. *Methods Enzymol.*, **257**, 209.
26. Wang, K., Huang, S., Kapoor-Munshi, A., and Nemerow, G. (1998) Adenovirus internalization and infection require dynamin. *J. Virol.*, **72**, 3455.
27. Bartlett, J. S., Wilcher, R., and Samulski, R. J. (2000) Infectious entry pathway of adeno-associated virus and adeno-associated virus vectors. *J. Virol.*, **74**, 2777.
28. DeTulleo, L. and Kirchhausen, T. (1998) The clathrin endocytic pathway in viral infection. *EMBO J.*, **17**, 4585.
29. Henley, J. R., Krueger, E. W., Oswald, B. J., and McNiven, M. A. (1998) Dynamin-mediated internalization of caveolae. *J. Cell Biol.*, **141**, 85.
30. Damke, H., Baba, T., Warnock, D. E., and Schmid, S. L. (1994) Induction of mutant dynamin specifically blocks endocytic coated vesicle formation. *J. Cell Biol.*, **127**, 915.
31. Werling, D., Hope, J. C., Chaplin, P., Collins, R. A., Taylor, G., and Howard, C. J. (1999) Involvement of caveolae in the uptake of respiratory syncytial virus antigen by dendritic cells. *J. Leukoc. Biol.*, **66**, 50.
32. Rodal, S. K., Skretting, G., Garred, O., Vilhardt, F., van Deurs, B., and Sandvig, K. (1999) Extraction of cholesterol with methyl-beta-cyclodextrin perturbs formation of clathrin-coated endocytic vesicles. *Mol. Biol. Cell*, **10**, 961.
33. Subtil, A., Gaidarov, I., Kobylarz, K., Lampson, M. A., Keen, J. H., and McGraw, T. E. (1999) Acute cholesterol depletion inhibits clathrin-coated pit budding. *Proc. Natl. Acad. Sci. USA*, **96**, 6775.
34. Mukherjee, S. and Maxfield, F. R. (2000) Role of membrane organization and membrane domains in endocytic lipid trafficking. *Traffic*, **1**, 203.

35. Kartenbeck, J., Stukenbrok, H., and Helenius, A. (1989) Endocytosis of simian virus 40 into the endoplasmic reticulum. *J. Cell Biol.*, **109**, 2721.
36. Roy, S., Luetterforst, R., Harding, A., Apolloni, A., Etheridge, M., Stang, E., *et al.* (1999) Dominant-negative caveolin inhibits H-Ras function by disrupting cholesterol-rich plasma membrane domains. *Nat. Cell Biol.*, **1**, 98.
37. Pho, M. T., Ashok, A., and Atwood, W. J. (2000) JC virus enters human glial cells by clathrin-dependent receptor- mediated endocytosis. *J. Virol.*, **74**, 2288.
38. Patterson, S., Oxford, J. S., and Dourmashkin, R. R. (1979) Studies on the mechanism of influenza virus entry into cells. *J. Gen. Virol.*, **43**, 223.
39. Cirone, M., Zompetta, C., Angeloni, A., Ablashi, D. V., Salahuddin, S. Z., Pavan, A., *et al.* (1992) Infection by human herpesvirus 6 (HHV-6) of human lymphoid T cells occurs through an endocytic pathway. *AIDS Res. Hum. Retroviruses*, **8**, 2031.
40. Tosteson, M. T. and Chow, M. (1997) Characterization of the ion channels formed by poliovirus in planar lipid membranes. *J. Virol.*, **71**, 507.
41. Belnap, D. M., Filman, D. J., Trus, B. L., Cheng, N., Booy, F. P., Conway, J. F., *et al.* (2000) Molecular tectonic model of virus structural transitions: the putative cell entry states of poliovirus. *J. Virol.*, **74**, 1342.
42. Neutra, M. R. and Kraehenbuhl, J.-P. (1992) Transepithelial transport and mucosal defence 1: The role of M cells. *Trends Cell Biol.*, **2**, 134.
43. Amerongen, H. M., Weltzin, R., Farnet, C. M., Michetti, P., Haseltine, W. A., and Neutra, M. R. (1991) Transepithelial transport of HIV-1 by intestinal M cells: A mechanism for transmission of AIDS. *J. AIDS*, **4**, 760.
44. Johannes, L. and Goud, B. (2000) Facing inward from compartment shores: How many pathways were we looking for? *Traffic*, **1**, 119.
45. Kielian, M. C., Marsh, M., and Helenius, A. (1986) Kinetics of endosome acidification detected by mutant and wild-type Semliki Forest virus. *EMBO J.*, **5**, 3103.
46. Marsh, M., Bolzau, E., and Helenius, A. (1983) Semliki Forest virus penetration occurs from acid prelysosomal vacuoles. *Cell*, **32**, 931.
47. Daro, E., Sheff, D., Gomez, M., Kreis, T., and Mellman, I. (1997) Inhibition of endosome function in CHO cells bearing a temperature- sensitive defect in the coatomer (COPI) component epsilon-COP. *J. Cell Biol.*, **139**, 1747.
48. Whitney, J. A., Gomez, M., Sheff, D., Kreis, T. E., and Mellman, I. (1995) Cytoplasmic coat proteins involved in endosome function. *Cell*, **83**, 703.
49. Aniento, F., Gu, F., Parton, R. G., and Gruenberg, J. (1996) An endosomal betaCOP is involved in the pH-dependent formation of transport vesicles destined for late endosomes. *J. Cell Biol.*, **133**, 29.
50. Clague, M. J., Urbe, S., Aniento, F., and Gruenberg, J. (1994) Vacuolar ATPase activity is required for endosomal carrier vesicle formation. *J. Biol. Chem.*, **269**, 21.
51. Bayer, N., Schober, D., Prchla, E., Murphy, R. F., Blaas, D., and Fuchs, R. (1998) Effect of bafilomycin A1 and nocodazole on endocytic transport in HeLa cells: implications for viral uncoating and infection. *J. Virol.*, **72**, 9645.
52. Rauma, T., Tuukkanen, J., Bergelson, J. M., Denning, G., and Hautala, T. (1999) rab5 GTPase regulates adenovirus endocytosis. *J. Virol.*, **73**, 9664.
53. Li, E., Stupack, D., Bokoch, G. M., and Nemerow, G. R. (1998) Adenovirus endocytosis requires actin cytoskeleton reorganization mediated by rho family GTPases. *J. Virol.*, **72**, 8806.
54. Dangoria, N. S., Breau, W. C., Anderson, H. A., Cishek, D. M., and Norkin, L. C. (1996) Extracellular simian virus 40 induces an ERK/MAP kinase-independent signalling pathway that activates primary response genes and promotes virus entry. *J. Gen. Virol.*, **77**, 2173.

55. Griffiths, G. and Rottier, P. (1992) Cell biology of viruses that assemble along the biosynthetic pathway. *Semin. Cell Biol.*, **3**, 367.
56. Smith, J. D. and de Harven, E. (1978) Herpes simplex virus and human cytomegalovirus replication in WI-38 cells. III. Cytochemical localization of lysosomal enzymes in infected cells. *J. Virol.*, **26**, 102.
57. Tooze, J., Hollinshead, M., Reis, B., Radsak, K., and Kern, H. (1993) Progeny vaccinia and human cytomegalovirus particles utilize early endosomal cisternae for their envelopes. *Eur. J. Cell Biol.*, **60**, 163.
58. Sauter, M. M., Pelchen-Matthews, A., Bron, R., Marsh, M., LaBranche, C. C., Vance, P., *et al.* (1996) An internalization signal in the simian immunodeficiency virus transmembrane protein cytoplasmic domain modulates expression of envelope glycoproteins on the cell surface. *J. Cell Biol.*, **132**, 795.
59. Berlioz-Torrent, C., Shacklett, B. L., Erdtmann, L., Delamarre, L., Bouchaert, I., Sonigo, P., *et al.* (1999) Interactions of the cytoplasmic domains of human and simian retroviral transmembrane proteins with components of the clathrin adaptor complexes modulate intracellular and cell surface expression of envelope glycoproteins. *J. Virol.*, **73**, 1350.
60. Ohno, H., Aguilar, R. C., Fournier, M. C., Hennecke, S., Cosson, P., and Bonifacino, J. S. (1997) Interaction of endocytic signals from the HIV-1 envelope glycoprotein complex with members of the adaptor medium chain family. *Virology*, **238**, 305.
61. Rowell, J. F., Stanhope, P. E., and Siliciano, R. F. (1995) Endocytosis of endogenously synthesized HIV-1 envelope protein—Mechanism and role in processing for association with Class-II MHC. *J. Immunol.*, **155**, 473.
62. Egan, M. A., Carruth, L. M., Rowell, J. F., Yu, X., and Siliciano, R. F. (1996) Human immunodeficiency virus type 1 envelope protein endocytosis mediated by a highly conserved intrinsic internalization signal in the cytoplasmic domain of gp41 is supressed in the presence of the Pr55gag precursor protein. *J. Virol.*, **70**, 6547.
63. Marsh, M., Pelchen-Matthews, A., and Hoxie, J. A. (1997) Roles for endocytosis in lentiviral replication. *Trends Cell Biol.*, **7**, 1.
64. Lee, C. A. (1996) Pathogenicity islands and the evolution of bacterial pathogens. *Infect. Agents Dis.*, **5**, 1.
65. Hacker, J., Blum-Oehler, G., Muhldorfer, I., and Tschape, H. (1997) Pathogenicity islands of virulent bacteria: structure, function and impact on microbial evolution. *Mol. Microbiol.*, **23**, 1089.
66. Leong, J. M., Morrissey, P. E., Marra, A., and Isberg, R. R. (1995) An aspartate residue of the Yersinia pseudotuberculosis invasin protein that is critical for integrin binding. *EMBO J.*, **14**, 422.
67. Galan, J. E. (1999) Interaction of Salmonella with host cells through the centisome 63 type III secretion system. *Curr. Opin. Microbiol.*, **2**, 46.
68. Iriarte, M. and Cornelis, G. R. (1998) YopT, a new Yersinia Yop effector protein, affects the cytoskeleton of host cells. *Mol. Microbiol.*, **29**, 915.
69. Bliska, J. B., Galan, J. E., and Falkow, S. (1993) Signal transduction in the mammalian cell during bacterial attachment and entry. *Cell*, **73**, 903.
70. Hardt, W. D., Urlaub, H., and Galan, J. E. (1998) A substrate of the centisome 63 type III protein secretion system of Salmonella typhimurium is encoded by a cryptic bacteriophage. *Proc. Natl. Acad. Sci. USA*, **95**, 2574.
71. Steele-Mortimer, O., Knodler, L. A., and Finlay, B. B. (2000) Poisons, ruffles and rockets: Bacterial pathogens and the host cell cytoskeleton. *Traffic*, **1**, 107.

72. Russell, D. G. (1983) Host cell invasion by Apicomplexa: an expression of the parasite's contractile system? *Parasitology*, **87**, 199.
73. Dobrowolski, J. and Sibley, L. D. (1997) The role of the cytoskeleton in host cell invasion by Toxoplasma gondii. *Behring Inst. Mitt.*, **99**, 90.
74. Mordue, D. G. and Sibley, L. D. (1997) Intracellular fate of vacuoles containing Toxoplasma gondii is determined at the time of formation and depends on the mechanism of entry. *J. Immunol.*, **159**, 4452.
75. Dubremetz, J. F., Garcia-Reguet, N., Conseil, V., and Fourmaux, M. N. (1998) Apical organelles and host-cell invasion by Apicomplexa. *Int. J. Parasitol.*, **28**, 1007.
76. Mordue, D. G., Desai, N., Dustin, M., and Sibley, L. D. (1999) Invasion by Toxoplasma gondii establishes a moving junction that selectively excludes host cell membrane proteins on the basis of their membrane anchoring. *J. Exp. Med.*, **190**, 1783.
77. Lingelbach, K. and Joiner, K. A. (1998) The parasitophorous vacuole membrane surrounding Plasmodium and Toxoplasma: an unusual compartment in infected cells. *J. Cell Sci.*, **111**, 1467.
78. Rodriguez, A., Martinez, I., Chung, A., Berlot, C. H., and Andrews, N. W. (1999) cAMP regulates Ca2+-dependent exocytosis of lysosomes and lysosome- mediated cell invasion by trypanosomes. *J. Biol Chem.*, **274**, 16754.
79. Burleigh, B. A. and Andrews, N. W. (1998) Signaling and host cell invasion by Trypanosoma cruzi. *Curr. Opin. Microbiol.*, **1**, 461.
80. Caler, E. V., Vaena de Avalos, S., Haynes, P. A., Andrews, N. W., and Burleigh, B. A. (1998) Oligopeptidase B-dependent signaling mediates host cell invasion by Trypanosoma cruzi. *EMBO J.*, **17**, 4975.
81. Canonne-Hergaux, F., Gruenheid, S., Govoni, G., and Gros, P. (1999) The Nramp1 protein and its role in resistance to infection and macrophage function. *Proc. Assoc. Am. Physicians*, **111**, 283.
82. Govoni, G. and Gros, P. (1998) Macrophage NRAMP1 and its role in resistance to microbial infections. *Inflamm. Res.*, **47**, 277.
83. Hackam, D. J., Rotstein, O. D., Zhang, W. J., Demaurex, N., Woodside, M., Tsai, O., *et al.* (1997) Regulation of phagosomal acidification. Differential targeting of Na^+/H^+ exchangers, Na^+/K^+-ATPases, and vacuolar-type H^+-ATPases. *J. Biol. Chem.*, **272**, 29810.
84. Underhill, D. M., Ozinsky, A., Hajjar, A. M., Stevens, A., Wilson, C. B., Bassetti, M., *et al.* (1999) The Toll-like receptor 2 is recruited to macrophage phagosomes and discriminates between pathogens. *Nature*, **401**, 811.
85. Falkow, S., Isberg, R. R., and Portnoy, D. A. (1992) The interaction of bacteria with mammalian cells. *Annu. Rev. Cell Biol.*, **8**, 333.
86. Portnoy, D. A., Tweten, R. K., Kehoe, M., and Bielecki, J. (1992) Capacity of listeriolysin O, streptolysin O, and perfringolysin O to mediate growth of Bacillus subtilis within mammalian cells. *Infect. Immun.*, **60**, 2710.
87. Goldfine, H., Knob, C., Alford, D., and Bentz, J. (1995) Membrane permeabilization by Listeria monocytogenes phosphatidylinositol-specific phospholipase C is independent of phospholipid hydrolysis and cooperative with listeriolysin O [retracted by Goldfine, H., Knob, C., Alford, D., and Bentz, J. (1997) *Proc. Natl. Acad. Sci. USA*, **94**, 2772]. *Proc. Natl. Acad. Sci. USA*, **92**, 2979.
88. Marquis, H., Doshi, V., and Portnoy, D. A. (1995) The broad-range phospholipase C and a metalloprotease mediate listeriolysin O-independent escape of Listeria monocytogenes from a primary vacuole in human epithelial cells. *Infect. Immun.*, **63**, 4531.

89. Bannam, T. and Goldfine, H. (1999) Mutagenesis of active-site histidines of Listeria monocytogenes phosphatidylinositol-specific phospholipase C: effects on enzyme activity and biological function. *Infect. Immun.*, **67**, 182.
90. Wadsworth, S. J. and Goldfine, H. (1999) Listeria monocytogenes phospholipase C-dependent calcium signaling modulates bacterial entry into J774 macrophage-like cells. *Infect. Immun.*, **67**, 1770.
91. Armstrong, J. A. and Hart, P. D. (1975) Phagosome-lysosome interactions in cultured macrophages infected with virulent tubercle bacilli. Reversal of the usual nonfusion pattern and observations on bacterial survival. *J. Exp. Med.*, **142**, 1.
92. Hart, P. D., Young, M. R., Gordon, A. H., and Sullivan, K. H. (1987) Inhibition of phagosome-lysosome fusion in macrophages by certain mycobacteria can be explained by inhibition of lysosomal movements observed after phagocytosis. *J. Exp. Med.*, **166**, 933.
93. Clemens, D. L. and Horwitz, M. A. (1995) Characterization of the Mycobacterium tuberculosis phagosome and evidence that phagosomal maturation is inhibited. *J. Exp. Med.*, **181**, 257.
94. Sturgill-Koszycki, S., Schlesinger, P. H., Chakraborty, P., Haddix, P. L., Collins, H. L., Fok, A. K., *et al.* (1994) Lack of acidification in Mycobacterium phagosomes produced by exclusion of the vesicular proton-ATPase [published erratum appears in *Science* (1994) **263**, 1359]. *Science*, **263**, 678.
95. Sturgill-Koszycki, S., Schaible, U. E., and Russell, D. G. (1996) Mycobacterium-containing phagosomes are accessible to early endosomes and reflect a transitional state in normal phagosome biogenesis. *EMBO J.*, **15**, 6960.
96. Russell, D. G., Dant, J., and Sturgill-Koszycki, S. (1996) Mycobacterium avium- and Mycobacterium tuberculosis-containing vacuoles are dynamic, fusion-competent vesicles that are accessible to glycosphingolipids from the host cell plasmalemma. *J. Immunol.*, **156**, 4764.
97. Clemens, D. L. and Horwitz, M. A. (1996) The Mycobacterium tuberculosis phagosome interacts with early endosomes and is accessible to exogenously administered transferrin. *J. Exp. Med.*, **184**, 1349.
98. Ullrich, H. J., Beatty, W. L., and Russell, D. G. (1999) Direct delivery of procathepsin D to phagosomes: implications for phagosome biogenesis and parasitism by Mycobacterium. *Eur. J. Cell Biol.*, **78**, 739.
99. McDonough, K. A., Kress, Y., and Bloom, B. R. (1993) The interaction of Mycobacterium tuberculosis with macrophages: a study of phagolysosome fusion. *Infect. Agents Dis.*, **2**, 232.
100. Teitelbaum, R., Cammer, M., Maitland, M. L., Freitag, N. E., Condeelis, J., and Bloom, B. R. (1999) Mycobacterial infection of macrophages results in membrane-permeable phagosomes. *Proc. Natl. Acad. Sci. USA*, **96**, 15190.
101. Oh, Y. K., Harding, C. V., and Swanson, J. A. (1997) The efficiency of antigen delivery from macrophage phagosomes into cytoplasm for MHC class I-restricted antigen presentation. *Vaccine*, **15**, 511.
102. Rodriguez, A., Regnault, A., Kleijmeer, M., Ricciardi-Castagnoli, P., and Amigorena, S. (1999) Selective transport of internalized antigens to the cytosol for MHC class I presentation in dendritic cells. *Nat. Cell Biol.*, **1**, 362.
103. Via, L. E., Deretic, D., Ulmer, R. J., Hibler, N. S., Huber, L. A., and Deretic, V. (1997) Arrest of mycobacterial phagosome maturation is caused by a block in vesicle fusion between stages controlled by rab5 and rab7. *J. Biol. Chem.*, **272**, 13326.

104. Ferrari, G., Langen, H., Naito, M., and Pieters, J. (1999) A coat protein on phagosomes involved in the intracellular survival of mycobacteria. *Cell*, **97**, 435.
105. Webster, P., JW, I. J., Chicoine, L. M., and Fikrig, E. (1998) The agent of human granulocytic ehrlichiosis resides in an endosomal compartment. *J. Clin. Invest.*, **101**, 1932.
106. Barnewall, R. E., Rikihisa, Y., and Lee, E. H. (1997) Ehrlichia chaffeensis inclusions are early endosomes which selectively accumulate transferrin receptor. *Infect. Immun.*, **65**, 1455.
107. Barnewall, R. E., Ohashi, N., and Rikihisa, Y. (1999) Ehrlichia chaffeensis and E. sennetsu, but not the human granulocytic ehrlichiosis agent, colocalize with transferrin receptor and up-regulate transferrin receptor mRNA by activating iron-responsive protein 1. *Infect. Immun.*, **67**, 2258.
108. Horwitz, M. A. (1983) The Legionnaires' disease bacterium (Legionella pneumophila) inhibits phagosome-lysosome fusion in human monocytes. *J. Exp. Med.*, **158**, 2108.
109. Horwitz, M. A. and Maxfield, F. R. (1984) Legionella pneumophila inhibits acidification of its phagosome in human monocytes. *J. Cell Biol.*, **99**, 1936.
110. Swanson, M. S. and Isberg, R. R. (1993) Formation of the Legionella pneumophila replicative phagosome. *Infect. Agents Dis.*, **2**, 224.
111. Swanson, M. S. and Isberg, R. R. (1995) Association of Legionella pneumophila with the macrophage endoplasmic reticulum. *Infect. Immun.*, **63**, 3609.
112. Vogel, J. P., Andrews, H. L., Wong, S. K., and Isberg, R. R. (1998) Conjugative transfer by the virulence system of Legionella pneumophila. *Science*, **279**, 873.
113. Vogel, J. P. and Isberg, R. R. (1999) Cell biology of Legionella pneumophila. *Curr. Opin. Microbiol.*, **2**, 30.
114. Pizarro-Cerda, J., Moreno, E., Sanguedolce, V., Mege, J. L., and Gorvel, J. P. (1998) Virulent Brucella abortus prevents lysosome fusion and is distributed within autophagosome-like compartments. *Infect. Immun.*, **66**, 2387.
115. Pizarro-Cerda, J., Meresse, S., Parton, R. G., van der Goot, G., Sola-Landa, A., Lopez-Goni, I., *et al.* (1998) Brucella abortus transits through the autophagic pathway and replicates in the endoplasmic reticulum of nonprofessional phagocytes. *Infect. Immun.*, **66**, 5711.
116. Escalante-Ochoa, C., Ducatelle, R., and Haesebrouck, F. (1998) The intracellular life of Chlamydia psittaci: how do the bacteria interact with the host cell? *FEMS Microbiol. Rev.*, **22**, 65.
117. Hackstadt, T. (1998) The diverse habitats of obligate intracellular parasites. *Curr. Opin. Microbiol.*, **1**, 82.
118. van Ooij, C., Apodaca, G., and Engel, J. (1997) Characterization of the Chlamydia trachomatis vacuole and its interaction with the host endocytic pathway in HeLa cells. *Infect. Immun.*, **65**, 758.
119. Scidmore, M. A., Rockey, D. D., Fischer, E. R., Heinzen, R. A., and Hackstadt, T. (1996) Vesicular interactions of the Chlamydia trachomatis inclusion are determined by chlamydial early protein synthesis rather than route of entry. *Infect. Immun.*, **64**, 5366.
120. Scidmore-Carlson, M. A., Shaw, E. I., Dooley, C. A., Fischer, E. R., and Hackstadt, T. (1999) Identification and characterization of a chlamydia trachomatis early operon encoding four novel inclusion membrane proteins. *Mol. Microbiol.*, **33**, 753.
121. Rockey, D. D., Grosenbach, D., Hruby, D. E., Peacock, M. G., Heinzen, R. A., and Hackstadt, T. (1997) Chlamydia psittaci IncA is phosphorylated by the host cell and is exposed on the cytoplasmic face of the developing inclusion. *Mol. Microbiol.*, **24**, 217.
122. Heinzen, R. A., Scidmore, M. A., Rockey, D. D., and Hackstadt, T. (1996) Differential interaction with endocytic and exocytic pathways distinguish parasitophorous vacuoles of Coxiella burnetii and Chlamydia trachomatis. *Infect. Immun.*, **64**, 796.

123. Boleti, H., Benmerah, A., Ojcius, D. M., Cerf-Bensussan, N., and Dautry-Varsat, A. (1999) Chlamydia infection of epithelial cells expressing dynamin and Eps15 mutants: clathrin-independent entry into cells and dynamin-dependent productive growth. *J. Cell Sci.*, **112**, 1487.
124. Joiner, K. A., Fuhrman, S. A., Miettinen, H. M., Kasper, L. H., and Mellman, I. (1990) Toxoplasma gondii: fusion competence of parasitophorous vacuoles in Fc receptor-transfected fibroblasts. *Science*, **249**, 641.
125. Sinai, A. P., Webster, P., and Joiner, K. A. (1997) Association of host cell endoplasmic reticulum and mitochondria with the Toxoplasma gondii parasitophorous vacuole membrane: a high affinity interaction. *J. Cell Sci.*, **110**, 2117.
126. Carruthers, V. B. and Sibley, L. D. (1997) Sequential protein secretion from three distinct organelles of Toxoplasma gondii accompanies invasion of human fibroblasts. *Eur. J. Cell Biol.*, **73**, 114.
127. Sibley, L. D. and Krahenbuhl, J. L. (1988) Modification of host cell phagosomes by Toxoplasma gondii involves redistribution of surface proteins and secretion of a 32 kDa protein. *Eur. J. Cell Biol.*, **47**, 81.
128. Mercier, C., Cesbron-Delauw, M. F., and Sibley, L. D. (1998) The amphipathic alpha helices of the toxoplasma protein GRA2 mediate post-secretory membrane association. *J. Cell Sci.*, **111**, 2171.
129. Schwab, J. C., Beckers, C. J., and Joiner, K. A. (1994) The parasitophorous vacuole membrane surrounding intracellular Toxoplasma gondii functions as a molecular sieve. *Proc. Natl. Acad. Sci. USA*, **91**, 509.
130. Mordue, D. G., Hakansson, S., Niesman, I., and Sibley, L. D. (1999) Toxoplasma gondii resides in a vacuole that avoids fusion with host cell endocytic and exocytic vesicular trafficking pathways. *Exp. Parasitol.*, **92**, 87.
131. Rabinovitch, M. and Veras, P. S. (1996) Cohabitation of Leishmania amazonensis and Coxiella burnetii. *Trends Microbiol.*, **4**, 158.
132. Lem, L., Riethof, D. A., Scidmore-Carlson, M., Griffiths, G. M., Hackstadt, T., and Brodsky, F. M. (1999) Enhanced interaction of HLA-DM with HLA-DR in enlarged vacuoles of hereditary and infectious lysosomal diseases. *J. Immunol.*, **162**, 523.
133. Antoine, J. C., Prina, E., Lang, T., and Courret, N. (1998) The biogenesis and properties of the parasitophorous vacuoles that harbour Leishmania in murine macrophages. *Trends Microbiol.*, **6**, 392.
134. Antoine, J. C., Prina, E., Jouanne, C., and Bongrand, P. (1990) Parasitophorous vacuoles of Leishmania amazonensis-infected macrophages maintain an acidic pH. *Infect. Immun.*, **58**, 779.
135. Prina, E., Antoine, J. C., Wiederanders, B., and Kirschke, H. (1990) Localization and activity of various lysosomal proteases in Leishmania amazonensis-infected macrophages. *Infect. Immun.*, **58**, 1730.
136. Russell, D. G., Xu, S., and Chakraborty, P. (1992) Intracellular trafficking and the parasitophorous vacuole of Leishmania mexicana-infected macrophages. *J. Cell Sci.*, **103**, 1193.
137. Desjardins, M. and Descoteaux, A. (1997) Inhibition of phagolysosomal biogenesis by the Leishmania lipophosphoglycan. *J. Exp. Med.*, **185**, 2061.
138. Desjardins, M. and Descoteaux, A. (1998) Survival strategies of Leishmania donovani in mammalian host macrophages. *Res. Immunol.*, **149**, 689.
139. Schaible, U. E., Schlesinger, P. H., Steinberg, T. H., Mangel, W. F., Kobayashi, T., and Russell, D. G. (1999) Parasitophorous vacuoles of Leishmania mexicana acquire macromolecules from the host cell cytosol via two independent routes. *J. Cell Sci.*, **112**, 681.

140. Finlay, B. B. (1997) Interactions of enteric pathogens with human epithelial cells. Bacterial exploitation of host processes. *Adv. Exp. Med. Biol.*, **412**, 289.
141. Garcia-del Portillo, F., Zwick, M. B., Leung, K. Y., and Finlay, B. B. (1993) Intracellular replication of Salmonella within epithelial cells is associated with filamentous structures containing lysosomal membrane glycoproteins. *Infect. Agents Dis.*, **2**, 227.
142. Garcia-del Portillo, F. and Finlay, B. B. (1995) Targeting of Salmonella typhimurium to vesicles containing lysosomal membrane glycoproteins bypasses compartments with mannose 6-phosphate receptors. *J. Cell Biol.*, **129**, 81.
143. Rathman, M., Sjaastad, M. D., and Falkow, S. (1996) Acidification of phagosomes containing Salmonella typhimurium in murine macrophages. *Infect. Immun.*, **64**, 2765.
144. Beuzon, C. R., Banks, G., Deiwick, J., Hensel, M., and Holden, D. W. (1999) pH-dependent secretion of SseB, a product of the SPI-2 type III secretion system of Salmonella typhimurium. *Mol. Microbiol.*, **33**, 806.
145. Meresse, S., Steele-Mortimer, O., Finlay, B. B., and Gorvel, J. P. (1999) The rab7 GTPase controls the maturation of Salmonella typhimurium- containing vacuoles in HeLa cells. *EMBO J.*, **18**, 4394.
146. Uchiya, K., Barbieri, M. A., Funato, K., Shah, A. H., Stahl, P. D., and Groisman, E. A. (1999) A Salmonella virulence protein that inhibits cellular trafficking. *EMBO J.*, **18**, 3924.
147. Antoine, J. C., Lang, T., Prina, E., Courret, N., and Hellio, R. (1999) H-2M molecules, like MHC class II molecules, are targeted to parasitophorous vacuoles of Leishmania-infected macrophages and internalized by amastigotes of L. amazonensis and L. mexicana. *J. Cell Sci.*, **112**, 2559.
148. Hmama, Z., Gabathuler, R., Jefferies, W. A., de Jong, G., and Reiner, N. E. (1998) Attenuation of HLA-DR expression by mononuclear phagocytes infected with Mycobacterium tuberculosis is related to intracellular sequestration of immature class II heterodimers. *J. Immunol.*, **161**, 4882.
149. Wolfram, M., Fuchs, M., Wiese, M., Stierhof, Y. D., and Overath, P. (1996) Antigen presentation by Leishmania mexicana-infected macrophages: activation of helper T cells by a model parasite antigen secreted into the parasitophorous vacuole or expressed on the amastigote surface. *Eur. J. Immunol.*, **26**, 3153.
150. Kima, P. E., Soong, L., Chicharro, C., Ruddle, N. H., and McMahon-Pratt, D. (1996) Leishmania-infected macrophages sequester endogenously synthesized parasite antigens from presentation to CD4+ T cells. *Eur. J. Immunol.*, **26**, 3163.
151. Ploegh, H. L. (1998) Viral strategies of immune evasion. *Science*, **280**, 248.
152. Piguet, V., Schwartz, O., Le Gall, S., and Trono, D. (1999) The downregulation of CD4 and MHC-I by primate lentiviruses: a paradigm for the modulation of cell surface receptors. *Immunol. Rev.*, **168**, 51.
153. Oldridge, J. and Marsh, M. (1998) Nef—an adaptor adaptor? *Trends Cell Biol.*, **8**, 302.
154. Piguet, V., Gu, F., Foti, M., Demaurex, N., Gruenberg, J., Carpentier, J. L., *et al.* (1999) Nef-induced CD4 degradation: a diacidic-based motif in Nef functions as a lysosomal targeting signal through the binding of beta-COP in endosomes. *Cell*, **97**, 63.
155. Piguet, V., Wan, L., Borel, C., Mangasarian, A., Demaurex, N., Thomas, G., *et al.* (2000) HIV-1 Nef protein binds to the cellular protein PACS-1 to downregulate class I major histocompatibility complexes. *Nat. Cell Biol.*, **2**, 163.
156. Lu, X., Yu, H., Liu, S. H., Brodsky, F. M., and Peterlin, B. M. (1998) Interactions between HIV1 Nef and vacuolar ATPase facilitate the internalization of CD4. *Immunity*, **8**, 647.

157. Sanfridson, A., Hester, S., and Doyle, C. (1997) Nef proteins encoded by human and simian immunodeficiency viruses induce the accumulation of endosomes and lysosomes in human T cells. *Proc. Natl. Acad. Sci. USA*, **94**, 873.
158. Koralnik, I. J., Mulloy, J. C., Andresson, T., Fullen, J., and Franchini, G. (1995) Mapping of the intermolecular association of human T cell leukaemia/lymphotropic virus type I p12I and the vacuolar H^+-ATPase 16 kDa subunit protein. *J. Gen. Virol.*, **76**, 1909.
159. Tong, X., Boll, W., Kirchhausen, T., and Howley, P. M. (1998) Interaction of the bovine papillomavirus E6 protein with the clathrin adaptor complex AP-1. *J. Virol.*, **72**, 476.
160. Reusch, U., Muranyi, W., Lucin, P., Burgert, H. G., Hengel, H., and Koszinowski, U. H. (1999) A cytomegalovirus glycoprotein re-routes MHC class I complexes to lysosomes for degradation. *EMBO J.*, **18**, 1081.
161. Overath, P. and Aebischer, T. (1999) Antigen presentation by macrophages harboring intravesicular pathogens. *Parasitol. Today*, **15**, 325.
162. Schaible, U. E., Sturgill-Koszycki, S., Schlesinger, P. H., and Russell, D. G. (1998) Cytokine activation leads to acidification and increases maturation of Mycobacterium avium-containing phagosomes in murine macrophages. *J. Immunol.*, **160**, 1290.
163. Via, L. E., Fratti, R. A., McFalone, M., Pagan-Ramos, E., Deretic, D., and Deretic, V. (1998) Effects of cytokines on mycobacterial phagosome maturation. *J. Cell Sci.*, **111**, 897.
164. Ting, L. M., Kim, A. C., Cattamanchi, A., and Ernst, J. D. (1999) Mycobacterium tuberculosis inhibits IFN-gamma transcriptional responses without inhibiting activation of STAT1. *J. Immunol.*, **163**, 3898.
165. Pourshafie, M., Ayub, Q., and Barrow, W. W. (1993) Comparative effects of Mycobacterium avium glycopeptidolipid and lipopeptide fragment on the function and ultrastructure of mononuclear cells. *Clin. Exp. Immunol.*, **93**, 72.
166. Barrow, W. W., de Sousa, J. P., Davis, T. L., Wright, E. L., Bachelet, M., and Rastogi, N. (1993) Immunomodulation of human peripheral blood mononuclear cell functions by defined lipid fractions of Mycobacterium avium. *Infect. Immun.*, **61**, 5286.
167. Moura, A. C. and Mariano, M. (1997) Lipids from Mycobacterium leprae cell wall suppress T-cell activation *in vivo* and *in vitro*. *Immunology*, **92**, 429.
168. Beatty, W. L., Rhoades, E. R., Ullrich, H. J., Chatterjee, D., and Russell, D. G. (2000) Trafficking and release of mycobacterial lipids from infected macrophages. *Traffic*, **1**, 235.
169. Chinen, T., Qureshi, M. H., Koguchi, Y., and Kawakami, K. (1999) Candida albicans suppresses nitric oxide (NO) production by interferon- gamma (IFN-gamma) and lipopolysaccharide (LPS)-stimulated murine peritoneal macrophages. *Clin. Exp. Immunol.*, **115**, 491.
170. Haslov, K., Fomsgaard, A., Takayama, K., Fomsgaard, J. S., Ibsen, P., Fauntleroy, M. B., *et al.* (1992) Immunosuppressive effects induced by the polysaccharide moiety of some bacterial lipopolysaccharides. *Immunobiology*, **186**, 378.
171. Issekutz, T. B. and Stoltz, J. M. (1985) Suppression of lymphocyte proliferation by Pseudomonas aeruginosa: mediation by Pseudomonas-activated suppressor monocytes. *Infect. Immun.*, **48**, 832.
172. Kawakami, K., Zhang, T., Qureshi, M. H., and Saito, A. (1997) Cryptococcus neoformans inhibits nitric oxide production by murine peritoneal macrophages stimulated with interferon-gamma and lipopolysaccharide. *Cell. Immunol.*, **180**, 47.
173. Petit, J. C., Richard, G., Burghoffer, B., and Daguet, G. L. (1985) Suppression of cellular immunity to Listeria monocytogenes by activated macrophages: mediation by prostaglandins. *Infect. Immun.*, **49**, 383.

174. Truitt, R. L., Hanke, C., Radke, J., Mueller, R., and Barbieri, J. T. (1998) Glycosphingolipids as novel targets for T-cell suppression by the B subunit of recombinant heat-labile enterotoxin. *Infect. Immun.*, **66**, 1299.
175. Yao, T., Mecsas, J., Healy, J. I., Falkow, S., and Chien, Y. (1999) Suppression of T and B lymphocyte activation by a Yersinia pseudotuberculosis virulence factor, yopH. *J. Exp. Med.*, **190**, 1343.
176. VanHeyningen, T. K., Collins, H. L., and Russell, D. G. (1997) IL-6 produced by macrophages infected with Mycobacterium species suppresses T cell responses. *J. Immunol.*, **158**, 330.
177. Castilla, V., Mersich, S. E., Candurra, N. A., and Damonte, E. B. (1994) The entry of junin virus into vero cells. *Arch. Virol.*, **136**, 363.
178. Castilla, V., Mersich, S. E., and Damonte, E. B. (1991) [Lysosomotropic compounds inhibiting the multiplication of Junin virus]. *Rev. Argent. Microbiol.*, **23**, 86.
179. Cash, P. (1982) Inhibition of La Crosse virus replication by monensin, monovalent ionophore. *J. Gen. Virol.*, **59**, 193.
180. Gonzalez-Scarano, F. (1985) La Crosse virus G1 glycoprotein undergoes a conformational change at the pH of fusion. *Virology*, **140**, 209.
181. Gonzalez-Scarano, F., Pobjecky, N., and Nathanson, N. (1984) La Crosse bunyavirus can mediate pH-dependent fusion from without. *Virology*, **132**, 222.
182. Pekosz, A. and Gonzalez-Scarano, F. (1996) The extracellular domain of la-crosse virus g1 forms oligomers and undergoes pH-dependent conformational-changes. *Virology*, **225**, 243.
183. Valdeira, M. L. and Geraldes, A. (1985) Morphological study on the entry of African swine fever virus into cells. *Biol. Cell*, **55**, 35.
184. Valdeira, M. L., Bernardes, C., Cruz, B., and Geraldes, A. (1998) Entry of African swine fever virus into Vero cells and uncoating. *Vet. Microbiol.*, **60**, 131.
185. Dourmashkin, R. R. and Tyrrell, D. A. J. (1974) Electron microscopic observations on the entry of influenza virus into susceptible cells. *J. Gen. Virol.*, **24**, 129.
186. Matlin, K. S., Reggio, H., Helenius, A., and Simons, K. (1981) Infectious entry pathway of influenza virus in a canine kidney cell line. *J. Cell Biol.*, **91**, 601.
187. Wiley, D. C., Skehel, J. J., and Waterfield, M. (1977) Evidence from studies with a cross-linking reagent that the haemagglutinin of influenza virus is a trimer. *Virology*, **79**, 446.
188. Skehel, J. J., Bayley, P. M., Brown, E. B., Martin, S. R., Waterfield, M. D., White, J. M., *et al.* (1982) Changes in the conformation of influenza virus hemagglutinin at the pH optimum of virus-mediated membrane fusion. *Proc. Natl. Acad. Sci. USA*, **79**, 968.
189. Guinea, R. and Carrasco, L. (1995) Requirement for vacuolar proton-ATPase activity during entry of influenza virus into cells. *J. Virol.*, **69**, 2306.
190. Redmond, S., Peters, G., and Dickson, C. (1984) Mouse mammary tumor virus can mediate cell fusion at reduced pH. *Virology*, **133**, 393.
191. Matlin, K. S., Reggio, H., Helenius, A., and Simons, K. (1982) Pathway of vesicular stomatitis virus entry leading to infection. *J. Mol. Biol.*, **156**, 609.
192. Gaudin, Y., Ruigrok, R., Knossow, M., and Flamand, A. (1993) Low-pH conformational changes of rabies virus glycoprotein and their role in membrane fusion. *J. Virol.*, **67**, 1365.
193. Superti, F., Derer, M., and Tsiang, H. (1988) Mechanism of Rabies virus entry into cells. *J. Gen. Virol.*, **65**, 781.
194. Doxsey, S. J., Brodsky, F. M., Blank, G. S., and Helenius, A. (1987) Inhibition of endocytosis by anti-clathrin antibodies. *Cell*, **50**, 453.

195. Guinea, R. and Carrasco, L. (1994) Concanamycin A: A powerful inhibitor of enveloped animal-virus entry into cells. *Biochem. Biophys. Res. Commun.*, **201**, 1270.
196. Kimura, T. and Ohyama, A. (1988) Association between the pH-dependent conformational change of West Nile flavivirus E protein and virus-mediated membrane fusion. *J. Gen. Virol.*, **69**, 124.
197. Gollins, S. W. and Porterfield, J. S. (1986) pH-dependent fusion between the Flavivirus West Nile and liposomal model membranes. *J. Gen. Virol.*, **67**, 157.
198. Kimura, T., Gollins, S. W., and Porterfield, J. S. (1986) The effect of pH on the early interaction of West Nile virus with P388D1 cells. *J. Gen. Virol.*, **67**, 2423.
199. Heinz, F. X., Auer, G., Stiasny, K., Holzmann, H., Mandl, C., Guirakhoo, F., *et al.* (1994) The interactions of the flavivirus envelope proteins: implications for virus entry and release. *Arch. Virol.*, **Suppl. 9**, 339.
200. Allison, S. L., Schalich, J., Stiasny, K., Mandl, C. W., Kunz, C., and Heinz, F. X. (1995) Oligomeric rearrangement of tick-borne encephalitis virus envelope proteins induced by an acidic pH. *J. Virol.*, **69**, 695.
201. Pastan, I., Seth, P., Fitzgerald, D., and Willingham, M. (1986) Entry of adenovirus. In *Virus attachment and entry into cells* (ed. R. L. Crowell and K. Lonberg-Holm), p. 141. American Society for Microbiology, Washington.
202. Varga, M. J., Weibull, C., and Everitt, E. (1991) Infectious entry pathway of adenovirus type 2. *J. Virol.*, **65**, 6061.
203. Kreutz, L. C. and Seal, B. S. (1995) The pathway of feline calicivirus entry. *Virus Res.*, **35**, 63.
204. Breiner, K. M. and Schaller, H. (2000) Cellular receptor traffic is essential for productive duck hepatitis B virus infection. *J. Virol.*, **74**, 2203.
205. Tanner, J., Weis, J., Fearon, D., Whang, Y., and Kieff, E. (1987) Epstein-Barr virus gp350/220 binding to the B lymphocyte C3d receptor mediates adsorption, capping, and endocytosis. *Cell*, **50**, 203.
206. Murphy, C., Zacchi, P., Parton, R. G., Zerial, M., and Lim, F. (1997) HSV infection of polarized epithelial cells on filter supports: implications for transport assays and protein localization. *Eur. J. Cell Biol.*, **72**, 278.
207. Parker, J. S. and Parrish, C. R. (2000) Cellular uptake and infection by canine parvovirus involves rapid dynamin-regulated clathrin-mediated endocytosis, followed by slower intracellular trafficking. *J. Virol.*, **74**, 1919.
208. Duan, D., Li, Q., Kao, A. W., Yue, Y., Pessin, J. E., and Engelhardt, J. F. (1999) Dynamin is required for recombinant adeno-associated virus type 2 infection. *J. Virol.*, **73**, 10371.
209. Prchla, E., Kuechler, E., Blaas, D., and Fuchs, R. (1994) Uncoating of human rhinovirus type 2 (HRV2) from isolated endosomes. *J. Virol.*, **68**, 3713.
210. Tai, J. H., Ong, S. J., Chang, S. C., and Su, H. M. (1993) Giardiavirus enters Giardia lamblia WB trophozoite via endocytosis. *Exp. Parasitol.*, **76**, 165.

Index